Prentice Hall

Biology

Chapter Tests: Levels A and B

(Includes Unit Tests and Final Exams)

PEARSON
Prentice Hall

Upper Saddle River, New Jersey
Needham, Massachusetts

To the Teacher

Chapter Tests: Levels A and B for *Prentice Hall Biology* is designed to provide teachers with material to assess students' knowledge of each chapter and unit. The material consists of blackline masters for two tests for each chapter and unit. In addition, two cumulative final exams are included. Test A is for all students, and Test B is for students who need additional help.

PEARSON
Prentice Hall

ISBN 0-13-115287-4

3 4 5 6 7 8 9 10 08 07 06 05 04

Contents

Unit 8 Invertebrates

Unit 9 Chordates

Unit 10 The Human Body

Name______________________________ Class________________ Date ______________

Multiple Choice

Write the letter that best answers the question or completes the statement on the line provided.

____ **1.** Which of the following is NOT a goal of science?
a. to investigate and understand the natural world
b. to explain events in the natural world
c. to use data to support a particular point of view
d. to use derived explanations to make useful predictions

____ **2.** Information gathered from observing a plant that grows 3 cm over a two-week period results in
a. inferences. c. hypotheses.
b. variables. d. data.

____ **3.** Which of the following is NOT a way that scientists generate hypotheses?
a. using informed, creative imagination
b. using logical inference
c. using prior knowledge
d. using a feeling about what should occur

____ **4.** Scientific hypotheses must be proposed in a way that
a. ensures that an experiment will be valid.
b. enables them to be proved valid.
c. enables them to be tested.
d. doesn't contradict previous hypotheses.

____ **5.** A controlled experiment allows the scientist to isolate and test
a. a conclusion.
b. a mass of information.
c. several variables.
d. a single variable.

____ **6.** The ability to reproduce results is an important part of any
a. hypothesis. c. law.
b. theory. d. experiment.

____ **7.** A theory
a. is always true.
b. is the opening statement of an experiment.
c. may be revised or replaced.
d. is a problem to be solved.

____ **8.** Which of the following is NOT a characteristic of all living things?

a. growth and development

b. ability to move

c. response to the environment

d. ability to reproduce

____ **9.** The amount of light and temperature are examples of

a. factors necessary for life.

b. methods of energy production.

c. factors to which living things respond.

d. factors that affect reproduction.

____**10.** Which of the following terms includes all the others?

a. biologist
c. zoologist

b. botanist
d. ethologist

____**11.** The basic unit of mass in the International System of Units, or SI, is the

a. meter.
c. liter.

b. ounce.
d. gram.

____**12.** To observe a small living organism, a scientist might use a(an)

a. electronic balance.

b. TEM.

c. compound light microscope.

d. electron microscope.

____**13.** An instrument used to separate cell parts according to density is the

a. compound light microscope.

b. electron microscope.

c. blender.

d. centrifuge.

____**14.** Which of the following is NOT considered a safety procedure?

a. Read all the steps in your activity before doing it.

b. If in doubt about any part of an activity, trust your instincts.

c. Follow your teacher's instructions.

d. Follow the textbook direction exactly.

____**15.** Because you may come in contact with organisms you cannot see, what safety procedure MUST be followed?

a. Read over your activity.

b. Open the windows of the laboratory.

c. Wash your hands thoroughly after completing the activity.

d. Do not wear long sleeves.

Name ______________________ Class ______________ Date ______________

Completion

Complete each statement on the line provided.

16. The information you gather during an experiment is called your ______________.

17. Based on his ______________, Redi made a prediction that keeping flies away from meat would prevent maggots appearing on the meat.

18. The name given to the idea that life could arise from nonliving matter is called ______________.

19. The chemical reactions through which an organism builds up or breaks down materials as it carries out its life processes is called ______________.

20. A variable that is deliberately changed in an experiment is the ______________ variable.

Short Answer

In complete sentences, write the answers to the questions on the lines provided.

21. What is the difference between a theory and a hypothesis?

____

____

____

____

22. What are the differences between sexual and asexual reproduction?

____

____

____

____

23. What structures are included in the level of organization called groups of cells?

____

____

24. What is the metric system?

____

____

____

25. How do light microscopes differ from electron microscopes?

____

____

____

____

Name______________________ Class________________ Date ____________

Figure 1-1

Using Science Skills

Use Figure 1-1 to answer the following questions on the lines provided.

26. Interpreting Graphics According to Figure 1-1, how many populations of animals are represented by this community?

____

____

27. Interpreting Graphics According to Figure 1-1, what are the nonliving elements in this ecosystem?

____

____

____

____

28. Applying Concepts The ecosystem shown in Figure 1-1 is part of what larger level of organisms?

____

____

____

____

Name______________________________ Class__________________ Date ______________

29. **Applying Concepts** Each population in Figure 1-1 is made up of what level of organisms?

____

____

30. **Applying Concepts** A population of rabbits lives on the other side of the prairie shown in the figure. Are the rabbits considered part of this community? Explain your answer.

____

____

____

Essay

Write the answer to each question in the space provided.

31. What is the difference between inference and observation?

32. Can a theory change over time? Explain your answer.

Name________________________ Class________________ Date ____________

33. How would one determine whether something was living or nonliving?

34. Under what conditions would you use a compound light microscope and each type of electron microscope?

35. Compare and contrast two common laboratory techniques: cell culturing and cell fractionation.

Name______________________________ Class_________________ Date _____________

Multiple Choice

Write the letter that best answers the question or completes the statement on the line provided.

____ **1.** The work of scientists begins with
a. testing a hypothesis.
b. careful observations.
c. creating experiments.
d. drawing conclusions.

____ **2.** Hypotheses may arise from
a. prior knowledge.
b. logical inferences.
c. imaginative guesses.
d. all of the above

____ **3.** A controlled experiment allows the scientist to isolate and test
a. a conclusion. c. several variables.
b. a mass of information. d. a single variable.

____ **4.** Scientists publish the details of important experiments so that
a. their work can be repeated.
b. their experimental procedures can be reviewed.
c. others can try to reproduce the results.
d. all of the above

____ **5.** A well-tested explanation that unifies a broad range of observations is a(an)
a. hypothesis. c. inference.
b. theory. d. controlled experiment.

____ **6.** All of the following are characteristics of all living things EXCEPT
a. growth. c. movement.
b. reproduction. d. use of energy.

____ **7.** Biology is the study of
a. the land, water, and air on Earth.
b. living things.
c. animals and plants only.
d. the environment.

____ **8.** The process by which organisms keep their internal conditions fairly constant is called
a. homeostasis. c. metabolism.
b. evolution. d. photosynthesis.

____ **9.** What is the term for a group of organisms of one type living in the same place?

a. biosphere
c. population
b. ecosystem
d. environment

____**10.** In the metric system, the basic unit of length is the

a. centimeter.
c. millimeter.
b. kilometer.
d. meter.

____**11.** Which is NOT a unit of measurement in the International System of Units, or SI?

a. meter
c. liter
b. ounce
d. gram

____**12.** An instrument that allows light to pass through the specimen and uses two lenses to form an image is a(an)

a. compound light microscope.
b. electron microscope.
c. TEM.
d. SEM.

____**13.** What is the term given to a group of cells that develop from a single original cell?

a. community
c. nutrient solution
b. cell culture
d. cell fractionation

____**14.** What technique is used to separate the different cell parts?

a. microscopy
c. cell fractionation
b. cell culture
d. all of the above

____**15.** Safety procedures are important when working

a. in a laboratory.
c. with animals.
b. in the field.
d. all of the above

Completion

Complete each statement on the line provided.

16. The information gathered from observation is called ____________________ .

17. The smallest units that are considered to be alive are called ____________________ .

18. A revised version of the original ____________________ system is called SI, an abbreviation for the Système International d'Unités.

19. The ____________________ microscope is generally used in high-school laboratories.

20. An instrument used to separate cell parts is called a(an) ____________________ .

Name______________________ Class________________ Date ____________

Short Answer

In complete sentences, write the answers to the questions on the lines provided.

21. What is an inference?

22. What is a hypothesis?

23. What are five of the levels of organization that biologists study?

24. What does cell fractionation allow a scientist to study?

25. What safety precaution must you carry out after every scientific activity?

Name________________________ Class__________________ Date ______________

Using Science Skills

Use the table below to answer the following questions on the lines provided.

Common Metric Units	
Length	**Mass**
1 meter (m) = 100 centimeters (cm)	1 kilogram (kg) = 1000 grams (g)
1 meter = 1000 millimeters (mm)	1 gram = 1000 milligrams (mg)
1000 meters = 1 kilometer (km)	1000 kilograms = 1 metric ton (t)
Volume	**Temperature**
1 liter (L) = 1000 milliliters (mL)	0°C = freezing point of water
1 liter = 1000 cubic centimeters (cm^3)	100°C = boiling point of water

Figure 1-1

26. **Using Tables and Graphs** What four common metric units in Figure 1-1 are used to measure length?

____

____

27. **Applying Concepts** Referring to Figure 1-1, why are conversions easier to do using the metric system rather than using traditional English units, such as inches, feet, and yards?

____

____

____

____

28. **Using Tables and Graphs** What is the boiling point of water, in degrees Celsius?

____

29. **Applying Concepts** Using Figure 1-1, what number does the prefix *kilo-* represent?

____

30. **Calculating** If you have 2 L of water, how many milliliters do you have?

____

Name________________________ Class________________ Date ____________

Multiple Choice

Write the letter that best answers the question or completes the statement on the line provided.

____ **1.** The space surrounding the nucleus of an atom contains

a. protons. c. neutrons.
b. electrons. d. ions.

____ **2.** Which of the following is a use of radioactive isotopes?

a. can determine the ages of rocks and fossils
b. can be used to treat cancer and kill bacteria that cause food to spoil
c. used as "tracers" to follow the movements of substances within organisms
d. all of the above

____ **3.** Which of the following makes up a molecule of water?

a. one atom of hydrogen and one atom of oxygen
b. one atom of sodium and one atom of chlorine
c. one atom of hydrogen and two atoms of oxygen
d. two atoms of hydrogen and one atom of oxygen

____ **4.** Which of the following statements about a compound is true?

a. The physical and chemical properties of a compound are usually very different from those of the elements from which it is formed.
b. Only the physical properties of a compound are usually the same as those of the elements from which it is formed.
c. Only the chemical properties of a compound are usually the same as those of the elements from which it is formed.
d. The physical and chemical properties of a compound are usually the same as those of the elements from which it is formed.

____ **5.** What type of electron is available to form bonds?

a. valence c. ionic
b. nucleus d. covalent

____ **6.** What type of ion forms when an atom loses electrons?

a. neutral c. negative
b. positive d. possibly positive or negative

____ **7.** Ice floats on water because

a. of cohesion.
b. ice has a higher density than water.
c. water shrinks when it freezes.
d. water expands when it freezes.

Name______________________________ Class__________________ Date ______________

____ 8. The most abundant compound in most living things is
a. carbon dioxide. c. sodium chloride.
b. water. d. sugar.

____ 9. When salt is dissolved in water, water is the
a. reactant. c. solute.
b. solution. d. solvent.

____10. A substance with a pH of 6 is called
a. an acid. c. both an acid and a base.
b. a base. d. neither an acid nor a base.

____11. A monosaccharide is a
a. carbohydrate. c. nucleic acid.
b. lipid. d. protein.

____12. Which statement is true?
a. Simple sugars are made of polysaccharides.
b. Glycerol is made of fatty acids.
c. RNA molecules are made of nucleotides.
d. Amino acids are made of proteins.

____13. When hydrogen and oxygen combine to form water, water would be
a. a product. c. both a product and a reactant.
b. a reactant. d. neither a product nor a reactant.

____14. If a reaction in one direction releases energy, the reaction in the opposite direction
a. also releases energy. c. destroys energy.
b. absorbs energy. d. cannot occur.

____15. Enzymes affect the reactions in living cells by changing the
a. products of the reaction.
b. speed of the reaction.
c. temperature of the reaction.
d. pH of the reaction.

Completion

Complete each statement on the line provided.

16. If an atom contains 15 protons, it must contain 15 ____________________.

17. Due to ____________________ forces, the design of a gecko's feet enables it to climb up vertical surfaces.

18. The stomach produces a(an) ____________________ to help digest food.

19. The pH scale is a measurement system that indicates the concentration of ____________________ in solution.

20. Chemical reactions that ____________________ energy will not occur without a source of energy.

Name________________________ Class________________ Date ____________

Short Answer

In complete sentences, write the answers to the questions on the lines provided.

21. What is mass number?

22. What accounts for water's properties of adhesion and cohesion?

23. Use the terms *solvent* and *solute* in describing how to prepare a salt solution.

24. Distinguish between RNA and DNA.

25. Name two essential roles that enzymes play in cells.

Name________________________ Class________________ Date ____________

Using Science Skills

Use the table below to answer the following questions on the lines provided.

pH Values of Some Common Substances

Substance	pH
hydrochloric acid	1.0
sulfuric acid	1.2
tomatoes	4.2
rainwater	6.2
pure water	7.0
sea water	8.5
ammonium chloride	11.1
sodium hydroxide	13.0

Figure 2-1

26. **Applying Concepts** What is the strongest acid listed in Figure 2-1?

____

27. **Applying Concepts** What is the pH of the weakest acid listed in Figure 2-1?

____

28. **Applying Concepts** What is the pH of the strongest base listed in Figure 2-1?

____

29. **Applying Concepts** According to the pH values of Figure 2-1, does a solution with a hydrogen ion concentration less than that of pure water have a pH greater or less than 7?

____

30. **Calculating** A change of one unit on the pH scale represents a tenfold increase in the concentration of hydrogen ions. According to the pH values listed in Figure 2-1, how much greater is the hydrogen ion concentration in tomatoes than in rainwater?

____

Name______________________________ Class_________________ Date _____________

Essay

Write the answer to each question in the space provided.

31. Compare protons, electrons, and neutrons with respect to location within atoms, electric charge, and mass.

32. What relationship exists between the mass number of an element and the isotopes of that element?

33. Explain the difference between ionic compounds and covalently bonded compounds.

Name________________ Class________________ Date ____________

34. Compare and contrast adhesion and cohesion, using capillary action as an example.

35. Compare enzyme activity to a lock and key.

Name______________________________ Class__________________ Date ______________

Multiple Choice

Write the letter that best answers the question or completes the statement on the line provided.

____ **1.** The three particles that make up an atom are
a. protons, neutrons, and isotopes.
b. neutrons, isotopes, and electrons.
c. positives, negatives, and electrons.
d. protons, neutrons, and electrons.

____ **2.** The nucleus is made of
a. protons and electrons.
b. electrons and neutrons.
c. protons and neutrons.
d. protons, neutrons, and electrons.

____ **3.** Isotopes are atoms of the same element with the same number of protons and
a. a different number of electrons.
b. a different number of molecules.
c. a different number of neutrons.
d. the same number of neutrons.

____ **4.** Which of the following terms describes a substance formed by the combination of two or more elements in definite proportions?
a. compound c. nucleus
b. isotope d. enzyme

____ **5.** A covalent bond is formed as the result of
a. transferring electrons.
b. sharing electrons.
c. transferring protons.
d. sharing protons.

____ **6.** Water molecules are polar, with
a. the oxygen side being slightly positive and the hydrogen side being slightly negative.
b. the oxygen and hydrogen sides being slightly positive.
c. the oxygen and hydrogen sides being slightly negative.
d. the oxygen side being slightly negative and the hydrogen side being slightly positive.

____ 7. A solution is a(an)
a. breaking of a chemical bond.
b. chemical reaction.
c. evenly distributed mixture of two or more substances.
d. combination of two or more liquids.

____ 8. Suspensions are mixtures
a. of water and nondissolved material.
b. in which the components are evenly distributed throughout the solution.
c. both a and b
d. neither a nor b

____ 9. Solutions that contain concentrations of H^+ ions lower than pure water
a. have pH values below 7.
b. are acids.
c. are bases.
d. are enzymes.

____10. Which of the following organic compounds is the main source of energy for living things?
a. carbohydrates
b. lipids
c. nucleic acids
d. proteins

____11. Which of the following is NOT a function of proteins?
a. store and transmit heredity
b. help to fight disease
c. control the rate of reactions and regulate cell processes
d. used to form bones and muscles

____12. What is the process that changes one set of chemicals into another set of chemicals?
a. cohesion
b. adhesion
c. chemical reaction
d. dissolving

____13. What is the term used to describe the energy needed to get a reaction started?
a. adhesion energy
b. activation energy
c. cohesion energy
d. chemical energy

____14. Chemical reactions that release energy
a. will not occur.
b. will never explode.
c. will always explode.
d. often occur spontaneously.

____15. A substance that speeds up the rate of a chemical reaction is called a(an)
a. catalyst.
b. lipid.
c. molecule.
d. element.

Name_______________________ Class_______________ Date ___________

Completion

Complete each statement on the line provided.

16. The subatomic particles that make up atoms are protons, neutrons, and _______________ .

17. Because they have the same number of protons and electrons, all isotopes of an element have the same _______________ properties.

18. A chemical bond formed by the transfer of electrons is a(an) _______________ bond.

19. A water molecule is polar because there is an uneven distribution of electrons between the oxygen and _______________ atoms.

20. The elements or compounds produced by a chemical reaction are known as _______________ .

Short Answer

In complete sentences, write the answers to the questions on the lines provided.

21. What is a molecule?

____

____

22. What are the main types of chemical bonds?

____

____

23. What is a mixture?

____

____

____

____

24. What are the four groups of organic compounds found in living things?

____

____

25. What is one of the most important factors in determining whether a chemical reaction will occur?

____

____

Name_________________ Class_________________ Date_________________

Using Science Skills

Use the table below to answer the following questions on the lines provided.

Element	Symbol	Protons	Neutrons	Electrons	Atomic Number	Mass Number
Hydrogen	H	1			1	
Helium	He	2				4
Carbon	C		6		6	
Oxygen	O		8	8		
Neon	Ne			10	10	20
Aluminum	Al	13				27
Zinc	Zn			30	30	65

Figure 2-1

26. **Calculating** Based on Figure 2-1, what is the mass number of carbon?

____

27. **Applying Concepts** Based on Figure 2-1, what is the atomic number of oxygen?

____

28. **Applying Concepts** Using Figure 2-1, how many electrons does an atom of aluminum contain?

____

29. **Applying Concepts** According to Figure 2-1, an atom of which element contains two neutrons?

____

30. **Applying Concepts** From Figure 2-1, which element has a mass number of 16?

____

Name______________________ Class______________ Date __________

Multiple Choice

Write the letter that best answers the question or completes the statement on the line provided.

____ **1.** Which of the following is NOT a goal of science?
a. to explain events in the natural world
b. to propose explanations that are proven correct
c. to use explanations to make useful predictions
d. to investigate and understand the natural world

____ **2.** In science, data are
a. always expressed in numbers.
b. gathered from observations.
c. logical interpretations based on experience.
d. usually descriptive.

____ **3.** In a controlled experiment in which the amount of water plants receive is the manipulated variable, which of the following could be the responding variable?
a. amount of sunlight plants receive
b. soil composition in which plants grow
c. time over which plants are observed
d. growth in centimeters over a certain period

____ **4.** Which of the following unifies a broad range of observations?
a. hypothesis
b. controlled experiment
c. theory
d. inference

____ **5.** Which of the following is NOT a characteristic of all living things?
a. ability to maintain a stable internal environment
b. ability to sexually reproduce
c. ability to move in response to the environment
d. ability to grow and develop

____ **6.** Why is biology divided into different fields?
a. No single biologist could study the diversity of life in the biosphere.
b. Populations are not considered part of ecosystems.
c. Living things are alike in essential ways.
d. No field of biology can focus on a certain level of organization.

Name________________________ Class________________ Date ____________

____ 7. Which of the following is NOT an advantage of using an electron microscope?

a. great magnification
b. ability to produce images of objects smaller than a cell
c. ability to see movement within a living cell
d. ability to observe objects smaller than 0.2 micrometers

____ 8. Why do all isotopes of an element have the same chemical properties?

a. They all have the same number of electrons.
b. They all have the same valence protons.
c. They all have a different number of neutrons.
d. They all have the same atomic mass.

____ 9. When atoms combine to form compounds, they do so by transferring or sharing

a. neutrons. c. protons.
b. electrons. d. ions.

____10. Which of the following statements about water is true?

a. Water is one of the few compounds that is a gas at the temperatures found over much of Earth's surface.
b. A water molecule is polar.
c. A water molecule has more protons than electrons.
d. Because of an uneven distribution of electrons between atoms, water molecules cannot form hydrogen bonds.

____11. When sugar is dissolved in water, the sugar is the

a. solute. c. buffer.
b. solvent. d. suspension.

____12. Which is NOT a group of organic compounds found in living things?

a. lipids c. proteins
b. nucleic acids d. macromolecules

____13. Which statement is true?

a. Monosaccharides are formed from polysaccharides.
b. Proteins are polymers of amino acids.
c. Nucleotides consist of fats, oils, and waxes.
d. Lipids are generally soluble in water.

____14. When carbon and oxygen combine to form carbon dioxide, what are the reactants?

a. carbon and oxygen
b. oxygen and carbon dioxide
c. energy, carbon, and oxygen
d. energy and carbon dioxide

Name______________________________ Class__________________ Date ____________

____15. What kind of chemical reactions often occur spontaneously?
  a. chemical reactions that absorb energy
  b. chemical reactions that use enzymes
  c. chemical reactions that require no activation energy
  d. chemical reactions that release energy

Completion

Complete each statement on the line provided.

16. A proposed scientific explanation for a set of observations is called a(an) ____________________.
17. A new organism has a single parent in ____________________ reproduction.
18. If an atom contains 7 electrons, it must contain 7 ____________________.
19. Living things use ____________________ as their main source of energy.
20. Chemical reactions always involve the formation of new ____________________ in products.

Short Answer

In complete sentences, write the answers to the questions on the lines provided.

21. How should a hypothesis be tested?

____

____

____

22. How can cell cultures be used in scientific studies?

____

____

____

____

23. What is the difference between adhesion and cohesion?

____

____

____

____

24. Describe a saltwater solution.

____

____

____

____

Name________________________ Class________________ Date ____________

25. How does an enzyme speed up a chemical reaction?

____

____

____

____

Using Science Skills

Use the table below to answer the following questions on the lines provided.

pH Values of Some Common Substances	
Substance	**pH**
Hydrochloric acid	1.0
Lemon juice	2.2
Carbonated beverages	3.0
Coffee	5.0
Pure water	7.0
Blood	7.4
Detergent	9.0
Bleach	12.0

Figure 1

26. **Applying Concepts** What is the weakest acid listed in Figure 1?

____

27. **Applying Concepts** What is the pH of the strongest base listed in Figure 1?

____

28. **Applying Concepts** According to Figure 1, is blood a base or an acid?

____

29. **Applying Concepts** According to Figure 1, does lemon juice have more or less H^+ ions than it has OH^- ions?

____

30. **Calculating** A change in one unit on the pH scale represents a tenfold increase in the concentration of hydrogen ions. According to the pH values listed in Figure 1, how much greater is the hydrogen ion concentration in coffee than in carbonated beverages?

____

Name________________________________ Class__________________ Date ______________

Essay

Write the answer to each question in the space provided.

31. What is the difference between a hypothesis and a theory?

32. Describe the process of differentiation in multicellular organisms.

33. Describe the images produced by a compound light microscope, a transmission electron microscope, and a scanning electron microscope.

34. Distinguish between the formation of an ionic bond and a covalent bond.

35. In an enzyme-catalyzed reaction, what role does the active site play in the reaction?

Name______________________________ Class __________________ Date ______________

Multiple Choice

Write the letter that best answers the question or completes the statement on the line provided.

____ **1.** Which of the following is a goal of science?
a. to investigate and understand the natural world
b. to explain events in the natural world
c. to use explanations of events to make useful predictions
d. all of the above

____ **2.** A logical interpretation based on prior knowledge or experience is a(an)
a. hypothesis.
b. observation.
c. inference.
d. theory.

____ **3.** In a controlled experiment, how many variables are changed at a time?
a. none
b. one
c. at least two
d. as many as possible

____ **4.** What are the smallest units that are considered to be alive?
a. molecules
b. cells
c. atoms
d. organisms

____ **5.** Which of the following levels of organization includes all of the others?
a. community
b. organism
c. population
d. biosphere

____ **6.** What does SI currently stand for?
a. International System of Units
b. System of Metrical Measurement
c. System of the Internet
d. Metric and English Units

____ **7.** What kind of microscope uses lenses to focus light that has passed through a specimen?
a. scanning electron microscope
b. compound light microscope
c. transmission electron microscope
d. all of the above

Name________________________________ Class__________________ Date ______________

____ 8. The negatively charged subatomic particles in atoms are
a. protons.
b. neutrons.
c. electrons.
d. ions.

____ 9. An ionic bond is formed as the result of
a. transferring electrons.
b. sharing electrons.
c. transferring protons.
d. sharing protons.

____10. What is the attraction between molecules of different substances?
a. cohesion
b. adhesion
c. polarity
d. polymerization

____11. What is a mixture in which all the components are evenly distributed throughout?
a. solution
b. solvent
c. solute
d. suspension

____12. The measurement system that indicates the concentration of H+ ions in solution is the
a. metric system.
b. SI scale.
c. pH scale.
d. responding variable.

____13. Starches and sugars are examples of what group of organic compounds?
a. nucleic acids
b. lipids
c. proteins
d. carbohydrates

____14. Which of the following is(are) a function of nucleic acids in living things?
a. stores and transmits genetic information
b. stores energy
c. transports substances into cells
d. serves as chemical messengers

____15. What kind of chemical reactions will not occur without a source of energy?

a. reactions that release energy

b. reactions that absorb energy

c. reactions that use enzymes

b. reactions that occur in living things

Completion

Complete each statement on the line provided.

16. A(an) ___________________ is a proposed scientific explanation for a set of observations.

17. In a controlled experiment, the variable that is deliberately changed is called the ___________________ variable.

18. In ___________________ reproduction, two cells from different parents unite to produce the first cell of the new organism.

19. A chemical ___________________ is a substance formed by the chemical combination of two or more elements.

20. A substance that speeds up the rate of a chemical reaction is called a(an) ___________________.

Short Answer

In complete sentences, write the answers to the questions on the lines provided.

21. What is a theory?

____

____

____

22. What are five of the eight characteristics that all living things share?

____

____

____

____

____

23. Why do all the isotopes of an element have the same chemical properties?

____

____

____

24. What is the difference between an acidic solution and a basic solution?

____

____

____

Name ______________________ Class ______________ Date ____________

25. What is a chemical reaction?

____

____

____

Using Science Skills

Use the table below to answer the following questions on the lines provided.

Common Metric Units	
Length	**Mass**
1 meter (m) = 100 centimeters (cm) 1 meter = 1000 millimeters (mm) 1000 meters = 1 kilometer (km)	1 kilogram (kg) = 1000 grams (g) 1 gram = 1000 milligrams (mg) 1000 kilograms = 1 metric ton (t)
Volume	**Temperature**
1 liter (L) = 1000 milliliters (mL) 1 liter = 1000 cubic centimeters (cm^3)	0°C = freezing point of water 100°C = boiling point of water

Figure 1

26. **Using Tables and Graphs** What four common metric units in Figure 1 are used to measure mass?

____

____

27. **Using Tables and Graphs** What is the freezing point of water, in degrees Celsius, shown in Figure 1?

____

28. **Applying Concepts** Using Figure 1, what number does the prefix *centi-* represent?

____

29. **Calculating** Based on Figure 1, if you drive 5 kilometers in your car, how many meters have you driven?

____

30. **Applying Concepts** Referring to Figure 1, why would it be easier to convert meters to centimeters than to convert yards to inches, as you would if you were using English units?

____

____

____

____

Name______________________________ Class__________________ Date ______________

Multiple Choice

Write the letter that best answers the question or completes the statement on the line provided.

____ **1.** Which of the following descriptions about the organization of an ecosystem is correct?

a. Communities make up species, which make up populations.

b. Populations make up species, which make up communities.

c. Species make up communities, which make up populations.

d. Species are grouped in populations, which make up communities.

____ **2.** The simplest grouping of more than one kind of organism in the biosphere is a(an)

a. population. c. ecosystem.

b. community. d. species.

____ **3.** Which ecological inquiry method is an ecologist using when he or she enters an area periodically to count the population numbers of a certain species?

a. questioning c. experimenting

b. observing d. modeling

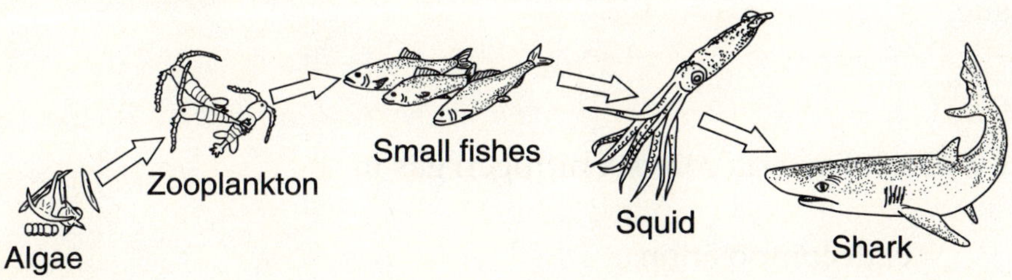

Figure 3-1

____**4.**The algae at the beginning of the food chain in Figure 3-1 are

a. consumers. c. producers.

b. decomposers. d. heterotrophs.

____ **5.** An organism that uses energy to produce its own food supply from inorganic compounds is called a(an)

a. heterotroph. c. detritivore.

b. consumer. d. autotroph.

____ **6.** Which of the following organisms does NOT require sunlight to live?

a. chemosynthetic bacteria

b. algae

c. trees

d. photosynthetic bacteria

____ **7.** The total amount of living tissue within a given trophic level is called the

a. organic mass. c. energy mass.
b. trophic mass. d. biomass.

____ **8.** What animals eat both producers and consumers?

a. herbivores c. chemotrophs
b. omnivores d. autotrophs

____ **9.** A snake that eats a frog that has eaten an insect that fed on a plant is a

a. first-level producer.
b. first-level consumer.
c. second-level producer.
d. third-level consumer.

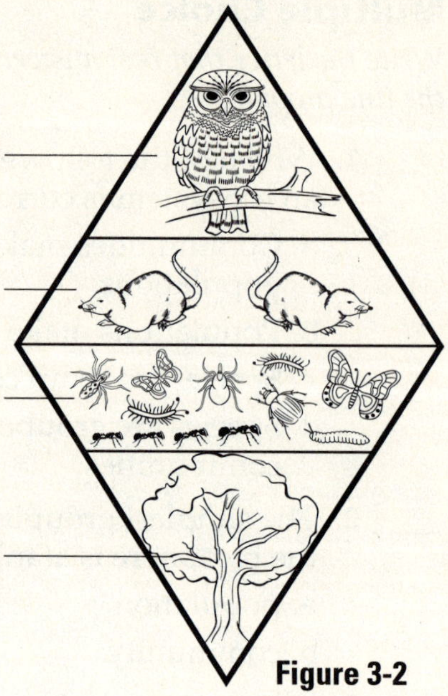

Figure 3-2

____**10.** The trophic levels in Figure 3-2 illustrate

a. the relative amount of energy.
b. the amount of living organic matter.
c. the relative number of individual organisms.
d. that producers outnumber first-level consumers.

____**11.** Only 10 percent of the energy stored in an organism can be passed on to the next trophic level. Of the remaining energy, some is used for the organism's life processes, and the rest is

a. used in reproduction.
b. stored as body tissue.
c. stored as fat.
d. eliminated as heat.

____**12.** What is the process by which bacteria convert nitrogen gas in the air to ammonia?

a. nitrogen fixation c. decomposition
b. excretion d. denitrification

____**13.** Carbon cycles through the biosphere in all of the following processes EXCEPT

a. photosynthesis. c. respiration.
b. transpiration. d. decomposition.

____**14.** The movements of energy and nutrients through living systems are different because

a. energy flows in one direction and nutrients recycle.
b. energy is limited in the biosphere and nutrients are always available.
c. nutrients flow in one direction and energy recycles.
d. energy forms chemical compounds and nutrients are lost as heat.

Name______________________________ Class__________________ Date ______________

____15. Which is most likely to be a limiting nutrient in a freshwater pond?

a. phosphorus c. carbon
b. nitrogen d. potassium

Completion

Complete each statement on the line provided.

16. Autotrophs capture energy from sunlight or ____________________ to produce food.

17. Organisms that break down organic matter and return it to the environment are called ____________________.

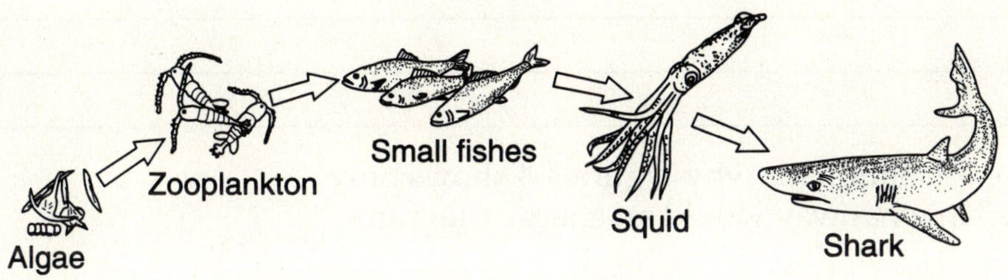

Figure 3-3

18. Of the organisms represented in Figure 3-3, the organisms in the oceans with the smallest total biomass are most likely the ____________________.

19. Water can enter the atmosphere through the processes of evaporation and ____________________.

20. Living organisms require nitrogen to make ____________________, which are used to build proteins.

Short Answer

In complete sentences, write the answers to the questions on the lines provided.

21. Describe the role of algae in ocean food chains.

Name________________________ Class________________ Date ____________

22. Describe the flow of energy among the following members of an ecosystem: decomposers, autotrophs, heterotrophs, and the sun.

____

____

____

____

23. Describe the flow of energy to the owl in Figure 3-4 if the tree provides 1500 calories of energy to the insects.

____

____

____

____

____

24. What is the most likely explanation for why Figure 3-4 shows only one organism at its base? In what way would an energy diagram be different?

____

____

____

____

25. Explain how seepage and transpiration in Figure 3-5 are related.

____

____

____

____

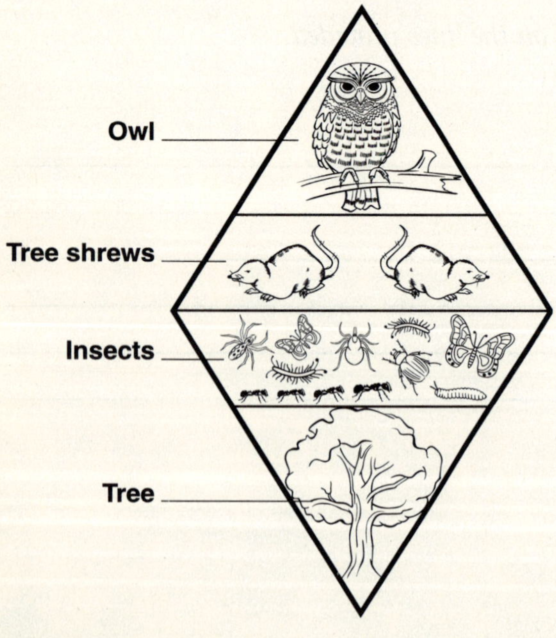

Figure 3-4

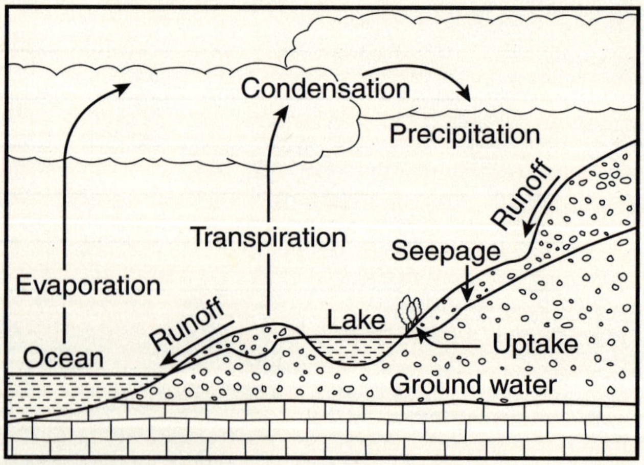

Figure 3-5

Name______________________________ Class__________________ Date ____________

Using Science Skills

Use the diagrams below to answer the following questions on the lines provided.

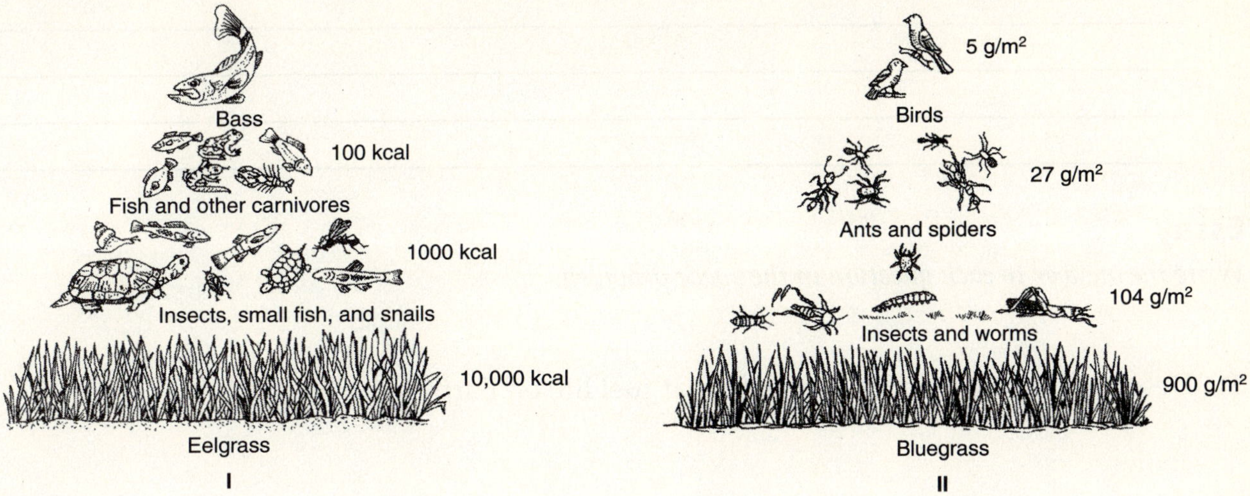

Figure 3-6

26. **Predicting** How might a large omnivore change the flow of energy in Figure 3-6, diagram II?

____

____

____

____

27. **Inferring** If a limiting nutrient is supplied to the producers in Figure 3-6, diagram II, what effect could it have on the birds?

____

____

____

____

28. **Interpreting Graphics** How many kilocalories (kcal) can the top carnivore in Figure 3-6, diagram I, store? Explain.

____

____

____

____

29. **Applying Concepts** What three scientific approaches do ecologists use to explain complex relationships, such as in the energy pyramid in Figure 3-6?

____

____

30. Applying Concepts Describe how chemosynthesis could contribute to the energy represented by Figure 3-6, diagram I.

____

____

____

____

____

Essay

Write the answer to each question in the space provided.

31. Describe the two sources of energy that fuel life on Earth.

32. How does a food web differ from a food chain?

33. Describe the three types of ecological pyramids.

34. Describe the roles of bacteria in the nitrogen cycle.

35. How might a large input of phosphorus affect a freshwater lake over time?

Name________________________ Class_________________ Date ___________

Multiple Choice

Write the letter that best answers the question or completes the statement on the line provided.

____ **1.** The branch of biology dealing with interactions among organisms and between organisms and their environment is called

a. economy. c. recycling.
b. modeling. d. ecology.

____ **2.** What is the combined portions of Earth called in which all living things exist?

a. biome c. ecosystem
b. community d. biosphere

____ **3.** All of the members of a particular species that live in the same area are called a(an)

a. biome. c. community.
b. population. d. ecosystem.

____ **4.** Which of the following is NOT a basic method used by ecologists to study the living world?

a. experimenting c. modeling
b. classifying d. observing

____ **5.** Green plants are

a. producers. c. herbivores.
b. consumers. d. omnivores.

____ **6.** What is the original source of almost all the energy in most ecosystems?

a. carbohydrates c. water
b. sunlight d. carbon

____ **7.** An organism that cannot make its own food is called a(an)

a. heterotroph. c. autotroph.
b. chemotroph. d. producer.

____ **8.** Organisms that obtain nutrients by breaking down dead and decaying plants and animals are called

a. decomposers. c. autotrophs.
b. omnivores. d. producers.

____ **9.** What is an organism that feeds only on plants called?

a. carnivore c. omnivore
b. herbivore d. detritivore

____ **10.** All the interconnected feeding relationships in an ecosystem make up a food

a. interaction. c. network.
b. chain. d. web.

Name________________________________ Class __________________ Date ______________

____11. Which type of pyramid shows the amount of living tissue at each trophic level in an ecosystem?

a. a numbers pyramid c. a biomass pyramid

b. an energy pyramid d. a food pyramid

____12. The repeated movement of water between Earth's surface and the atmosphere is called

a. the water cycle.

b. the condensation cycle.

c. precipitation.

d. evaporation.

____13. Nitrogen fixation is carried out primarily by

a. humans. c. bacteria.

b. plants. d. consumers.

____14. Organisms need nutrients in order to

a. utilize hydrogen and oxygen.

b. carry out essential life functions.

c. recycle chemical compounds.

d. carry out nitrogen fixation.

____15. The rate at which organic matter is created by producers in an ecosystem is called

a. a limiting nutrient. c. an algal bloom.

b. fertilization. d. primary productivity.

Completion

Complete each statement on the line provided.

16. A group of ecosystems that have the same climate and dominant communities is called a(an) __________________ .

17. Scientists classify the nitrogen, carbon, and water cycles as __________________ cycles.

18. Plant-eating animals such as cows are called __________________ .

19. Only about 10 percent of the __________________ in a trophic level is available to organisms at the next level.

20. The chemical substances that an organism requires to live are called __________________ .

Name______________________ Class______________ Date __________

Short Answer

In complete sentences, write the answers to the questions on the lines provided.

21. What is the biosphere? Where is it located?

____

____

____

____

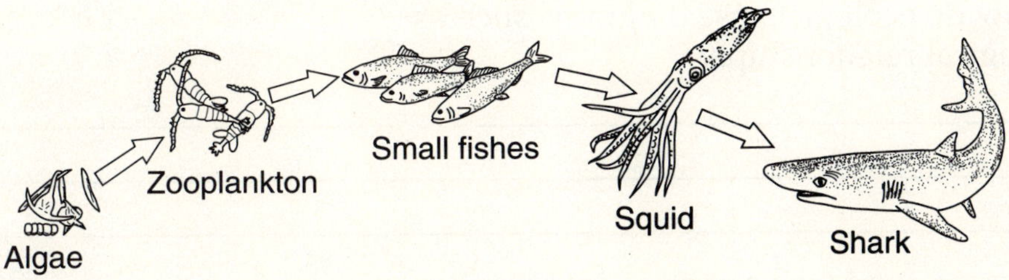

Figure 3-1

22. Using Figure 3-1, explain the relationship between sharks and the sun.

____

____

____

____

23. Compare the movement of energy in the biosphere with the movement of matter through the biosphere.

____

____

____

____

24. Using Figure 3-2, trace the path of water that leaves a lake through evaporation, and describe how it might return to the lake.

____

____

____

____

____

____

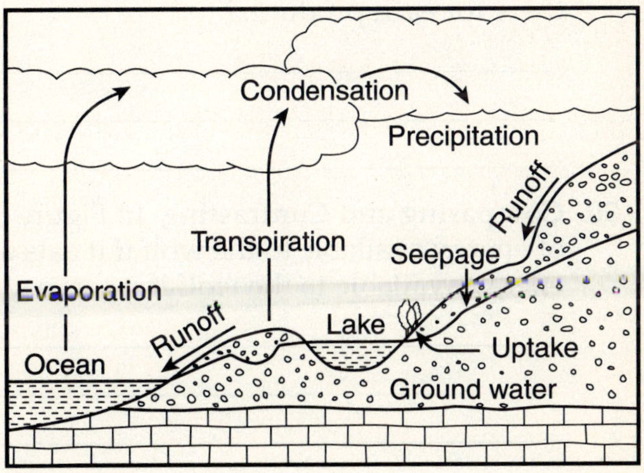

Figure 3-2

Name________________________________ Class__________________ Date ______________

25. What events typically contribute to an algal bloom in a lake or an ocean?

____

____

____

Using Science Skills

Use the diagram to answer the following questions on the lines provided.

26. **Applying Concepts** How do ecologists use diagrams such as Figure 3-3 to study ecological relationships?

____

____

____

27. **Inferring** Figure 3-3 shows a food web arranged into trophic levels. How many energy-transferring steps away from the sun is the wolf? How do you know?

____

____

____

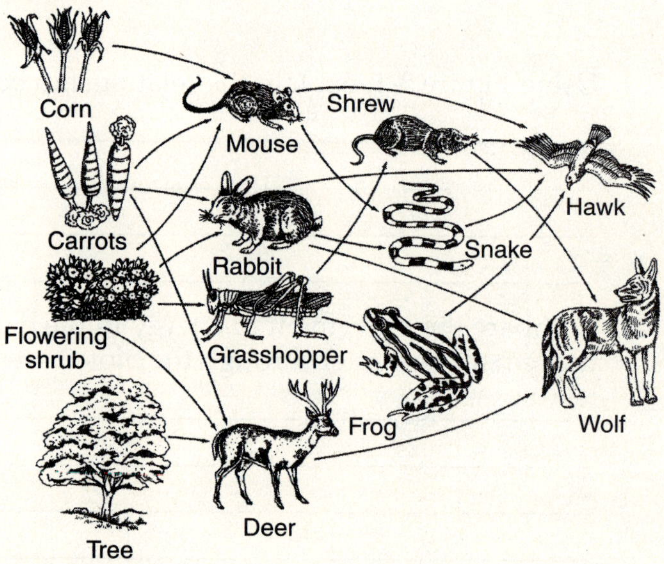

Figure 3-3

28. **Inferring** A food web, such as the one in Figure 3-3, is a model of the feeding relationships in an ecosystem. What makes this model representative of an ecosystem?

____

____

____

29. **Interpreting Graphics** In Figure 3-3, how many first-level consumers are there for each producer?

____

____

____

30. **Comparing and Contrasting** In Figure 3-3, compare the amount of energy available to the wolf if it eats a rabbit with the amount of energy available to the wolf if it eats a shrew.

____

____

____

Name________________________ Class________________ Date ____________

Multiple Choice

Write the letter that best answers the question or completes the statement on the line provided.

____ 1. Climate is a global factor that produces
a. Earth's unique ocean and atmosphere.
b. the shape and elevation of landmasses.
c. a wide range of environmental conditions that shapes communities.
d. solar energy within the atmosphere.

____ 2. The loss of heat to space is slowed by
a. radiation entering the atmosphere.
b. atmospheric gases.
c. solar energy.
d. the biosphere.

____ 3. Earth has three main climate zones because of the differences in latitude and
a. amount of solar energy received.
b. angle of heating.
c. ocean currents.
d. prevailing winds.

____ 4. The unequal heating of Earth's surface
a. drives wind and ocean currents.
b. causes winds that transport heat throughout the biosphere.
c. has important effects on Earth's climate regions.
d. all of the above

____ 5. Which is a biotic factor that affects the size of a population in a specific ecosystem?
a. average temperature of the ecosystem
b. type of soil in the ecosystem
c. number and kinds of predators in the ecosystem
d. concentration of oxygen in the ecosystem

____ 6. Several species of warblers can live in the same spruce tree ONLY because they
a. have different habitats within the tree.
b. eat different foods within the tree.
c. occupy different niches within the tree.
d. can find different temperatures within the tree.

____ 7. Different species can share the same habitat, but competition among them is reduced if they

a. reproduce at different times.

b. eat less.

c. increase their populations.

d. occupy different niches.

____ 8. The symbiotic relationship between a flower and the insect that feeds on its nectar is an example of

a. mutualism because the flower provides the insect with food, and the insect pollinates the flower.

b. parasitism because the insect lives off the nectar from the flower.

c. commensalism because the insect doesn't harm the flower, and the flower doesn't benefit from the relationship.

d. predation because the insect feeds on the flower.

____ 9. What is one difference between primary and secondary succession?

a. Primary succession is slow, and secondary succession is rapid.

b. Secondary succession begins on soil, and primary succession begins on newly exposed surfaces.

c. Primary succession modifies the environment, and secondary succession does not.

d. Secondary succession begins with lichens, and primary succession begins with trees.

____10. Which factor can influence continual change in an ecosystem?

a. further disturbances

b. long-term climate changes

c. introduction of nonnative species

d. all of the above

____11. An example of a place with a microclimate is

a. a mountain range capped with ice.

b. a forested park in a desert city.

c. an orchid growing in a rain forest.

d. coniferous trees in a temperate forest.

____12. Which two biomes have the least amount of precipitation?

a. tropical rain forest and temperate grassland

b. tropical savanna and tropical dry forest

c. tundra and desert

d. boreal forest and temperate woodland and shrubland

Name________________________ Class________________ Date ____________

____**13.** The chemistry of aquatic ecosystems is determined by the
- a. amount of salts, nutrients, and oxygen dissolved in the water.
- b. number of other organisms present in the water.
- c. amount of rainfall the water receives.
- d. biotic and abiotic factors in the water.

____**14.** Which is NOT an adaptation that organisms have for living in flowing water?
- a. hooks
- b. tentacles
- c. streamlined bodies
- d. suckers

____**15.** Which of the following statements is NOT true about the oceanic zone?
- a. The open ocean has very low levels of nutrients.
- b. Organisms in the deep oceanic zone are exposed to frigid temperatures and total darkness.
- c. The oceanic zone begins at the low-tide mark and extends to the end of the continental shelf.
- d. Most of the photosynthetic activity on Earth occurs in the open ocean within the photic zone.

Completion

Complete each statement on the line provided.

16. Weather differs from ________________ in that weather can change on a daily basis and is not considered a description of year-to-year conditions of temperature and precipitation.

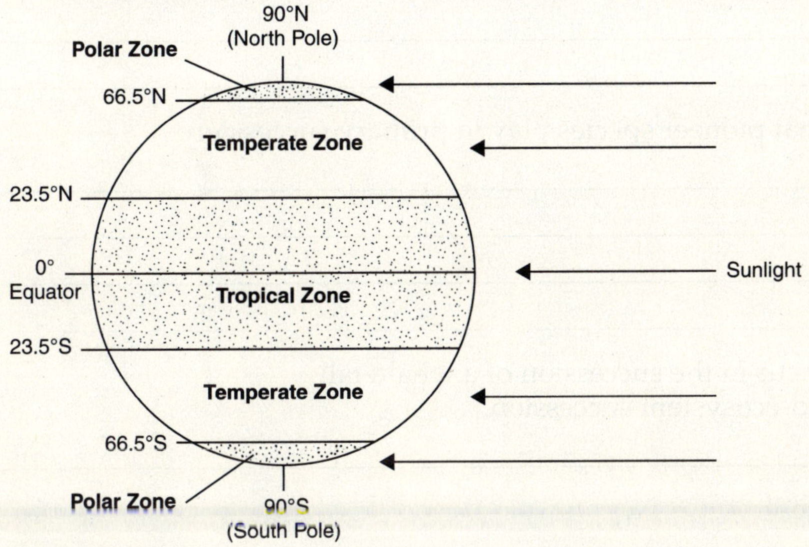

Figure 4-1

17. According to Figure 4-1, the north polar zone lies between ________________ latitude and the North Pole.

18. As shown in Figure 4-1, the climate zone that receives the most direct sunlight throughout the year lies between _____________ and _____________ latitudes.

19. An aquatic environment's water _____________ refers to the amount of dissolved salts, nutrients, and oxygen in the water.

20. A swamp is different from a lake as a freshwater ecosystem because a swamp is a(an) _____________ ecosystem.

Short Answer

In complete sentences, write the answers to the questions on the lines provided.

21. List five factors that contribute to global climate patterns.

22. Using Figure 4-1 on the previous page, explain why average temperatures decrease with increasing distance from the equator.

23. Explain why the food that a bullfrog eats is part of its niche.

24. Describe an important role that pioneer species play in primary succession.

25. Explain why plants do not occur in the succession of a whale-fall community as they do in land-ecosystem succession.

Using Science Skills

Use the diagram below to answer the following questions on the lines provided.

26. Applying Concepts Describe some of the biotic and abiotic factors you might find at point A and in region B of the flowing-water ecosystem shown in Figure 4-2.

27. Interpreting Graphics The boreal forest and river valley depicted in Figure 4-2 were swept by fire 20 years ago. The forest on the hills on each side of the river valley were destroyed. Does this illustration show an earlier or later stage of succession? Which kind of succession has taken place? Explain what will happen in this ecosystem if there are no more disturbances.

28. Predicting Examine Figure 4-2. Predict what might happen to this flowing-water ecosystem in a boreal forest biome if a dam was built on the river in region C.

29. Inferring Which regions of the river in Figure 4-2 contain more nutrients than the other regions? Explain.

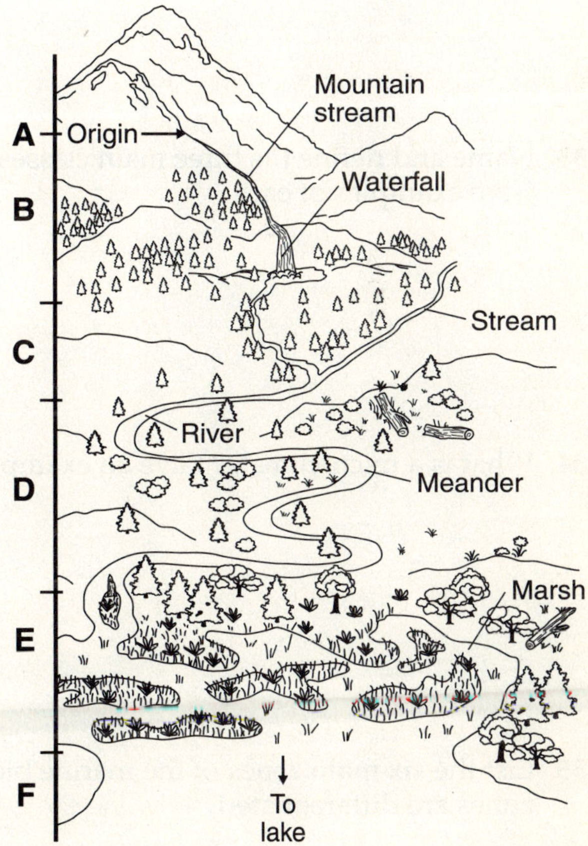

Figure 4-2

Name______________________________ Class________________ Date ____________

30. **Comparing and Contrasting** Compare and contrast region E in Figure 4-2 with region D. Describe the physical characteristics of each region.

____

____

____

Essay

Write the answer to each question in the space provided.

31. Explain how ocean currents originate and affect Earth's climates.

32. Describe the greenhouse effect and explain how it maintains Earth's temperature range.

33. Name and define the three main classes of symbiotic relationships. Give examples of each.

34. What is a microclimate? Give an example.

35. List the six main zones of the marine biome. Explain how these zones are differentiated.

Name________________________ Class________________ Date ____________

Multiple Choice

Write the letter that best answers the question or completes the statement on the line provided.

____ **1.** The average year-after-year conditions of temperature and precipitation in a particular region is the region's
a. weather. c. ecosystem.
b. latitude. d. climate.

____ **2.** Temperatures on Earth remain within a suitable range for life because of the
a. unequal heating of Earth's surface.
b. loss of heat to space.
c. radiation of sunlight back into the atmosphere.
d. greenhouse effect.

____ **3.** Cool air over the poles will
a. rise. c. absorb heat from the equator.
b. sink. d. flow parallel to Earth's surface.

____ **4.** The tendency for warm air to rise and cool air to sink results in
a. global wind patterns.
b. ocean upwelling.
c. unequal heat distribution.
d. climate zones.

____ **5.** Each of the following is an abiotic factor in the environment EXCEPT
a. plant life. c. rainfall.
b. soil type. d. temperature.

____ **6.** An organism's niche is
a. the way the organism uses the range of physical and biological conditions in which it lives.
b. all the physical factors in the organism's environment.
c. the range of temperatures that the organism needs to survive.
d. a full description of the place an organism lives.

____ **7.** An interaction in which one organism captures and feeds on another organism is called
a. competition. c. mutualism.
b. symbiosis. d. predation.

____ **8.** Symbiosis in which both species benefit is called
a. commensalism. c. predation.
b. mutualism. d. parasitism.

____ 9. The series of predictable changes that occur in a community over time is called
a. population growth.
b. ecological succession.
c. climax community.
d. climate change.

____10. Primary succession can begin after
a. a forest fire.
b. a lava flow.
c. farmland is abandoned.
d. a severe storm.

____11. Climate conditions in a small area that differ from the climate of the surrounding area are called
a. natural features.
b. microclimates.
c. biomes.
d. ecosystems.

____12. Which biome is characterized by very low temperatures, little precipitation, and permafrost?
a. desert
b. temperate forest
c. tundra
d. tropical dry forest

____13. Aquatic ecosystems are classified by all of the following EXCEPT
a. depth and flow of the water.
b. organisms that live there.
c. temperature of the water.
d. chemistry of the water.

____14. Ponds and lakes are
a. wetlands.
b. estuaries.
c. standing-water ecosystems.
d. flowing-water ecosystems.

____15. The photic zone
a. extends to the bottom of the open ocean.
b. extends to a depth of about 200 meters.
c. is deep, cold, and permanently dark.
d. is where chemosynthetic bacteria are the producers.

Completion

Complete each statement on the line provided.

16. The energy of incoming ____________________ drives Earth's weather and helps to determine climate.

17. Atmospheric gases that trap heat inside Earth's atmosphere are called ____________________ gases.

18. Organisms within an ecosystem are ____________________ factors in that ecosystem.

19. A small city park that is sunnier and windier than the climate of the surrounding areas has it own ____________________ .

20. Single-celled algae that grow in lakes and ponds are called ____________________ .

Short Answer

In complete sentences, write the answers to the questions on the lines provided.

21. How are atmospheric gases like the glass in a greenhouse?

____

____

22. What are greenhouse gases?

____

____

23. Describe a climate you might find in the tropical zone.

____

____

24. What is an ecological resource?

____

____

25. What abiotic factor determines the amount of light received by an aquatic ecosystem?

____

____

Using Science Skills

Use the table on the next page to answer the following questions on the lines provided.

26. **Comparing and Contrasting** In Figure 4-1, compare the average daytime temperature of the tropical rain forest with the average summer temperature of the tundra. How can you account for the difference?

____

____

____

27. **Calculating** What is the highest average temperature shown in Figure 4-1? the lowest average temperature? How many degrees of temperature are there between the highest and the lowest temperatures? What global phenomenon maintains this range of temperatures?

____

____

Name______________________ Class________________ Date ____________

Average Rainfall and Temperature of Earth's Biomes

Biome	Average Yearly Rainfall	Average Temperature	Climate Zone
Tropical Rain Forest	400 cm	Daytime: 34°C Nighttime: 20°C	Tropical
Tropical Dry Forest	250–300 cm	Dry season: 32°C Wet season: 20°C	Tropical
Temperate Forest	75–125 cm	Summer: 28°C Winter: 6°C	Mostly temperate
Boreal Forest	35–75 cm	Summer: 14°C Winter: –10°C	Mostly temperate
Tropical Savanna	150 cm	Dry season: 34°C Wet season: 16°C	Mostly tropical
Desert	Less than 25 cm	Summer: 38°C Winter: 7°C	Tropical and temperate
Temperate Grassland	25–75 cm	Summer: 30°C Winter: 0°C	Temperate
Temperate Woodland and Scrubland	Less than 50 cm	Summer: 20°C Winter: 10°C	Temperate
Tundra	30–50 cm	Summer: 12°C Winter: –26°C	Temperate and polar

Figure 4-1

28. Using Tables and Graphs Which climate zone listed in Figure 4-1 includes the most biomes?

____

____

____

29. Using Tables and Graphs Which biomes listed in Figure 4-1 include areas that have an average yearly rainfall of less than 75 cm? Which biomes include areas that have an average yearly rainfall of more than 200 cm?

____

____

____

30. Analyzing Data Which two biomes shown in Figure 4-1 have seasons determined by the amount of precipitation they receive at different times of the year?

____

____

____

Name______________________________ Class __________________ Date ______________

Multiple Choice

Write the letter that best answers the question or completes the statement on the line provided.

____ **1.** There are 150 Saguaro cacti plants per square kilometer in a certain area of Arizona desert. To which population characteristic does this information refer?

a. growth rate
b. geographic distribution
c. age structure
d. population density

____ **2.** What does the range of a population tell you that density does not?

a. the number that live in an area
b. the areas inhabited by a population
c. the births per unit area
d. the deaths per unit area

____ **3.** What can cause a population to grow?

a. The birthrate becomes higher than the death rate.
b. The birthrate stays the same, and the death rate increases.
c. The birthrate becomes lower than the death rate.
d. The birthrate and the death rate remain the same.

____ **4.** When individuals in a population reproduce at a constant rate, it produces a growth pattern called

a. logistic growth.
b. growth density.
c. demographic growth.
d. exponential growth.

____ **5.** Which are two ways a population can decrease in size?

a. immigration and emigration
b. increased death rate and immigration
c. decreased birthrate and emigration
d. emigration and increased birthrate

____ **6.** As resources in a population become less available, population growth

a. becomes negative.
b. increases slowly.
c. reaches carrying capacity.
d. enters a phase of exponential growth.

____ **7.** Which of the following is not likely to be a limiting factor on the sea otter population?

a. disease c. drought
b. competition d. predation

____ **8.** Which will reduce competition within a species' population?

a. fewer individuals c. fewer resources

b. higher birthrate d. higher population density

____ **9.** If a population grows larger than the carrying capacity of the environment, the

a. death rate may rise. c. population will grow faster.

b. birthrate may rise. d. carrying capacity will change.

____ **10.** Which would be least likely to be affected by a density-dependent limiting factor?

a. a small, scattered population

b. a population with a high birthrate

c. a large, dense population

d. a population with a high immigration rate

____ **11.** Which density-dependent factors, other than the predator/prey relationship, affected the populations of moose and wolves on Isle Royale?

a. extreme temperatures for the moose and flooding for the wolves

b. parasitic wasps for the wolves and clear-cut forest for the moose

c. a hurricane followed by drought for both moose and wolves

d. food availability for the moose and disease for the wolf

____ **12.** Demographic transition is change from high birthrates and high death rates to

a. exponential growth.

b. a low birthrate and a low death rate.

c. a low birthrate and a high death rate.

d. indefinite growth.

____ **13.** About 500 years ago, the world's population started

a. decreasing.

b. to reach carrying capacity.

c. growing more rapidly.

d. to level off.

____ **14.** In Rwanda, there are more young children than teenagers, and more teenagers than adults. This age structure indicates a population that

a. has stopped growing.

b. will double in 30 years.

c. has a steady growth rate.

d. will decrease in 30 years.

____ **15.** In countries like India, the human population is growing

a. exponentially. c. logistically.

b. transitionally. d. demographically.

Name________________________________ Class____________________ Date ________________

Completion

Complete each statement on the line provided.

16. When an individual moves into a population from a different population, it is called ____________________ .

17. Zero population growth is a characteristic of ____________________ growth.

18. Under conditions of logistic growth, population size will rise and fall around an average point called a(an) ____________________ .

19. Resource shortages triggered by increasing population size are density-____________________ limiting factors.

20. Social and ____________________ factors explain why some countries have high growth rates while other countries grow slowly or not at all.

Short Answer

In complete sentences, write the answers to the questions on the lines provided.

21. How does emigration affect population size?

____

____

____

22. What is a density-independent limiting factor? Give two examples.

____

____

____

23. What kept the human population growth rate low before the Industrial Revolution?

____

____

____

24. How can a demographer, or scientist who studies demography, predict how a population will change in the future?

____

____

____

25. How does population growth in the United States compare with that of China and India?

____

____

____

Name ______________________ Class ______________ Date ____________

Using Science Skills

Use the graphs to answer the following questions on the lines provided.

Graph I shows the curve for a culture of *Paramecium aurelia*. Graph II shows the growth curve for a culture of *Paramecium caudatum*, a larger species. Graph III shows the growth curves of each species when they are grown together.

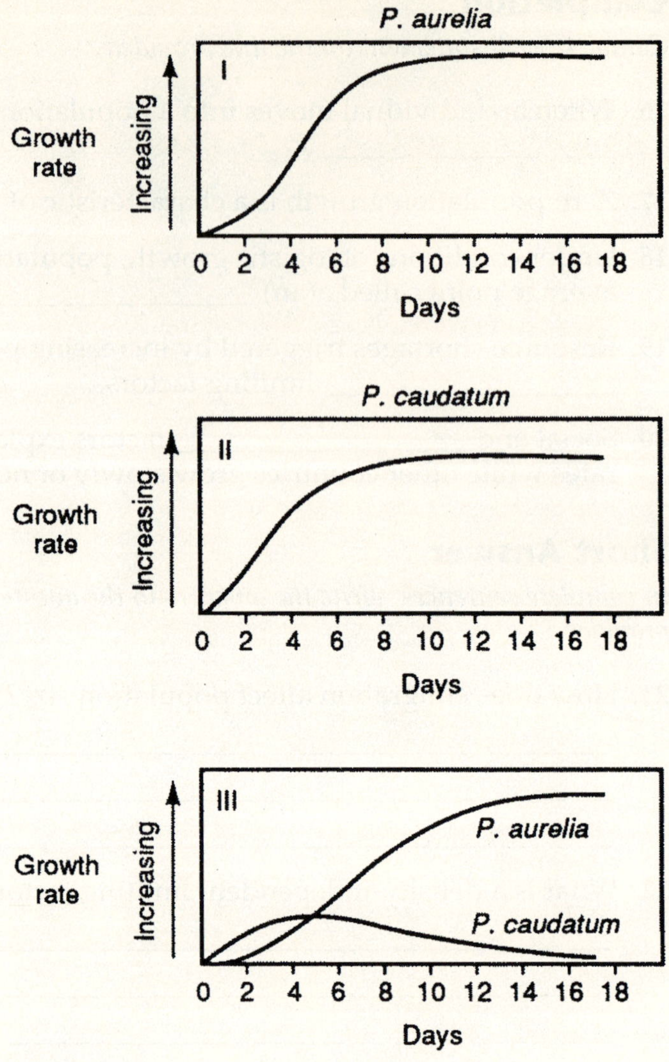

Figure 5-1

26. Interpreting Data According to Figure 5-1, which species has the greater initial growth rate when they are grown in separate cultures?

____

27. Interpreting Graphics From Figure 5-1, which species has the greater growth rate overall when grown together? Describe the growth curve of *P. caudatum* in graph III.

____

28. Observing What type of population growth curve can be observed in graphs I and II of Figure 5-1?

____

29. Drawing Conclusions What is the most likely explanation for the decline of the *P. caudatum* shown in graph III of Figure 5-1?

____

30. Drawing Conclusions Are the limiting factors for all cultures in Figure 5-1 density-dependent or density-independent? Explain.

____

Name______________________________ Class__________________ Date ______________

Essay

Write the answer to each question in the space provided.

31. Discuss four ways population size can change.

32. Differentiate between exponential and logistic growth.

33. Explain the limiting factors that control population growth.

Name______________________________ Class__________________ Date ______________

34. Differentiate between density-dependent and density-independent limiting factors.

35. Sequence the historical events that contributed to the current human population growth.

Name________________________________ Class__________________ Date ______________

Multiple Choice

Write the letter that best answers the question or completes the statement on the line provided.

____ **1.** Which of the following is NOT one of the factors that play a role in population growth rate?

a. immigration c. emigration
b. death rate d. demography

____ **2.** One of the main characteristics of a population is its

a. change over time.
b. geographic distribution.
c. dynamics.
d. habitat.

____ **3.** Sea otters are important to the populations of

a. kelp. c. killer whales.
b. sea urchins. d. all of the above

____ **4.** The movement of organisms into a given area from another area is called

a. immigration. c. population shift.
b. emigration. d. carrying capacity.

____ **5.** When the birthrate in a population becomes higher than the death rate, the population growth rate

a. increases. c. levels off.
b. decreases. d. decreases, then levels off.

____ **6.** When individuals in a population reproduce at a constant rate, it produces a pattern of growth called

a. logistic growth. c. exponential growth.
b. growth density. d. carrying capacity.

____ **7.** The various growth phases through which most populations go are represented by a(an)

a. logistic growth curve.
b. exponential growth curve.
c. demographic curve.
d. age-structure curve.

____ **8.** Any factor in the environment that causes population growth to decrease is a

a. carrying capacity. c. limiting factor.
b. limiting nutrient. d. growth factor.

Name_______________________________ Class_________________ Date ______________

____ **9.** Which of the following describes the largest number of individuals that an environment can support?

a. carrying capacity c. emigration
b. immigration d. exponential growth

____**10.** Each of the following is a density-dependent limiting factor EXCEPT

a. competition. c. crowding.
b. unusual weather. d. disease.

____**11.** Which of the following is a density-independent factor?

a. earthquake c. emigration
b. disease d. parasitism

____**12.** Human population growth has slowed down in

a. China. c. India.
b. the United States. d. Africa.

____**13.** Demography is the scientific study of

a. parasitism and disease.
b. modernized countries.
c. human populations.
d. economic transitions.

____**14.** Which country has not yet completed the demographic transition?

a. United States c. Great Britain
b. India d. Japan

____**15.** The anticipated human population by the year 2050 may be about

a. 7.8 billion. c. 9 trillion.
b. 9 billion. d. 78 billion.

Completion

Complete each statement on the line provided.

16. The number of individuals per unit area is a population's ________________.

17. Emigration can cause a population to ________________ in size.

18. Under ideal conditions with unlimited resources, a population will grow ________________.

19. One of the best known mechanisms of population control is the ________________ relationship.

20. A diagram that graphs the numbers of people of different age groups in the population is called a(an) ________________.

Name ______________________ Class ______________ Date ____________

Short Answer

In complete sentences, write the answers to the questions on the lines provided.

21. What are the four main characteristics of a population?

22. Describe the graph of exponential growth.

23. What is a limiting factor? Give two examples.

24. Why did the human population begin growing exponentially about 500 years ago?

25. What are some of the factors that economists believe will overcome the negative impacts of the growth of the human population?

Name______________ Class______________ Date______________

Using Science Skills

Use the table below to answer the following questions on the lines provided.

Population Statistics in the United States From 1900 to 1990

	1900	1910	1920	1930	1940	1950	1960	1970	1980	1990
Live births per thousand	32.3	30.1	27.7	21.3	19.4	24.1	23.7	18.4	15.9	15.6
Deaths per thousand	17.1	14.7	13.0	11.3	10.8	9.6	9.5	9.5	8.8	8.7
Number of births over deaths	15.2	15.4	14.7	10.0	8.6	14.5	14.2	8.9	7.1	6.9
Life expectation at birth (yrs)	47.3	50.0	54.1	59.7	62.9	68.2	69.7	70.8	73.7	74.8
Increase in longevity since 1900 (yrs)		2.7	6.8	12.4	15.6	20.9	22.4	23.5	26.4	27.5

Figure 5-1

26. Using Tables and Graphs According to Figure 5-1, which year had the highest number of live births per thousand of population?

____

27. Analyzing Data In Figure 5-1, what is unique about the number of deaths per thousand in 1960 and 1970?

____

____

28. Interpreting Graphics In which year was the number of births over deaths at its lowest point, according to Figure 5-1?

____

29. Drawing Conclusions Examine the data in Figure 5-1. Does the number of births over deaths since 1950 increase or decrease? Explain your answer.

____

____

30. Applying Concepts Examine the data for life expectancy at birth in Figure 5-1. Over the period of time covered by the data, the death rate decreased steadily, and the birthrate has declined since 1950. What factors are responsible for the case that our population is not decreasing?

____

____

____

____

Name______________________________ Class__________________ Date ______________

Multiple Choice

Write the letter that best answers the question or completes the statement on the line provided.

____ **1.** The arrival of Europeans in the Hawaiian Islands changed the islands by introducing

a. new crops. c. pigs and rats.
b. new ways of farming. d. all of the above

____ **2.** Early hunters and gatherers in North America are responsible for

a. natural plant and animal populations.
b. a major extinction event of large animals.
c. producing new varieties of crops.
d. inventing technology such as manufactured tools.

____ **3.** A resource, such as air or parts of the oceans, that is open to anyone is often preserved and protected by

a. national organizations.
b. everyone.
c. the country near which it is located.
d. no one.

____ **4.** Which is a way to limit deforestation?

a. use more wood products
b. plant and harvest trees on tree farms
c. increase carbon dioxide levels
d. fertilize the soil

____ **5.** The sulfur and nitrogen compounds in smog combine with water to form

a. ozone. c. acid rain.
b. ammonia. d. chlorofluorocarbons.

____ **6.** An example of a sustainable-use practice is the use of beneficial insects like ladybugs to

a. harm natural resources.
b. pollinate plants.
c. control unwanted pests.
d. eat unwanted plants.

____ **7.** The number of different species in the biosphere is called

a. biodiversity. c. genetic diversity.
b. ecosystem diversity. d. species diversity.

____ 8. Biodiversity is valuable in the biosphere because it
a. gives us interesting things to look at.
b. tells us about many other species.
c. is the biological life-support system of our planet.
d. provides humans with resistance to disease.

____ 9. One property that makes DDT hazardous over time is that DDT is
a. an insecticide.
b. a perfect pesticide.
c. nonbiodegradable.
d. deadly to herbivores.

____10. All of the following are threats to biodiversity EXCEPT
a. biological magnification of toxic compounds.
b. habitat fragmentation.
c. invasive species.
d. species preservation.

____11. As DDT moves up the trophic levels in food chains, its concentration
a. stays the same.
b. decreases.
c. increases.
d. is eliminated.

____12. The goals of conservation biology include all of the following EXCEPT
a. wise management of natural resources.
b. protection and management of individual species.
c. preservation of habitats and wildlife.
d. introducing foreign species into new environments.

____13. Protecting ecosystem diversity is a goal of
a. the green revolution.
b. conservation biology.
c. the captive breeding program.
d. the United Nations.

____14. Some scientists think that global warming is
a. a natural variation in climate.
b. a result of human activity.
c. melting the polar ice caps.
d. all of the above

____15. The major cause of ozone depletion is
a. nitric acid.
b. sulfuric acid.
c. chlorofluorocarbons.
d. ultraviolet light.

Name_______________ Class_______________ Date_______________

Completion

Complete each statement on the line provided.

16. To prevent certain fish populations from collapsing, certain _______________ have been temporarily closed.

17. A location that has 170 species of birds has greater _______________ than a location that has 100 species of birds.

18. Excessive levels of DDT in top-level consumers are a result of _______________ .

19. Zoos have established _______________ programs in which young animals are raised in protected surroundings until the population is stable and are later returned to the wild.

20. Intense UV radiation due to ozone depletion can damage plant leaves and _______________ in the oceans.

Short Answer

In complete sentences, write the answers to the questions on the lines provided.

21. How do modern subsistence hunters differ from ancient hunter-gatherers?

22. What is the "tragedy of the commons"?

23. What are the the goals of sustainable development, and upon what principles are sustainable practices based?

24. What is a biodiversity "hot spot"? Where is such a hot spot in the United States?

25. Describe three services that are provided by a healthy biosphere.

____

____

____

Using Science Skills

Use the graph below to answer the following questions on the lines provided.

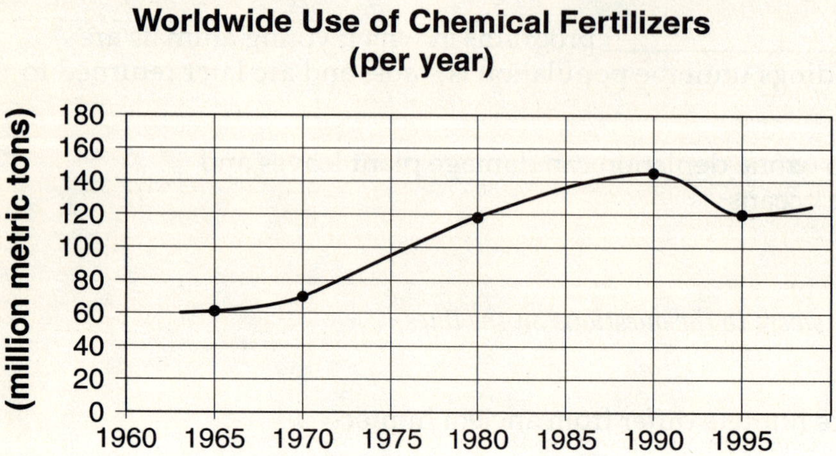

Figure 6-1

26. **Using Charts and Graphs** The Green Revolution has doubled food production over the last 50 years. Starting in 1965, in what year did the amount of chemical fertilizers used worldwide double, according to Figure 6-1?

____

27. **Analyzing Data** Describe the data in Figure 6-1. Does it indicate that the Green Revolution was meeting its goal? Explain.

____

____

____

____

____

____

28. **Applying Concepts** How might the trend in Figure 6-1 up to 1990 relate to the practice of monoculture?

____

____

____

____

____

____

29. **Formulating Hypotheses** Formulate a hypothesis that, if supported by testing, would explain the trend shown in Figure 6-1 after 1990.

30. **Making Judgments** Examine the trend in chemical fertilizer use from 1965 to 1990 shown in Figure 6-1. Is this a sustainable use of resources? Explain.

Essay

Write the answer to each question in the space provided.

31. How is Earth like an island?

32. What causes acid rain? What is the effect of acid rain on natural habitats?

Name____________________ Class______________ Date__________

33. Discuss biodiversity as a source of medicines.

34. What are introduced species? How are they a threat to biodiversity?

35. Discuss one of the challenges to conservation efforts in developing countries.

Name______________________ Class______________ Date ___________

Multiple Choice

Write the letter that best answers the question or completes the statement on the line provided.

____ **1.** In the very distant past, most people
a. lived in small groups.
b. lived in permanent settlements.
c. did not gather plants.
d. did not hunt animals.

____ **2.** Imported plants in Hawaii have
a. crowded out many native species.
b. reduced the native bird species.
c. introduced diseases.
d. depleted natural resources.

____ **3.** Land is a resource that provides
a. space for cities and suburbs.
b. raw materials for industry.
c. soil for growing crops.
d. all of the above

____ **4.** An old-growth forest is
a. a renewable resource.
b. a nonrenewable resource.
c. a microclimate.
d. poor in biodiversity.

____ **5.** The wearing away of surface soil by water and wind is known as
a. deforestation. c. overgrazing.
b. desertification. d. soil erosion.

____ **6.** Farmers can reduce soil erosion by
a. increasing irrigation.
b. contour plowing.
c. grazing cattle on the land.
d. plowing up roots.

____ **7.** Using renewable resources while ensuring that they are not depleted is a practice called
a. sustainable development. c. biological magnification.
b. monoculture. d. subsistence hunting.

Name________________________ Class ________________ Date ____________

____ 8. Biodiversity is important to human society because it
a. is a natural resource.
b. provides food and goods.
c. provides medicines.
d. all of the above

____ 9. The sum total of the genetically based variety of living organisms in the biosphere is called
a. species diversity. c. biodiversity.
b. sustainable use. d. conservation biology.

____10. An endangered species is
a. a diseased animal.
b. a dangerous predator.
c. a group of organisms near extinction.
d. all organisms at the top of a food chain.

____11. DDT was used to
a. fertilize soil. c. form ozone.
b. kill insects. d. feed animals.

____12. One of the goals of conservation biology is to
a. enforce environmental laws.
b. protect habitats.
c. manage natural resources.
d. all of the above

____13. An ecological "hot spot" is an area where
a. habitats and species are healthy.
b. hunting is encouraged.
c. habitats and species are in the most danger of extinction.
d. species diversity is too high.

____14. Overexposure to UV radiation can
a. cause cancer.
b. decrease organisms' resistance to disease.
c. damage eyes.
d. all of the above

____15. An increase in Earth's average temperature from the buildup of carbon dioxide and other gases in the atmosphere is called
a. the greenhouse effect.
b. ozone depletion.
c. global warming.
d. particulate dispersal.

Name______________________ Class______________ Date ___________

Completion

Complete each statement on the line provided.

16. Because the gases that make up the air can be replaced by natural processes, air is classified as a(an) _______________ resource.

17. Microscopic particles, such as dust and ash, that can enter the nose, mouth, and lungs are called _______________ .

18. Sustainable practices are based on principles of ecology as well as _______________ .

19. Substances that cannot be broken down into their elements by metabolic processes are _______________ .

20. The _______________ acts as Earth's sunscreen.

Short Answer

In complete sentences, write the answers to the questions on the lines provided.

21. What is suburban sprawl?

____

____

____

____

22. What development in human history led to humans gathering in settlements, towns, and cities?

____

____

____

____

23. Why is biodiversity one of Earth's greatest natural resources?

____

____

____

24. How is hunting a threat to biodiversity?

____

____

____

____

25. What is ozone? Why is it important?

____

____

____

Name______________________ Class________________ Date ____________

Using Science Skills

Use the tables to answer the following questions on the lines provided.

Existing Species and Endangered Organisms

I. How Many Species Are There?

Groups of Species	Number Identified	Possible Total
Viruses	5,000	perhaps 500,000
Bacteria	4,000	400,000 to 3 million
Fungi	70,000	1 to 1.5 million
Protozoans	40,000	100,000 to 200,000
Algae	40,000	200,000 to 10 million
Plants	250,000	300,000 to 500,000
Vertebrates	45,000	50,000
Roundworms	15,000	500,000 to 1 million
Mollusks	50,000	200,000
Crustaceans	40,000	150,000
Spiders, mites	75,000	750,000 to 1 million
Insects	950,000	8 to 100 million

II. Organisms Listed as Endangered in the United States

Type of Organism	Number of Endangered Species
Mammals	63
Birds	78
Reptiles	14
Amphibians	10
Fishes	70
Snails	20
Clams	61
Crustaceans	18
Insects	33
Spiders	12
Flowering plants	565
Conifers	2
Ferns and other plants	24

Figure 6-1

26. Calculating According to Figure 6-1, table I, what is the identified species diversity in the biosphere?

____

____

____

____

27. Comparing and Contrasting Which group of species listed in table I, Figure 6-1, is the most diverse?

____

____

____

____

28. Using Tables and Graphs Which group of species has the highest number on the list of endangered species in table II, Figure 6-1?

____

____

____

____

29. Analyzing Data Which table in Figure 6-1 tells you about the numbers of species whose population size is declining in a way that places it in danger of extinction?

____

30. Predicting After ten years, if conservation efforts are supported to protect entire ecosystems, how would table II in Figure 6-1 likely change?

____

____

Multiple Choice

Write the letter that best answers the question or completes the statement on the line provided.

____ **1.** When a new volcanic island appears, which of the following will probably grow first?

a. mosses
b. lichens
c. grasses
d. ferns

____ **2.** Assemblages of different populations that live together in a defined area are called

a. ecosystems.
b. habitats.
c. communities.
d. species.

____ **3.** Which of the following processes release(s) carbon dioxide into the atmosphere?

a. volcanic activity
b. respiration
c. burning of fossil fuels
d. all of the above

____ **4.** Climate zones are the result of differences in

a. angle of the sun's rays.
b. heat transport.
c. thickness of the ozone layer.
d. greenhouse gases.

____ **5.** An example of a biotic factor is

a. average temperature.
b. competing species.
c. soil type.
d. sunlight.

____ **6.** One type of symbiosis is

a. succession.
b. competition.
c. predation.
d. parasitism.

____ **7.** Which biome contains scorpions and drought-resistant shrubs?

a. tundra
b. desert
c. taiga
d. tropical rain forest

____ **8.** An ecosystem in which water either covers the soil or is present at or near the surface of the soil for at least part of the year is called a

a. mangrove swamp.
b. pond.
c. wetland.
d. salt marsh.

Name________________________________ Class____________________ Date _______________

____ **9.** Which two factors increase population size?

a. deaths and immigration

b. births and emigration

c. deaths and emigration

d. births and immigration

____ **10.** Density-dependent limiting factors include

a. disease.

b. earthquakes.

c. blizzards.

d damming of rivers.

____ **11.** Which age structure is most likely for a population that has not completed the demographic transition?

a. 50 percent of people below age 15

b. 15 percent of people below age 15

c. 5 percent of people aged 10–14

d. 10 percent of people aged 50–54

____ **12.** All renewable resources

a. are unlimited in supply.

b. can regenerate or be replenished.

c. are living or once were living.

d. can be recycled or reused.

____ **13.** Compounds that contribute to the formation of acid rain contain

a. calcium and phosphorus.

b. carbon dioxide and oxygen.

c. ammonia and nitrates.

d. nitrogen and sulfur.

____ **14.** Ecosystem services include

a. solar energy.

b. production of oxygen.

c. food production.

d. all of the above

____ **15.** Which of the following organisms is a detritivore?

a. caterpillar   c. snail

b. crow   d. fungus

Name______________________________ Class__________________ Date ______________

Completion

Complete each statement on the line provided.

16. Only about __________________ percent of the energy available within one trophic level is transferred to organisms at the next trophic level.

17. The process in which soil bacteria convert nitrates into nitrogen gas is called __________________.

18. The biome that is characterized by permafrost is the __________________.

19. Under ideal conditions with unlimited resources, a population will have a(an) __________________ pattern of growth.

20. Planting large fields with a single variety of crop year after year is a practice called __________________.

Short Answer

In complete sentences, write the answers to the questions on the lines provided.

21. What is the difference between photosynthesis and chemosynthesis?

____

____

____

22. How does energy flow through an ecosystem?

____

____

____

23. Why do foreign species introduced into new habitats often become invasive?

____

____

____

24. What are sustainable strategies for protecting land, forest, and fishery resources?

____

____

____

25. What is biological magnification?

____

____

____

Using Science Skills

Use the graph to answer the following questions on the lines provided.

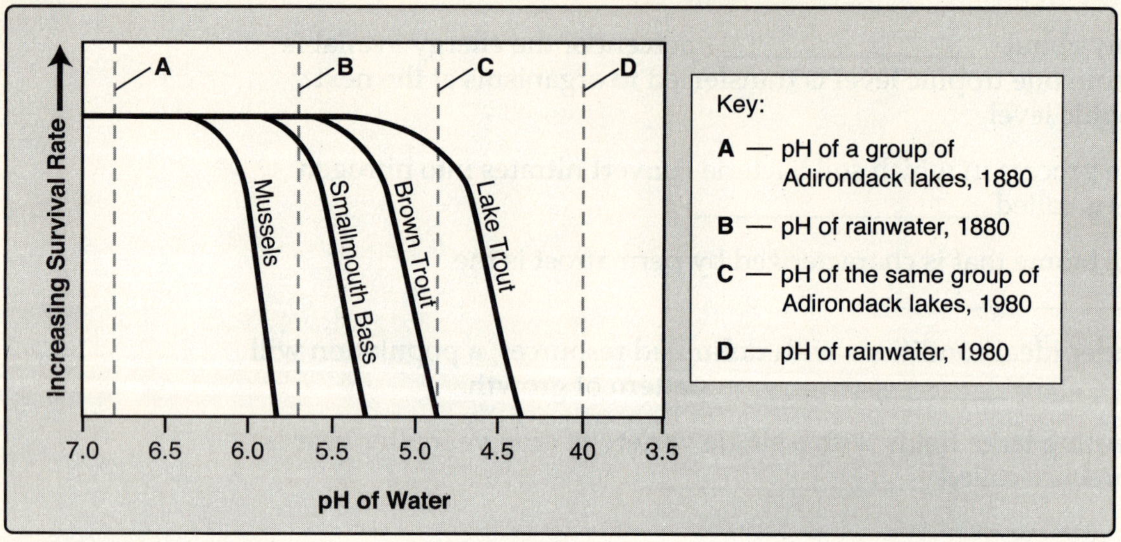

Figure 1

26. According to Figure 1, how did the acidity of Adirondack lake water change from 1880 to 1980, and what is the likely cause of the change?

____

27. Based on Figure 1, which species is least tolerant of increased acidity of its environment? Which species is most tolerant of increased acidity?

____

____

28. Based on Figure 1, at what pH value is survival of smallmouth bass first affected by increasing acidity of water? At what pH value is survival of brown trout first affected?

____

____

29. Based on Figure 1, which species is most likely to be threatened by extinction if the above trend in lake water acidity continues?

____

30. According to Figure 1, is the change in acidity a density-dependent or density-independent limiting factor? Explain your answer.

____

____

____

Name____________________________ Class __________________ Date ______________

Multiple Choice

Write the letter that best answers the question or completes the statement on the line provided.

____ 1. A collection of all the organisms that live in a particular place, together with their nonliving environment is a(an)
a. population. c. ecosystem.
b. community. d. biome.

____ 2. Organisms that can capture energy and produce food are known as
a. consumers. c. heterotrophs.
b. autotrophs. d. omnivores.

____ 3. The rate at which organic matter is created by producers is called
a. nutrient limit. c. nitrogen fixation.
b. primary succession. d. primary productivity.

____ 4. The average, year-after-year conditions of temperature and precipitation in a particular region are referred to as
a. climate. c. microclimate.
b. weather. d. zonation.

____ 5. The type of interaction in which one organism captures and feeds on another is
a. competition. c. predation.
b. parasitism. d. symbiosis.

____ 6. Boreal forest biomes are
a. hot and wet year-round.
b. home to more species than all other biomes combined.
c. also known as taiga.
d. found near the equator.

____ 7. What type of organism forms the base of many aquatic food webs?
a. plants c. mangrove trees
b. zooplankton d. phytoplankton

____ 8. The permanently dark zone of the ocean is called the
a. photic zone. c. coastal zone.
b. aphotic zone. d. intertidal zone.

____ 9. Population density refers to the
a. number of individuals per unit area.
b. area inhabited by a population.
c. number of people in each age group.
d. carrying capacity.

Name________________ Class________________ Date ____________

____10. An example of a density-independent limiting factor is
a. competition. c. predation.
b. unusual weather. d. disease.

____11. Demography is the scientific study of
a. limiting factors. c. human populations.
b. biological diversity. d. urban development.

____12. Human activities that affect the biosphere include
a. hunting and gathering.
b. agriculture.
c. industry.
d. all of the above

____13. An example of a nonrenewable resource is
a. trees. c. fish.
b. water. d. coal.

____14. The sum total of the genetically based variety of all organisms in the biosphere is called
a. genetic diversity. c. biodiversity.
b. species diversity. d. ecosystem diversity.

____15. The increase in average temperature of the biosphere over the past century is referred to as
a. global warming. c. secondary succession.
b. tolerance. d. greenhouse effect.

Completion

Complete each statement on the line provided.

16. The combined portion of the planet in which all of life exists is the ________________.

17. The main source of energy for life on Earth is ________________.

18. The full range of physical and biological conditions in which an organism lives and the way in which the organism uses those conditions is the organism's ________________.

19. A growth pattern represented by an S-shaped curve is called ________________ growth.

20. The process in which concentrations of a harmful substance increase in organisms at higher trophic levels is called ________________.

Name________________________ Class__________________ Date ______________

Short Answer

In complete sentences, write the answers to the questions on the lines provided.

21. What factors determine aquatic ecosystems?

____

____

____

22. How is latitude related to climate?

____

____

____

23. What are four important characteristics of populations?

____

____

____

24. What is the green revolution?

____

____

____

25. What are three different types of ecological pyramids?

____

____

____

Name________________________ Class________________ Date ____________

Using Science Skills

Use the graph to answer the following questions on the lines provided.

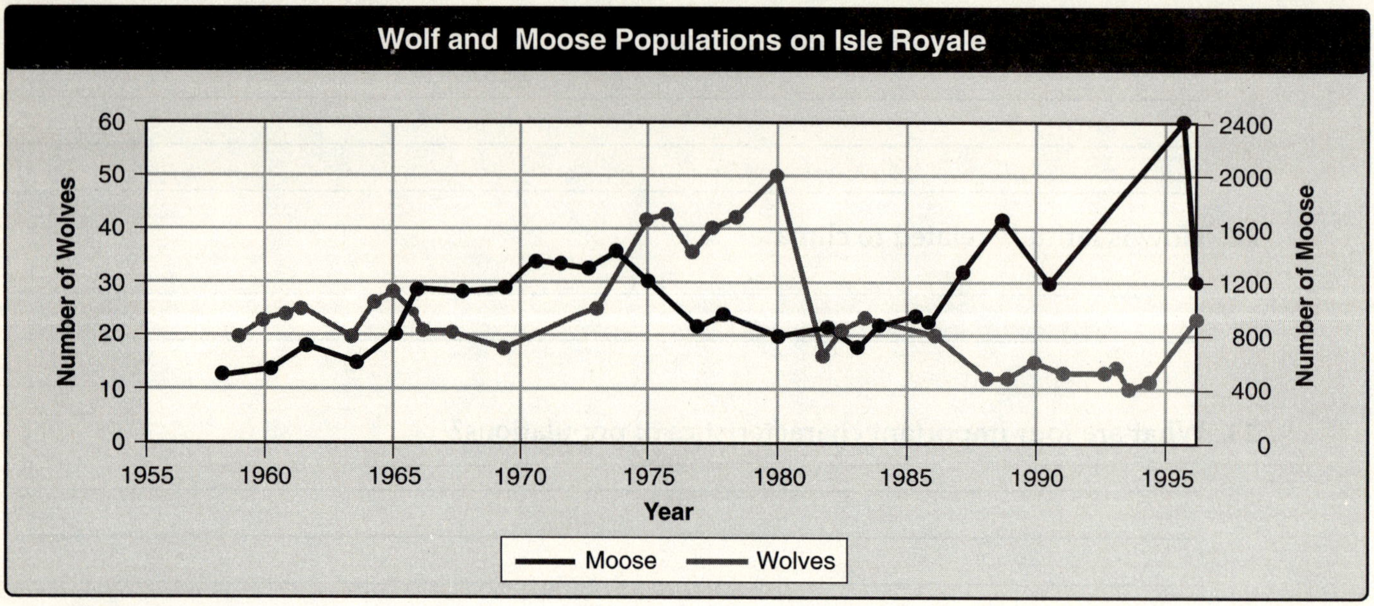

Figure 1

26. What type of interaction does the graph in Figure 1 represent?

27. In Figure 1, why does the wolf population generally rise whenever the moose population increases?

28. In Figure 1, why is an increase in the number of wolves usually followed by a decline in the moose population?

29. In Figure 1, why are there almost always fewer wolves than moose?

30. Does the graph in Figure 1 show the action of density-dependent or density-independent limiting factors? Explain your answer.

Name______________________________ Class__________________ Date ______________

Multiple Choice

Write the letter that best answers the question or completes the statement on the line provided.

____ **1.** The work of Schleiden and Schwann can be summarized by saying that
a. all plants are made of cells.
b. all animals are made of cells.
c. plants and animals have specialized cells.
d. all plants and animals are made of cells.

____ **2.** Which cell structure contains the cell's genetic material and controls many of the cell's activities?
a. organelle c. cell envelope
b. nucleus d. cytoplasm

____ **3.** Cells fall into two broad categories, depending on whether they
a. have a cell wall.
b. contain genetic material.
c. have a nucleus.
d. contain chloroplasts.

____ **4.** Eukaryotes usually contain
a. a nucleus.
b. specialized organelles.
c. genetic material.
d. all of the above

____ **5.** Which of the following is NOT found in the nucleus?
a. cytoplasm c. chromatin
b. nucleolus d. DNA

____ **6.** Which structures carry out cell movement?
a. cytoplasm and ribosomes
b. nucleolus and nucleus
c. microtubules and microfilaments
d. chromosomes

____ **7.** Which organelle breaks down compounds into small particles that the cell can use?
a. Golgi apparatus c. endoplasmic reticulum
b. lysosome d. mitochondrion

____ **8.** Which organelle makes proteins using coded instructions that come from the nucleus?

a. Golgi apparatus
b. mitochondrion
c. vacuole
d. ribosome

____ **9.** Which organelle converts the chemical energy stored in food into compounds that are more convenient for the cell to use?

a. chloroplast
b. Golgi apparatus
c. endoplasmic reticulum
d. mitochondrion

____ **10.** Which of the following is a function of the cell membrane?

a. breaks down lipids, carbohydrates, and proteins from foods
b. stores water, salt, proteins, and carbohydrates
c. keeps the cell wall in place
d. regulates which materials enter and leave the cell

____ **11.** Diffusion occurs because

a. molecules constantly move and collide with one another.
b. the concentration of a solution is never the same throughout a solution.
c. the concentration of a solution is always the same throughout a solution.
d. molecules never move or collide with one another.

____ **12.** An animal cell that is surrounded by fresh water will burst because the osmotic pressure causes

a. water to move into the cell.
b. water to move out of the cell.
c. solutes to move into the cell.
d. solutes to move out of the cell.

____ **13.** The cells of multicellular organisms are

a. smaller than those of unicellular organisms.
b. simpler than those of unicellular organisms.
c. specialized to perform particular functions.
d. not dependent on one another.

____ **14.** Which of the following is an organ of the digestive system?

a. stomach
b. nerve tissue
c. muscle cell
d. epithelial tissue

____ **15.** An organ system is a group of organs that

a. are made up of similar cells.
b. are made up of similar tissues.
c. work together to perform a specific function.
d. work together to perform all the functions in a multicellular organism.

Name________________________ Class________________ Date____________

Completion

Complete each statement on the line provided.

16. According to the cell theory, new cells are produced from existing ________________ .

17. In a eukaryote, the portion of the cell outside the nucleus is called the ________________ .

18. During cell division, chromatin condenses to form ________________ , which are threadlike structures containing genetic information.

19. Unlike smooth endoplasmic reticulum, rough endoplasmic reticulum has ________________ attached to it.

20. The cell takes in food and water and eliminates wastes through the ________________ .

Short Answer

In complete sentences, write the answers to the questions on the lines provided.

21. How do prokaryotes and eukaryotes differ?

____

____

22. Using Figure 7-1, give the letter of the structure in drawing II that corresponds to structure H in drawing I. Name the structure and state what process occurs in it.

____

____

____

23. Using Figure 7-1, give the label letters and full names of the three structures that are found in drawing II but not in drawing I. State the function of each of these structures.

____

____

____

____

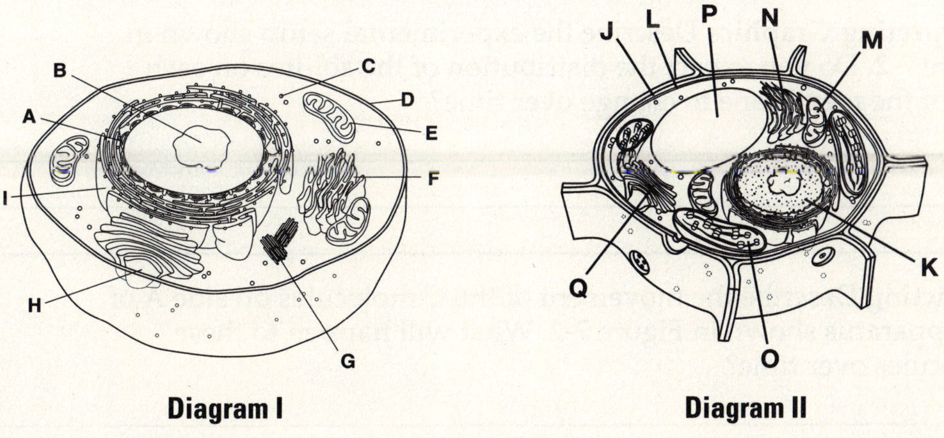

Figure 7-1

24. Define diffusion.

____

25. A hypertonic salt solution has a higher concentration of solutes than a blood cell. Explain what happens when a blood cell is placed in a hypertonic salt solution.

____

Using Science Skills

Use the diagram below to answer the following questions on the lines provided.

A student put together the experimental setup shown below. The selectively permeable membrane is permeable to both types of solute molecules shown.

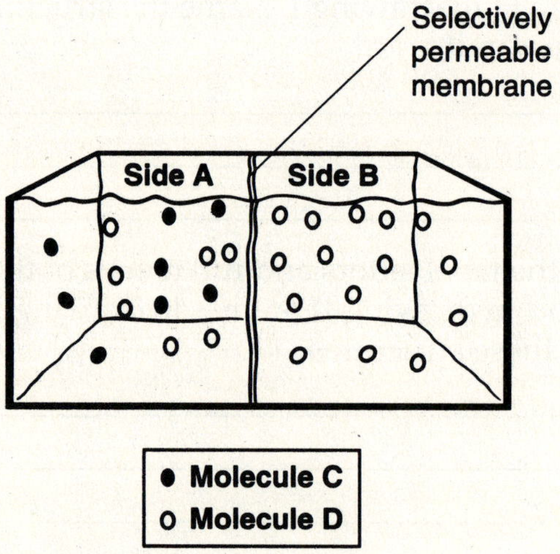

Figure 7-2

26. **Interpreting Graphics** Describe the experimental setup shown in Figure 7-2. Do you expect the distribution of the solutes on each side of the membrane to change over time?

____

27. **Predicting** Describe the movement of the C molecules on side A of the apparatus shown in Figure 7-2. What will happen to these molecules over time?

____

Name______________________________ Class________________ Date ____________

28. **Predicting** Look at Figure 7-2. Describe the movement of the D molecules on side A. What will happen to these molecules over time?

____

____

____

29. **Predicting** What will the apparatus shown in Figure 7-2 look like when equilibrium is reached?

____

____

____

30. **Predicting** Once equilibrium is reached in the apparatus shown in Figure 7-2, will the molecules continue to move? Explain your answer.

____

____

____

Essay

Write the answer to each question in the space provided.

31. Summarize three statements from the cell theory. Explain the significance of the cell theory to biology.

32. Compare the cell membrane to a mosaic.

33. How do facilitated diffusion and active transport differ?

34. Compare a cell from a unicellular organism with a cell from a multicellular organism in terms of cell specialization.

35. Discuss the levels of organization in multicellular organisms, and explain why these levels are not used to describe unicellular organisms.

Name______________________________ Class __________________ Date ______________

Multiple Choice

Write the letter that best answers the question or completes the statement on the line provided.

____ **1.** Who was the first person to identify and see cells?
a. Anton van Leeuwenhoek
b. Robert Hooke
c. Matthias Schleiden
d. Rudolf Virchow

____ **2.** The thin, flexible barrier around a cell is called the
a. cell membrane.
b. cell wall.
c. cell envelope.
d. cytoplasm.

____ **3.** Prokaryotes lack
a. cytoplasm.
b. a cell membrane.
c. a nucleus.
d. genetic material.

____ **4.** Which of the following contains a nucleus?
a. prokaryotes c. eukaryotes
b. bacteria d. organelles

____ **5.** The main function of the cell wall is to
a. support and protect the cell.
b. store DNA.
c. direct the activities of the cell.
d. help the cell move.

____ **6.** Which of the following is a function of the nucleus?
a. stores DNA
b. controls most of the cell's processes
c. contains the information needed to make proteins
d. all of the above

____ **7.** Which of the following is a function of the cytoskeleton?
a. helps a cell keep its shape
b. contains DNA
c. surrounds the cell
d. helps make proteins

____ 8. Which of the following is an organelle found in the cytoplasm?

a. nucleolus
c. chromatin
b. ribosome
d. cell wall

____ 9. Which organelle would you expect to find in plant cells?

a. mitochondrion
c. chloroplast
b. ribosome
d. smooth endoplasmic reticulum

____10. Which of the following structures serves as the cell's boundary from its environment?

a. mitochondrion
b. cell membrane
c. chloroplast
d. channel proteins

____11. Diffusion is the movement of particles from

a. an area of low concentration to an area of high concentration.
b. an area of high concentration to an area of low concentration.
c. an area of equilibrium to an area of high concentration.
d. all of the above

____12. The diffusion of water across a selectively permeable membrane is called

a. osmotic pressure.
b. osmosis.
c. facilitated diffusion.
d. active transport.

____13. Which term refers to cells having different tasks in an organism?

a. multicellular
c. levels of organization
b. cell specialization
d. unicellular

____14. Which of the following is an example of an organ?

a. heart
c. digestive system
b. epithelial tissue
d. nerve cell

____15. A group of similar cells that perform a particular function is called a(an)

a. organ.
c. tissue.
b. organ system.
d. division of labor.

Name________________________________ Class____________________ Date ______________

Completion

Complete each statement on the line provided.

16. The portion of the cell outside the nucleus is called the ____________________.

17. Eukaryotes contain structures that act as if they are specialized organs. These structures are called ____________________.

18. Molecules tend to move from an area where they are more concentrated to an area where they are less concentrated. This process is called ____________________.

19. The cells in a multicellular organism have specific jobs. This is called cell ____________________.

20. The levels of organization in a multicellular organism are ____________________, tissues, ____________________, and organ systems.

Short Answer

In complete sentences, write the answers to the questions on the lines provided.

21. What does the cell theory state?

____

____

____

22. What are two functions of the nucleus?

____

____

23. List two functions of the cytoskeleton.

____

____

____

24. Explain, in terms of osmosis, why a raisin placed in a cup of pure water overnight will puff up with water.

____

____

____

25. List the four levels of organization in order from simplest to most complex.

____

____

Name____________________ Class______________ Date ____________

Using Science Skills

Use the diagrams below to answer the following questions on the lines provided.

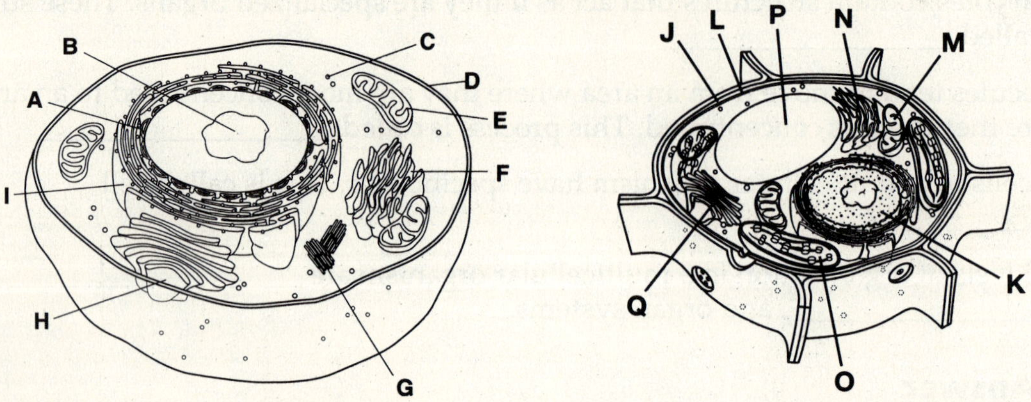

Figure 7-1

Diagram I

Diagram II

26. **Interpreting Graphics** Which drawing in Figure 7-1 contains a structure that captures sunlight and converts it into chemical energy? What is the name of the structure described, and what is it labeled in the diagram?

____

____

27. **Comparing and Contrasting** Look at Figure 7-1. Give the letter of the structure in drawing I that corresponds to structure M in drawing II. What is the name of this structure?

____

____

28. **Comparing and Contrasting** Give the letter of the structure in drawing I of Figure 7-1 that corresponds to structure L in drawing II. What is the name of this structure?

____

29. **Interpreting Graphics** Which organelle is labeled P in Figure 7-1? What is the function of this organelle?

____

____

30. **Interpreting Graphics** Do the drawings in Figure 7-1 represent prokaryotes or eukaryotes? How do you know?

____

____

Name______________________ Class________________ Date ____________

Multiple Choice

Write the letter that best answers the question or completes the statement on the line provided.

____ 1. Which of the following is an autotroph?
a. mushroom
b. dog
c. monkey
d. tree

____ 2. Which of the following is NOT an example of a heterotroph?
a. mushroom
b. leopard
c. grass
d. human

____ 3. Energy is released from ATP when
a. a phosphate group is added.
b. adenine bonds to ribose.
c. ATP is exposed to sunlight.
d. a phosphate group is removed.

____ 4. Look at Figure 8-1. All of the following are parts of an ADP molecule EXCEPT
a. structure A.
b. structure B.
c. structure C.
d. structure D.

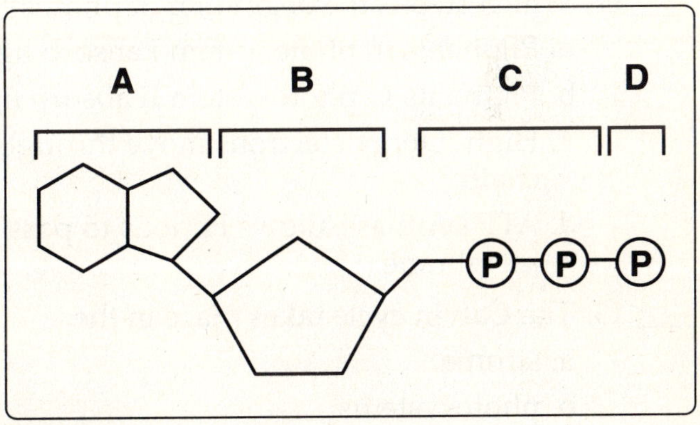

Figure 8-1

____ 5. Which structures shown in Figure 8-1 make up an ATP molecule?
a. A and B
b. A, B, and C
c. A, B, C, and D
d. C and D

____ 6. Which scientists showed that plants need light to grow?
a. van Helmont and Calvin
b. Priestley and Ingenhousz
c. van Helmont and Priestley
d. Priestley and Calvin

____ 7. Which of the following are used in the overall reactions for photosynthesis?
a. carbon dioxide
b. water
c. light
d. all of the above

____ 8. Most plants appear green because chlorophyll
   a. does not absorb green light.
   b. reflects violet light.
   c. absorbs green light.
   d. none of the above

____ 9. A granum is a(an)
   a. stack of chloroplasts.
   b. stack of thylakoids.
   c. membrane enclosing a thylakoid.
   d. photosynthetic pigment molecule.

____10. The light-collecting units of a chloroplast are the
   a. electron carriers.
   b. photosystems.
   c. stroma.
   d. high-energy sugars.

____11. What are the products of the light-dependent reactions?
   a. oxygen gas
   b. ATP
   c. NADPH
   d. all of the above

____12. Which step is the beginning of photosynthesis?
   a. Pigments in photosystem I absorb light.
   b. Pigments in photosystem II absorb light.
   c. High-energy electrons move through the electron transport chain.
   d. ATP synthase allows H^+ ions to pass through the thylakoid membrane.

____13. The Calvin cycle takes place in the
   a. stroma.
   b. photosystems.
   c. thylakoid membranes.
   d. chlorophyll molecules.

____14. If carbon dioxide is removed from a plant's environment, what would you expect to happen to its production of high-energy sugars?
   a. More sugars will be produced.
   b. No sugars will be produced.
   c. The same number of sugars will be produced but without carbon dioxide.
   d. Carbon dioxide does not affect the production of high-energy sugars in plants.

____15. If you continue to increase the intensity of light that a plant receives, what happens?
   a. The rate of photosynthesis increases with light intensity.
   b. The rate of photosynthesis decreases with light intensity.
   c. The rate of photosynthesis increases and then levels off.
   d. The rate of photosynthesis does not change.

Name______________________ Class______________ Date ____________

Completion

Complete each statement on the line provided.

16. Photosynthesis requires light, water, carbon dioxide, and ______________ .

17. If you separate the pigments found in a typical plant cell's chloroplasts, you will find ______________ , orange, and red pigments.

18. Thylakoids are a(an) ______________ color because they contain chlorophyll.

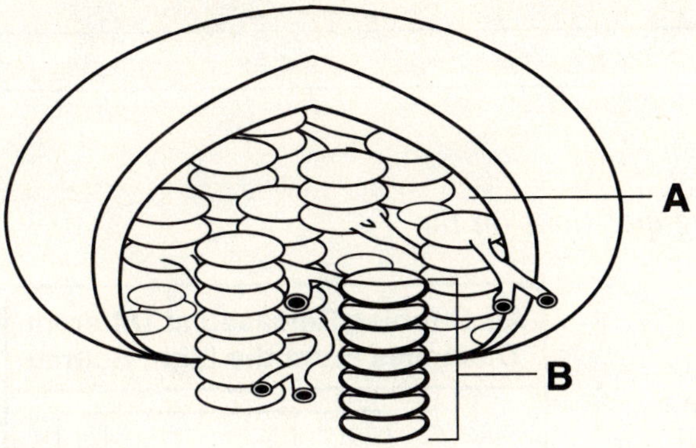

Figure 8-2

19. Photosystems I and II are found in the structure labeled ______________ in Figure 8-2.

20. In many plants, the rate of photosynthesis ______________ when the weather becomes very cold.

Short Answer

In complete sentences, write the answers to the questions on the lines provided.

21. How do heterotrophs obtain energy?

____

____

____

22. What is ATP, and when is energy released from it?

____

____

____

23. Write the overall equation for photosynthesis in both symbols and words.

____

____

24. Photosystems I and II are both located in the thylakoid membrane. What advantage does their proximity provide?

25. What does the Calvin cycle do?

Using Science Skills

Use the diagram and graph below to answer the following questions on the lines provided.

A student prepared two beakers with identical sprigs of a water plant as shown below. She placed one beaker in the shade and the other beaker beside a fluorescent lamp. She then systematically changed the distance of the beaker from the lamp. She counted the bubbles given off by each sprig of the water plant. Shown here is the graph of the data for the beaker she placed in the light.

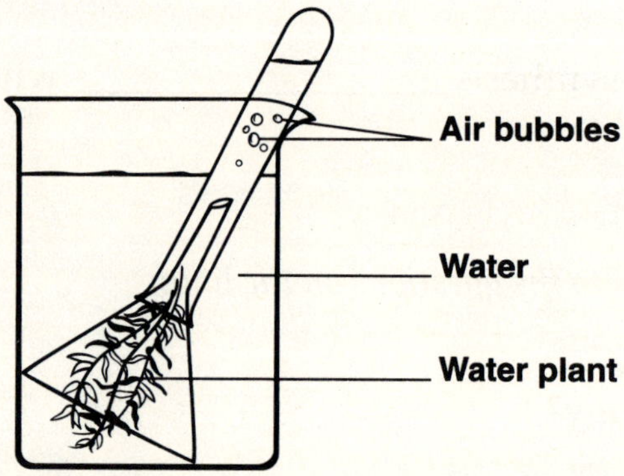

Figure 8-3

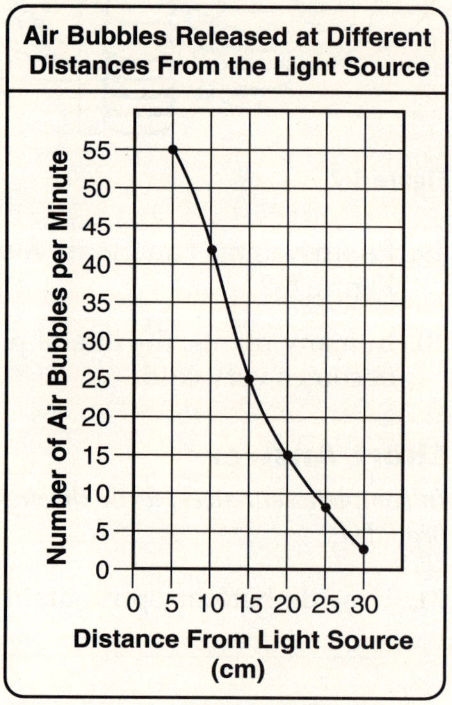

26. Controlling Variables Which beaker is the student's control beaker, the one in the shade or the one in the light?

27. Applying Concepts Look at Figure 8-3. If the student later tested the air bubbles collected in the test tube, what would she find they are made of? How do you know?

28. **Using Tables and Graphs** Look at the graph in Figure 8-3. At what distance from the light source was the greatest number of bubbles produced?

____

29. **Analyzing Data** Look at the graph in Figure 8-3. What do the student's data show?

____

____

30. **Predicting** If the lamp was placed closer than 5 centimeters from the water plant, would the plant give off many more bubbles? Why or why not?

____

____

____

Essay

Write the answer to each question in the space provided.

31. Discuss the relationship between autotrophs and heterotrophs. Do heterotrophs depend on autotrophs for their survival? Explain your answer.

32. Compare the storage capacity of ATP and glucose. How does the cell use each of these molecules to store energy?

33. Describe how pigments obtain energy from light. Use chlorophyll as an example of the process you describe.

34. Identify three factors that affect the rate of photosynthesis, and explain the effect of each.

35. Trace the events that occur in the thylakoid membrane during the light-dependent reactions.

Multiple Choice

Write the letter that best answers the question or completes the statement on the line provided.

____ **1.** Organisms, such as plants, that make their own food are called

a. autotrophs. c. thylakoids.
b. heterotrophs. d. pigments.

____ **2.** Organisms that cannot make their own food and must obtain energy from the foods they eat are called

a. autotrophs. c. thylakoids.
b. heterotrophs. d. plants.

____ **3.** Which of the following is NOT a part of an ATP molecule?

a. adenine c. chlorophyll
b. ribose d. phosphate

____ **4.** In Figure 8-1, between which parts of the molecule must the bonds be broken to form an ADP molecule?

a. A and B
b. B and C
c. C and D
d. all of the above

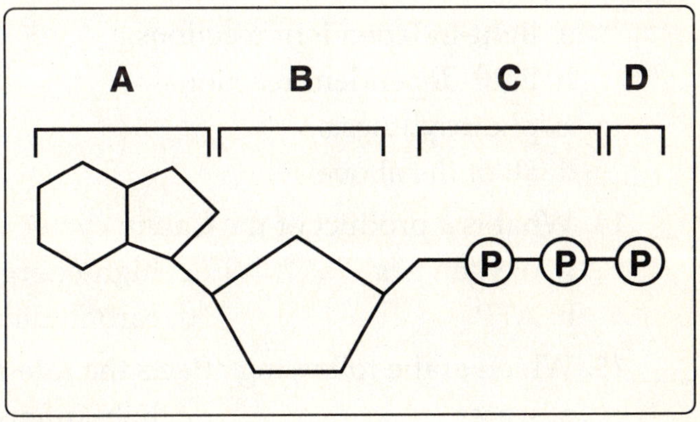

Figure 8-1

____ **5.** Jan van Helmont concluded that plants gain most of their mass from

a. water.
b. the soil.
c. carbon dioxide in the air.
d. oxygen in the air.

____ **6.** Ingenhousz showed that plants produce oxygen bubbles when exposed to

a. ATP. c. light.
b. carbon dioxide. d. a burning candle.

____ **7.** Photosynthesis uses sunlight to convert water and carbon dioxide into

a. oxygen. c. ATP and oxygen.
b. high-energy sugars. d. oxygen and high-energy sugars.

____ **8.** Plants gather the sun's energy with light-absorbing molecules called

a. pigments. c. chloroplasts.
b. thylakoids. d. glucose.

_____ **9.** Plants take in the sun's energy by absorbing
a. high-energy sugars. c. chlorophyll *b*.
b. chlorophyll *a*. d. sunlight.

_____ **10.** The stroma is the space that surrounds
a. thylakoids. c. plant cells.
b. chloroplasts. d. all of the above

_____ **11.** Where do the light-dependent reactions take place?
a. in the stroma
b. outside the chloroplasts
c. in the thylakoid membranes
d. only in chlorophyll molecules

_____ **12.** Where are photosystems I and II found?
a. in the stroma
b. in the thylakoid membrane
c. in the Calvin cycle
d. all of the above

_____ **13.** The Calvin cycle is another name for
a. light-independent reactions.
b. light-dependent reactions.
c. photosynthesis.
d. all of the above

_____ **14.** What is a product of the Calvin cycle?
a. oxygen gas c. high-energy sugars
b. ATP d. carbon dioxide

_____ **15.** Which of the following affects the rate of photosynthesis?
a. water c. light intensity
b. temperature d. all of the above

Completion

Complete each statement on the line provided.

16. Organisms, such as hawks and leopards, that obtain energy from the foods they consume are called ____________________ .

17. Ingenhousz found that plants produce oxygen bubbles only in the presence of ____________________ .

18. Photosynthesis uses the energy of sunlight to convert water and carbon dioxide into oxygen and ____________________ .

Name________________________ Class________________ Date ____________

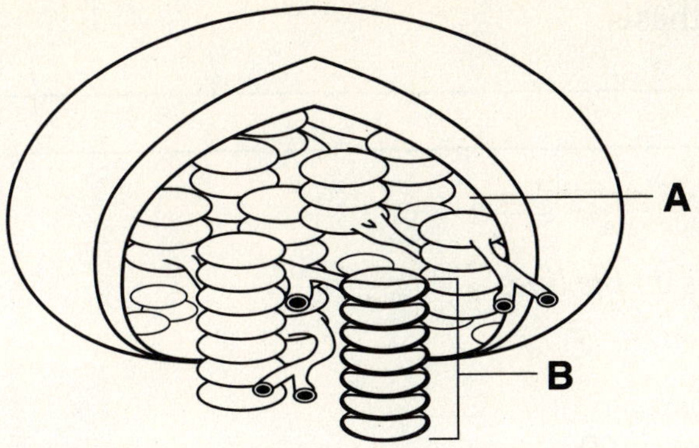

Figure 8-2

19. The area indicated in Figure 8-2 by the letter A is called ________________.

20. The light-dependent reactions convert $NADP^+$ and ADP into the enery carriers NADPH and ________________.

Short Answer

In complete sentences, write the answers to the questions on the lines provided.

21. What is the difference between an autotroph and a heterotroph? Give an example of each type of organism.

22. Explain how heterotrophs get their energy from the sun even though they cannot make their own food.

23. What happens when a phosphate group is removed from an ATP molecule?

24. Identify the structures labeled B in Figure 8-2. What is their function?

Name________________________ Class________________ Date ____________

25. List two factors that affect the rate of photosynthesis.

____

____

Using Science Skills

Use the diagram below to answer the following questions on the lines provided.

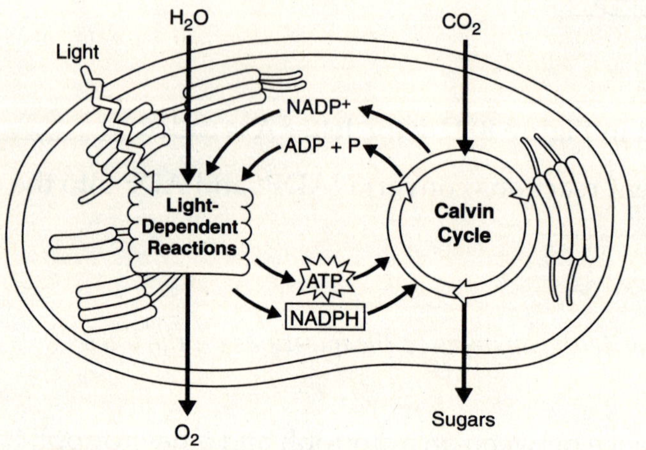

Figure 8-3

26. **Interpreting Graphics** What process is shown in Figure 8-3?

____

27. **Interpreting Graphics** What structure is shown in Figure 8-3?

____

28. **Interpreting Graphics** Look at Figure 8-3. What are the products of the light-dependent reactions?

____

____

____

____

29. **Interpreting Graphics** What are the products of the Calvin cycle shown in Figure 8-3?

____

30. **Interpreting Graphics** In Figure 8-3, what chemical from the atmosphere is used in the Calvin cycle to produce sugars?

____

Name____________________ Class____________________ Date __________

Multiple Choice

Write the letter that best answers the question or completes the statement on the line provided.

____ **1.** Which of the following is NOT a stage of cellular respiration?
a. fermentation c. glycolysis
b. electron transport d. Krebs cycle

____ **2.** What are the reactants in the equation for cellular respiration?
a. oxygen and lactic acid
b. carbon dioxide and water
c. glucose and oxygen
d. water and glucose

____ **3.** The starting molecule for glycolysis is
a. ADP. c. citric acid.
b. pyruvic acid. d. glucose.

____ **4.** One cause of muscle soreness is
a. alcoholic fermentation.
b. glycolysis.
c. lactic acid fermentation.
d. the Krebs cycle.

____ **5.** Which process is used to produce beer and wine?
a. lactic acid fermentation
b. glycolysis
c. alcoholic fermentation
d. the Krebs cycle

____ **6.** The conversion of pyruvic acid into lactic acid requires
a. alcohol. c. ATP.
b. oxygen. d. NADH.

____ **7.** Which organism is NOT likely to carry out cellular respiration?
a. tree c. anaerobic bacterium
b. mushroom d. tiger

____ **8.** During one turn, the Krebs cycle produces
a. oxygen. c. electron carriers.
b. lactic acid. d. glucose.

____ **9.** Which of the following passes high-energy electrons into the electron transport chain?
a. NADH and $FADH_2$ c. citric acid
b. ATP and ADP d. acetyl-CoA

Name______________________ Class______________ Date ____________

_____10. Cellular respiration uses one molecule of glucose to produce
- a. 2 ATP molecules.
- b. 34 ATP molecules.
- c. 36 ATP molecules.
- d. 38 ATP molecules.

_____11. Breathing heavily after running a race is your body's way of
- a. making more citric acid.
- b. repaying an oxygen debt.
- c. restarting glycolysis.
- d. recharging the electron transport chain.

_____12. All of the following are sources of energy during exercise EXCEPT
- a. stored ATP.
- b. alcoholic fermentation.
- c. lactic acid fermentation.
- d. cellular respiration.

_____13. Which process does NOT release energy from glucose?
- a. glycolysis
- b. photosynthesis
- c. fermentation
- d. cellular respiration

_____14. Photosynthesis is to chloroplasts as cellular respiration is to
- a. chloroplasts.
- b. cytoplasm.
- c. mitochondria.
- d. nucleus.

_____15. Plants cannot release energy from glucose using
- a. glycolysis.
- b. photosynthesis.
- c. the Krebs cycle.
- d. cellular respiration.

Completion

Complete each statement on the line provided.

16. Cellular respiration occurs only in the presence of ________________.
17. Without oxygen, a cell can extract a net gain of only ________________ molecules of ATP from each glucose molecule.
18. The pathway labeled B in Figure 9-1 is called ________________ fermentation.
19. In Figure 9-1, only the pathway labeled ________________ requires oxygen.
20. A high level of lactic acid in the blood is a sign that ________________ fermentation has occurred.

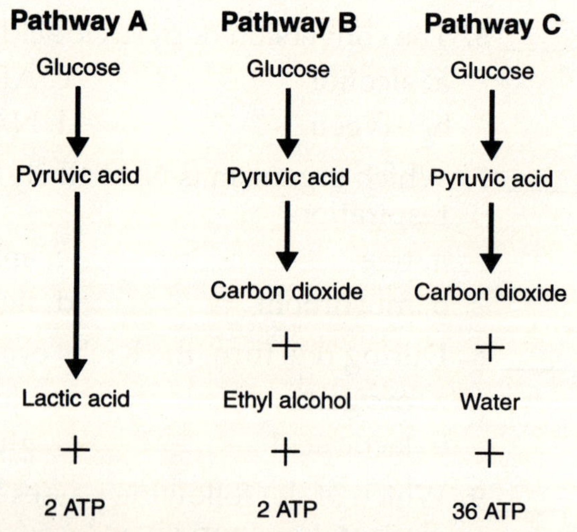

Figure 9-1

Name______________________ Class________________ Date ____________

Short Answer

In complete sentences, write the answers to the questions on the lines provided.

21. List the three main stages of cellular respiration in order. Where does each stage take place in the cell?

____

____

____

22. What are the two types of fermentation? How do their products differ?

____

____

____

23. Based on Figure 9-2, which type of fermentation does NOT give off carbon dioxide? Explain your answer.

____

____

____

____

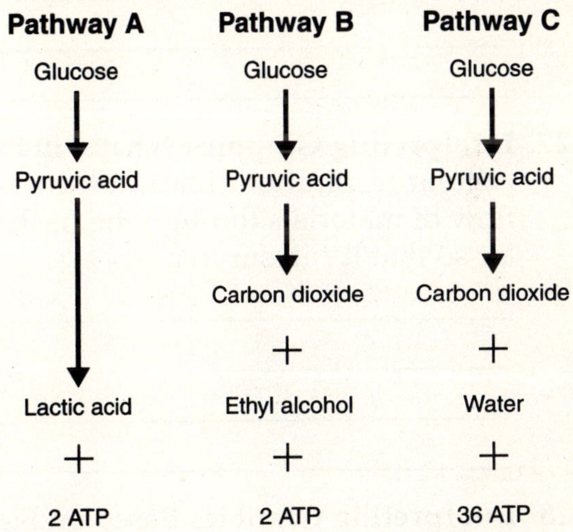

Figure 9-2

24. What role does oxygen play in the electron transport chain?

____

____

____

____

25. What three sources of ATP does your body use during a long aerobic exercise session?

____

____

____

Using Science Skills

Use the diagram on the next page to answer the following questions on the lines provided.

A scientist set up a respiration chamber as shown in Figure 9-3. She placed a mouse in flask B. Into flasks A, C, and D, she poured distilled water mixed with the acid-base indicator phenolphthalein. In the presence of CO_2, phenolphthalein turns from pink to clear. She allowed the mouse to stay in the chamber for about an hour.

Name______________________ Class______________ Date____________

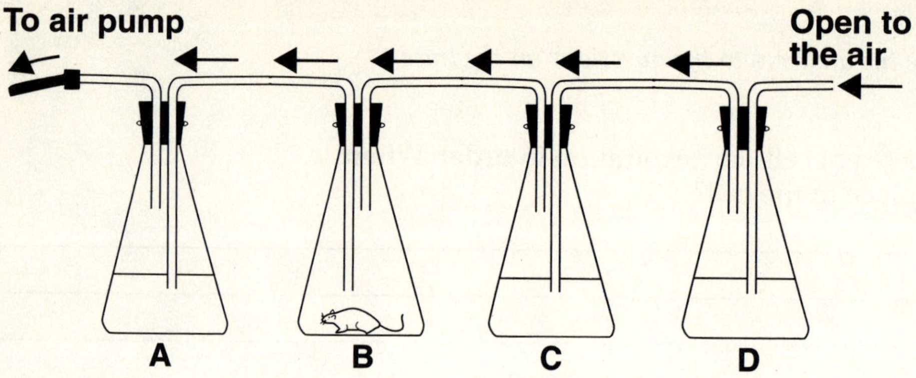

Figure 9-3

26. Inferring Write the equation for cellular respiration. Based on this equation and the setup shown in Figure 9-3, what substance(s) would you expect the mouse in flask B to give off?

____
____
____

27. Interpreting Graphics What will the mouse require to carry out cellular respiration? Look at the flasks in Figure 9-3. Describe the flow of materials through the flasks. Will the mouse receive fresh air so that it can survive?

____
____
____
____

28. Interpreting Graphics Based on Figure 9-3, how will the scientist be able to detect whether the mouse is carrying out cellular respiration?

____
____
____

29. Applying Concepts Assume that the scientist set up an identical respiration chamber, except that in this setup she placed a cricket in flask B instead of a mouse. At the end of one hour, she measured the amount of CO_2 given off by the cricket and the mouse. A small amount of CO_2 had been given off by the mouse, but little to no CO_2 had been given off by the cricket. Is the cricket undergoing cellular respiration? Explain these results.

____
____
____
____

Name________________________ Class________________ Date ____________

30. **Predicting** Assume that the scientist set up an identical respiration chamber, except that in this setup she placed a mouse that had been exercising on a hamster wheel. Then, the scientist measured the amount of CO_2 given off by both mice at the end of 15 minutes. Predict which setup produced the most CO_2. Explain your answer.

____

____

____

Essay

Write the answer to each question in the space provided.

31. List the main events of glycolysis. How many ATP molecules are produced and consumed by glycolysis? What effect does the presence of oxygen have on the events that follow glycolysis?

32. Compare lactic acid fermentation with alcoholic fermentation. Where does each process occur? What are the products of each process?

33. Identify the electron carriers of cellular respiration. Discuss the relationship between the electron carriers and the electron transport chain.

34. Explain how high-energy electrons are used by the electron transport chain.

35. Which pathways does the body use to release energy during exercise? Discuss these pathways in terms of a footrace.

Name________________________________ Class________________ Date ____________

Multiple Choice

Write the letter that best answers the question or completes the statement on the line provided.

____ **1.** Which of the following is released during cellular respiration?

a. oxygen c. energy
b. air d. lactic acid

____ **2.** Cellular respiration releases energy by breaking down

a. food molecules. c. carbon dioxide.
b. ATP. d. water.

____ **3.** Which of these is a product of cellular respiration?

a. oxygen c. glucose
b. water d. all of the above

____ **4.** Which of these processes takes place in the cytoplasm of a cell?

a. glycolysis c. Krebs cycle
b. electron transport d. all of the above

____ **5.** Glycolysis provides a cell with a net gain of

a. 2 ATP molecules. c. 18 ATP molecules.
b. 4 ATP molecules. d. 36 ATP molecules.

____ **6.** Lactic acid fermentation occurs in

a. bread dough.
b. any environment containing oxygen.
c. muscle cells.
d. mitochondria.

____ **7.** The two main types of fermentation are called

a. alcoholic and aerobic.
b. aerobic and anaerobic.
c. alcoholic and lactic acid.
d. lactic acid and anaerobic.

____ **8.** In the presence of oxygen, glycolysis is followed by

a. lactic acid fermentation.
b. alcoholic fermentation.
c. photosynthesis.
d. the Krebs cycle.

____ **9.** Cellular respiration is called an aerobic process because it requires

a. light. c. oxygen.
b. exercise. d. glucose.

____**10.** The starting molecule for the Krebs cycle is
a. glucose. c. pyruvic acid.
b. NADH. d. coenzyme A.

____**11.** In eukaryotes, electron transport occurs in the
a. mitochondria. c. cell membrane.
b. chloroplasts. d. cytoplasm.

____**12.** The energy of the electrons passing along the electron transport chain is used to make
a. lactic acid. c. alcohol.
b. citric acid. d. ATP.

____**13.** When the body needs to exercise for longer than 90 seconds, it generates ATP by carrying out
a. lactic acid fermentation.
b. alcoholic fermentation.
c. cellular respiration.
d. glycolysis.

____**14.** Unlike photosynthesis, cellular respiration occurs in
a. animal cells only. c. all but plant cells.
b. plant cells only. d. all eukaryotic cells.

____**15.** The products of photosynthesis are the
a. products of cellular respiration.
b. reactants of cellular respiration.
c. products of glycolysis.
d. reactants of fermentation.

Completion

Complete each statement on the line provided.

16. Glycolysis converts glucose into two molecules of ________________.

17. Based on Figure 9-1, ________________ ATP molecules are formed by fermentation.

18. The ________________ is a series of carrier proteins.

19. The body gets rid of lactic acid in a chemical pathway that requires ________________.

20. Based on Figure 9-1, the complete breakdown of glucose through cellular respiration results in the production of ________________ ATP molecules.

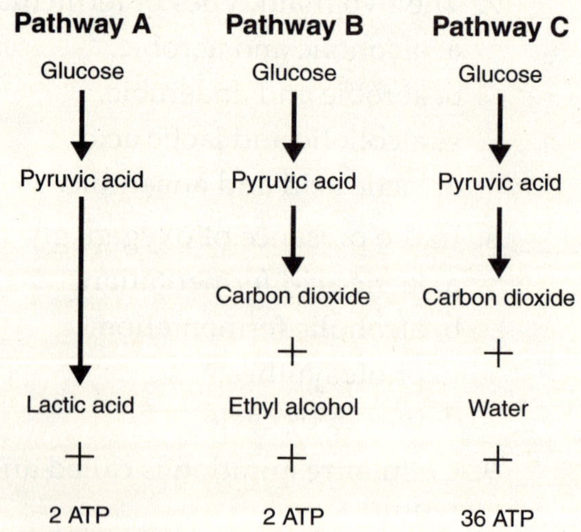

Figure 9-1

Name________________________ Class________________ Date ____________

Short Answer

In complete sentences, write the answers to the questions on the lines provided.

21. What is cellular respiration?

22. What happens during glycolysis?

23. Why is the Krebs cycle also known as the citric acid cycle?

24. What is the main function of the electron transport chain?

25. What roles does oxygen play in photosynthesis and in cellular respiration?

(Test B continues on page 116)

Name______________________________ Class __________________ Date ____________

Using Science Skills

Use the diagram below to answer the following questions on the lines provided.

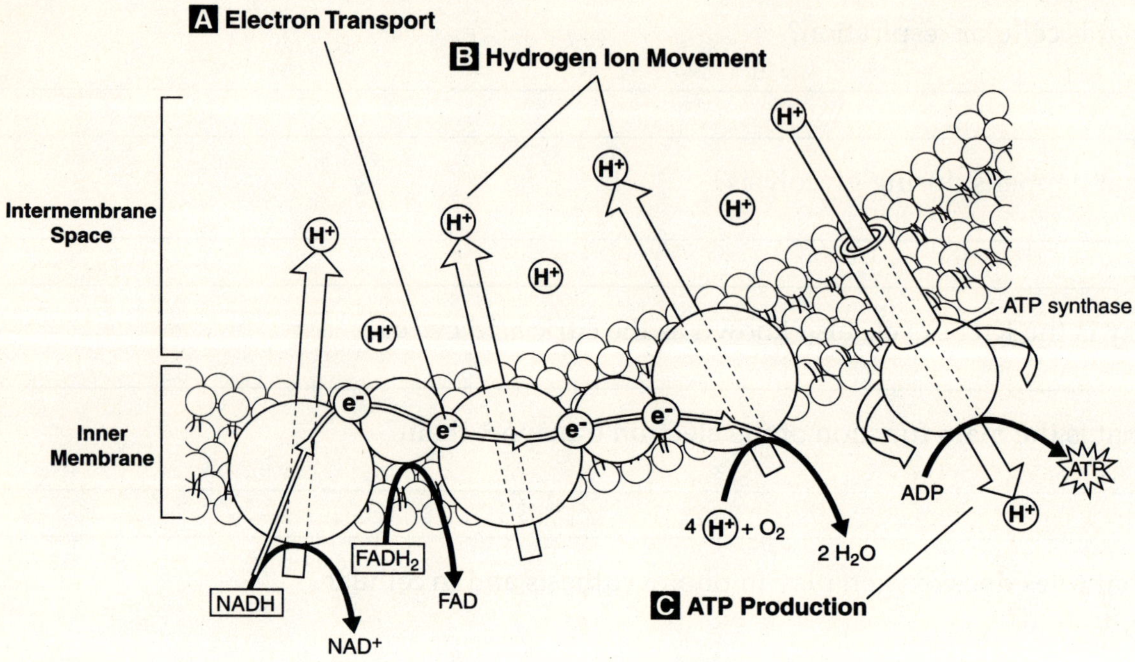

Figure 9-2

26. Interpreting Graphics What process does Figure 9-2 show?

____

27. Interpreting Graphics Look at Figure 9-2. Where do the electrons moving along the inner membrane come from?

____

____

28. Interpreting Graphics Where do the electrons moving along the inner membrane in Figure 9-2 end up?

____

29. Inferring Look at the arrows and H^+ ions in Figure 9-2. Which direction do most of the H^+ ions move in? What is the result of this movement?

____

____

____

30. Interpreting Graphics ATP synthase is an enzyme. Find ATP synthase in Figure 9-2. What reaction does ATP synthase catalyze when an H^+ ion passes through its channel?

____

____

Name______________________________ Class__________________ Date ______________

Multiple Choice

Write the letter that best answers the question or completes the statement on the line provided.

____ **1.** As a cell becomes larger, its
a. volume increases faster than its surface area.
b. surface area increases faster than its volume.
c. volume increases, but its surface area stays the same.
d. surface area stays the same, but its volume increases.

____ **2.** All of the following are problems that growth causes for cells EXCEPT
a. DNA overload.
b. excess oxygen.
c. obtaining enough food.
d. expelling wastes.

____ **3.** Which of the following is NOT a way that cell division solves the problems of cell growth?
a. Cell division provides each daughter cell with its own copy of DNA.
b. Cell division increases the mass of the original cell.
c. Cell division increases the surface area of the original cell.
d. Cell division reduces the original cell's volume.

____ **4.** When during the cell cycle are chromosomes visible?
a. only during interphase
b. only when they are being replicated
c. only during the M phase
d. only during the G_1 phase

____ **5.** Which pair is correct?
a. G_1 phase, DNA replication
b. G_2 phase, preparation for mitosis
c. S phase, cell division
d. M phase, cell growth

____ **6.** When during the cell cycle is a cell's DNA replicated?
a. G_1 phase
b. G_2 phase
c. S phase
d. M phase

____ **7.** Which event occurs during interphase?
a. The cell grows.
b. Centrioles appear.
c. Spindle fibers begin to form.
d. Centromeres divide.

____ **8.** During which phase of mitosis do the chromosomes line up along the middle of the dividing cell?
a. prophase
b. telophase
c. metaphase
d. anaphase

____ 9. Which of the following represents the phases of mitosis in their proper sequence?

a. prophase, metaphase, anaphase, telophase

b. interphase, prophase, metaphase, anaphase, telophase

c. interphase, prophase, metaphase, telophase

d. prophase, metaphase, anaphase, telophase, cytokinesis

____10. What is the role of the spindle during mitosis?

a. It helps separate the chromosomes.

b. It breaks down the nuclear membrane.

c. It duplicates the DNA.

d. It divides the cell in half.

____11. The two main stages of cell division are called

a. mitosis and interphase.

b. synthesis and cytokinesis.

c. the M phase and the S phase.

d. mitosis and cytokinesis.

____12. Which of the following is a factor that can stop normal cells from growing?

a. contact with other cells

b. growth factors

c. a cut in the skin

d. cyclin that has been taken from a cell in mitosis

____13. Which of the following explains why normal cells grown in a petri dish tend to stop growing once they have covered the bottom of the dish?

a. The cells lack cyclin.

b. The petri dish inhibits cell growth.

c. Contact with other cells stops cell growth.

d. Most cells grown in petri dishes have a defective p53.

____14. Cyclins are a family of closely related proteins that

a. regulate the cell cycle.

b. produce p53.

c. cause cancer.

d. work to heal wounds.

____15. Cancer is a disorder in which some cells have lost the ability to control their

a. size.

b. spindle fibers.

c. growth rate.

d. surface area.

Name________________________ Class________________ Date____________

Completion

Complete each statement on the line provided.

16. The larger a cell becomes, the ________________ efficiently it is able to function.

17. Before a normal cell becomes too large to carry out normal activities, it will usually divide to form two ________________ cells.

18. The structure labeled A in Figure 10-1 is called the ________________ .

19. Proteins that regulate the cell cycle based on events inside the cell are called ________________ regulators.

20. In all forms of ________________ , certain cells fail to respond to the signals that regulate the cell cycle of most cells.

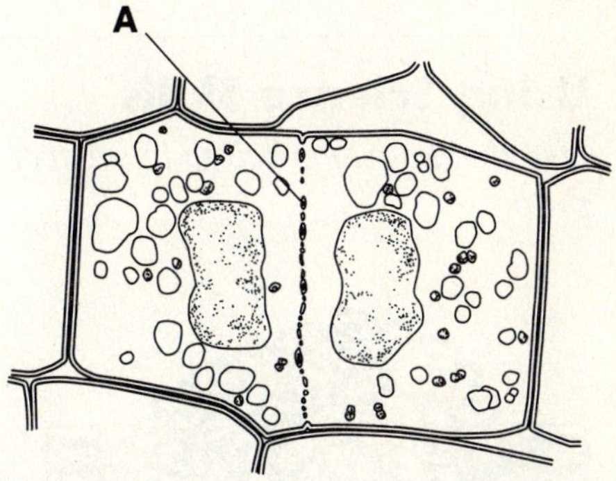

Figure 10-1

Short Answer

In complete sentences, write the answers to the questions on the lines provided.

21. What effect does cell size have on a cell's ability to efficiently carry out its activities? Give an example.

22. How does an increase in cell size affect the ratio of surface area to volume?

23. What is interphase, and which phases of the cell cycle does interphase include?

24. Describe how a plant cell produces a new cell wall during cytokinesis.

Name________________ Class________________ Date________________

25. Identify a factor that can stop cells from growing. Give an example of how this factor affects cell growth.

____

____

____

____

Using Science Skills

Use the diagrams below to answer the following questions on the lines provided.

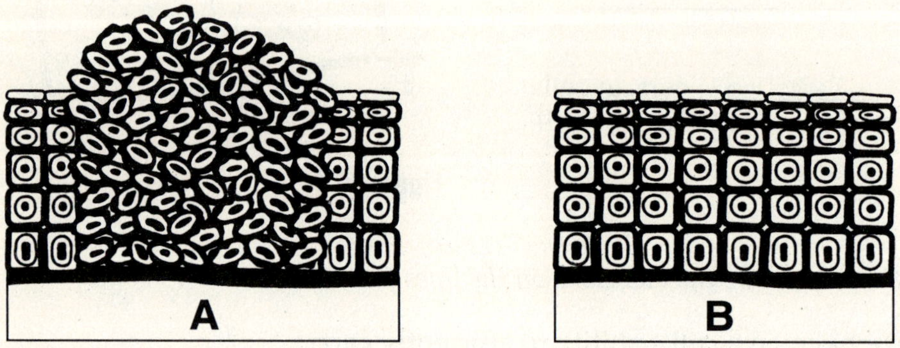

Figure 10-2

26. **Interpreting Graphics** Look at Figure 10-2. Which diagram shows cancer cells? How do you know?

____

____

27. **Comparing and Contrasting** Explain how cancer cells are different from normal cells. Then, relate these characteristics to the diagrams in Figure 10-2.

____

____

____

____

____

____

____

28. **Predicting** Look at the cancer cells shown in Figure 10-2. What can happen if these cells are left untreated?

____

____

____

Name______________________________ Class__________________ Date ______________

29. Applying Concepts Explain the role that p53 might have had in the growth and division of the cells shown in each diagram in Figure 10-2.

____

____

____

____

____

____

30. Problem Solving How might the cancer cells shown in Figure 10-2 be prevented from doing more harm to the organism of which they are a part?

____

____

____

____

____

Essay

Write the answer to each question in the space provided.

31. What kinds of problems does growth cause for cells? How does cell division help a cell solve these problems?

32. List and describe the main events of the cell cycle. Illustrate your description with a diagram of the cell cycle.

Name_____________________________ Class__________________ Date ______________

33. Describe what happens during the four phases of mitosis.

34. Describe how the skin cells near a cut behave. What role does contact with other cells have in the behavior of cells near a cut?

35. Describe how cancer cells are different from other cells. Based on these differences, explain why cancer has been such a difficult condition to cure.

Name________________________________ Class___________________ Date ______________

Multiple Choice

Write the letter that best answers the question or completes the statement on the line provided.

____ **1.** As a cell grows, it
a. places more demands on its DNA.
b. uses up food and oxygen more quickly.
c. has more trouble moving enough materials across its cell membrane.
d. all of the above

____ **2.** Compared with small cells, large cells have more trouble
a. dividing.
b. producing daughter cells.
c. moving needed materials in and waste products out.
d. making copies of their DNA.

____ **3.** The process by which a cell divides into two daughter cells is called
a. cell division. c. interphase.
b. metaphase. d. mitosis.

____ **4.** Which of the following happens when a cell divides?
a. The cell's volume increases.
b. It becomes more difficult for the cell to get enough oxygen and nutrients.
c. The cell has DNA overload.
d. Each daughter cell receives its own copy of the parent cell's DNA.

____ **5.** Which of the following is a phase in the cell cycle?
a. G_1 phase c. M phase
b. G_2 phase d. all of the above

____ **6.** Cell division is represented in Figure 10-1 by the letter
a. A. c. C.
b. B. d. D.

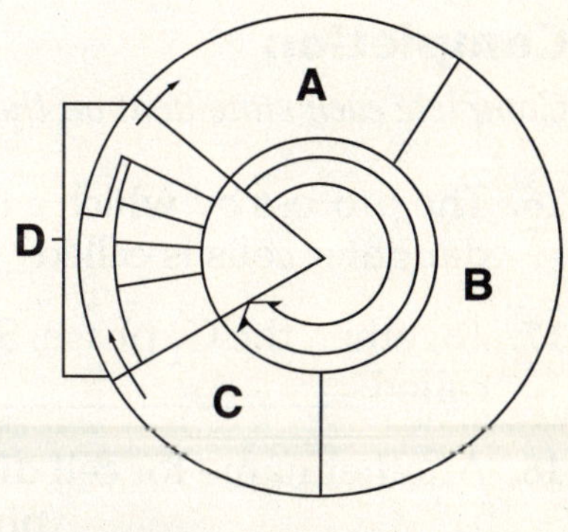

Figure 10-1

____ **7.** The cell cycle is the
a. series of events that cells go through as they grow and divide.
b. period of time between the birth and the death of a cell.
c. time from prophase until cytokinesis.
d. time it takes for one cell to undergo mitosis.

____ 8. The structure labeled A in Figure 10-2 is called the
   a. centromere.   c. sister chromatid.
   b. centriole.   d. spindle.

____ 9. The structures labeled B in Figure 10-2 are called
   a. centromeres.   c. sister chromatids.
   b. centrioles.   d. spindles.

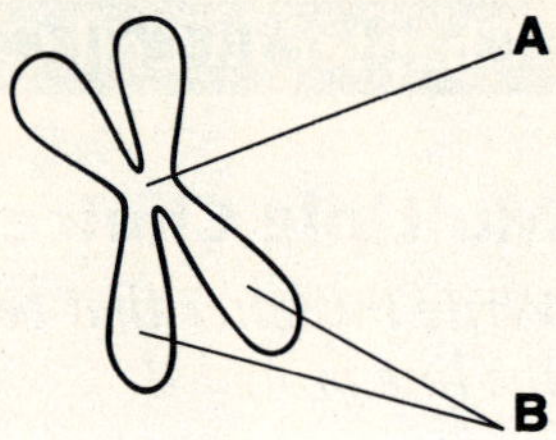

Figure 10-2

____10. Which of the following is a phase of mitosis?
   a. cytokinesis   c. anaphase
   b. interphase   d. S phase

____11. The first phase of mitosis is called
   a. prophase.   c. metaphase.
   b. anaphase.   d. interphase.

____12. In which phase of mitosis do chromosomes become visible?
   a. prophase   c. metaphase
   b. interphase   d. telophase

____13. What happens when cells come into contact with other cells?
   a. They divide more quickly.
   b. They stop growing.
   c. They produce cyclins.
   d. They produce p53.

____14. In eukaryotic cells, the timing of the cell cycle is regulated by
   a. the centrioles.   c. the spindle.
   b. cyclins.   d. all of the above

____15. What is a tumor?
   a. an accumulation of cyclins
   b. a mass of cancer cells
   c. the rapidly dividing cells found at the site of a wound
   d. a defective p53 gene

Completion

Complete each statement on the line provided.

16. The process by which a cell divides into two daughter cells is called __________________ .

17. Together, the G_1 phase, S phase, and G_2 phase are called __________________ .

18. Another name for cell division is the __________________ phase.

19. Look at Figure 10-3. The process shown occurs directly following mitosis. This process is called __________________ .

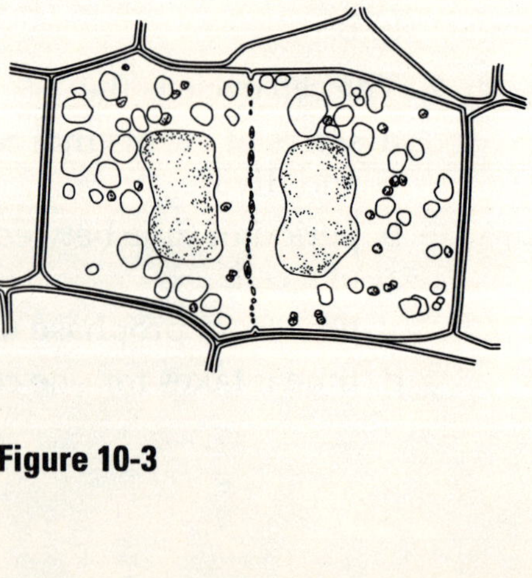

Figure 10-3

20. Proteins called __________________ regulate the timing of the cell cycle in eukaryotic cells.

Short Answer

In complete sentences, write the answers to the questions on the lines provided.

21. List two problems that growth causes for cells.

____

____

____

22. Why are chromosomes not visible in most cells except during cell division?

____

____

____

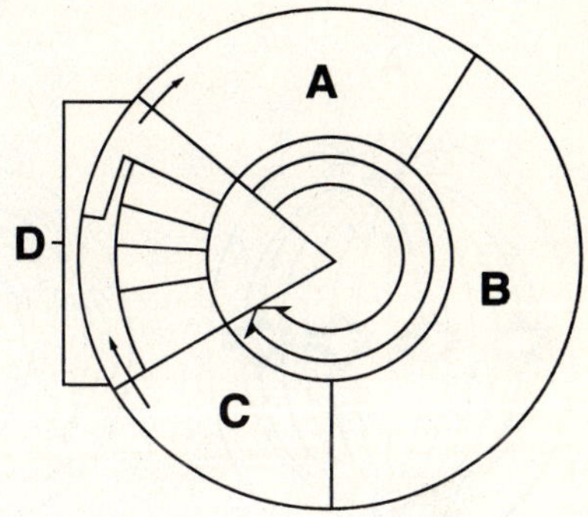

Figure 10-4

23. The main events of the cell cycle are labeled A, B, C, and D in Figure 10-4. Name these events. Then, briefly state what happens during each event.

____

____

____

____

____

____

____

____

24. Name two factors that help regulate the timing of the cell cycle.

____

____

____

25. How do cancer cells differ from normal cells?

____

____

____

____

____

Name________________________ Class________________ Date____________

Using Science Skills

Use the diagrams below to answer the following questions on the lines provided.

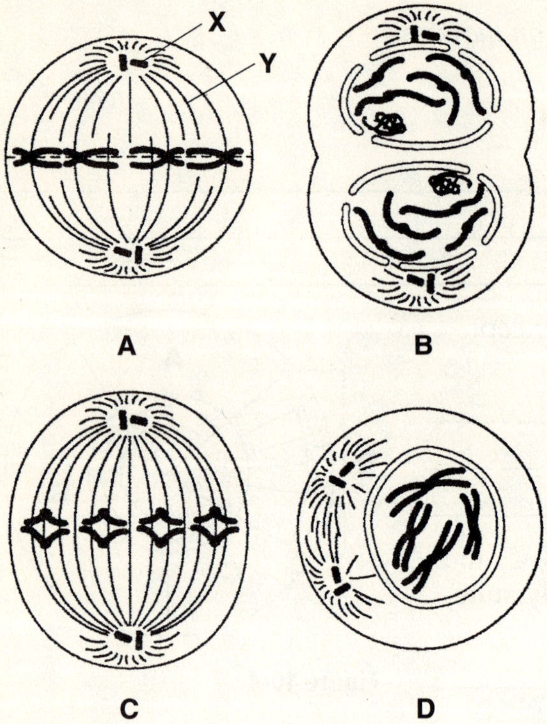

Figure 10-5

26. **Interpreting Graphics** What does Figure 10-5 represent? How do you know if this is an animal cell or a plant cell?

____

____

____

27. **Inferring** What is the chromosome number of the cell shown in Figure 10-5?

____

28. **Inferring** Identify the structures labeled X and Y in Figure 10-5.

____

29. **Applying Concepts** List the correct order for the diagrams in Figure 10-5.

____

30. **Predicting** After the steps shown in Figure 10-5 have been arranged in the correct order, what would a diagram of a final step show?

____

____

Name______________________________ Class________________ Date ____________

Multiple Choice

Write the letter that best answers the question or completes the statement on the line provided.

____ **1.** In a eukaryote, the portion of the cell outside the nucleus is the

a. nuclear envelope.
b. epithelial tissue.
c. cytoplasm.
d. organelle.

____ **2.** Which of the following is NOT true about the cell nucleus?

a. assembles lipid components of the cell membrane
b. contains nearly all the cell's DNA
c. contains a region where the assembly of ribosomes begins
d. controls many of the cell's activities

____ **3.** The main function of the cell wall is to

a. provide a lipid bilayer for the cell membrane.
b. contain the nucleus outside the cytoplasm.
c. provide support and protection for the cell.
d. prevent facilitated diffusion.

____ **4.** A group of similar cells that perform a particular function is called a(an)

a. organ.
b. tissue.
c. organ system.
d. multicellular organism.

____ **5.** Organisms that use light energy to produce food are called

a. heterotrophs.
b. autotrophs.
c. consumers.
d. pigments.

____ **6.** Photosynthesis begins when

a. pigments in photosystem II absorb light.
b. ATP and NADPH enter the stroma.
c. pigments in photosytem I absorb light.
d. high-energy electrons move from photosystem II to photosystem I.

____ **7.** Which of the following is NOT a factor that affects photosynthesis?

a. temperature
b. water availability
c. light intensity
d. heterotroph availability

____ **8.** Which of the following is NOT a main stage of cellular respiration?

a. Krebs cycle
b. glycolysis
c. electron transport
d. fermentation

____ **9.** The products of lactic acid fermentation are

a. lactic acid and NAD^+.
b. pyruvic acid and NADH.
c. CO_2 and NAD^+.
d. alcohol and lactic acid.

____**10.** In the Krebs cycle, pyruvic acid is

a. generated so that glycolysis can continue.
b. broken down into carbon dioxide.
c. used to convert ATP into ADP.
d. broken down into glucose.

____**11.** In eukaryotes, where is the electron transport chain located?

a. in the inner membrane of the mitochondrion
b. in the cytoplasm
c. in the stroma of the chloroplast
d. at the center of the mitochondrion

____**12.** What is both a reactant of cellular respiration and a product of photosynthesis?

a. carbon dioxide
b. oxygen
c. citric acid
d. water

____**13.** Which of the following is NOT a problem as a cell becomes larger?

a. More demands are placed on the cell's DNA.
b. The cell has more trouble moving nutrients into the cell.
c. Cell volume decreases as cell surface area increases.
d. The cell has more trouble moving wastes out of the cell.

____14. During which phase of the cell cycle are chromosomes replicated?

a. M phase

b. S phase

c. G_1 phase

d. G_2 phase

____15. Cancer cells do not respond to signals that

a. cause cells to continue to grow.

b. regulate the growth of most cells.

c. produce a defect in a gene called p53.

d. produce contact with other cells.

Completion

Complete each statement on the line provided.

16. The diffusion of water through a selectively permeable membrane is called ____________________.

17. Organisms that obtain energy from the foods they consume are ____________________.

18. The process that uses the energy of sunlight to convert water and carbon dioxide into high-energy sugars is ____________________.

19. In ____________________, one molecule of glucose is broken in half, producing two molecules of pyruvic acid.

20. In the process of cell division, the division of the cell nucleus is called ____________________.

Short Answer

In complete sentences, write the answers to the questions on the lines provided.

21. What are the three statements of the cell theory?

____

____

____

22. What is ATP, and how is energy released from it?

____

____

____

____

Name________________________________ Class__________________ Date ______________

23. Write the overall equation for photosynthesis in both symbols and words.

____

____

24. Write the overall equation for cellular respiration in both symbols and words.

____

____

25. How does cell division affect a cell's ratio of surface area to volume?

____

____

____

Using Science Skills

Use the drawings below to answer the following questions on the lines provided.

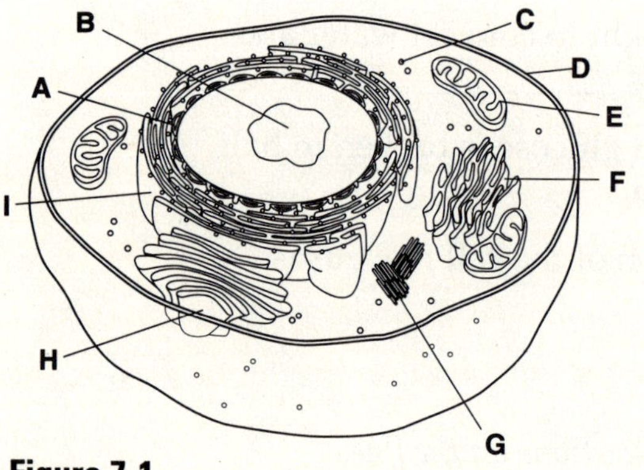

Figure 7-1

Diagram I

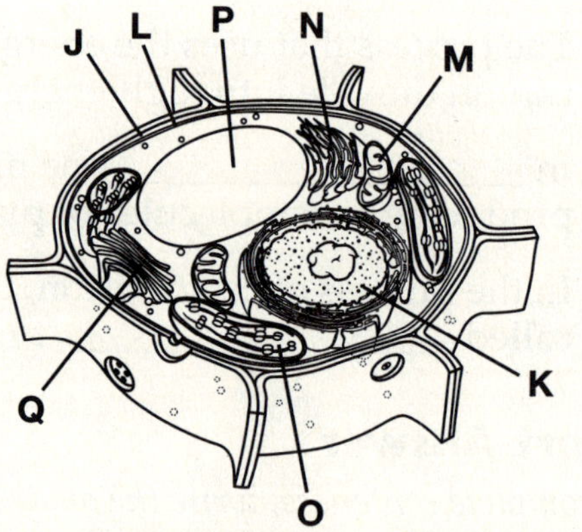

Diagram II

26. **Classifying** Would both drawings in Figure 1 be classified as eukaryotes or prokaryotes? Explain how you know.

____

____

27. **Comparing and Contrasting** Give the letter of the structure in drawing II of Figure 1 that corresponds with structure D in drawing I. Identify the structure by name and explain in what way the two examples differ.

____

____

Name______________________________ Class__________________ Date ______________

28. Comparing and Contrasting Give the letter of the structure in drawing I of Figure 1 that corresponds to structure O in drawing II. Identify that structure and the process that occurs there.

____

____

29. Comparing and Contrasting Give the letter of the structure in drawing II of Figure 1 that corresponds to structure E in drawing I. What is that structure, and what process occurs there?

____

____

30. Drawing Conclusions From the structures you can identify in the two cells shown in drawings I and II of Figure 1, what do you conclude about the type of cell each drawing depicts?

____

____

____

Name______________________________ Class __________________ Date ______________

Essay

Write the answer to each question in the space provided.

31. How are facilitated diffusion and active transport similar, and how do they differ?

32. What happens when pigments obtain energy from light? Use details from the light-dependent reactions in your response.

33. Explain how much energy a cell can produce from a single molecule of glucose when oxygen is available and when oxygen is not available.

34. How are electron carriers essential in both photosynthesis and cellular respiration?

35. What happens during each of the four phases of mitosis in an animal cell?

Multiple Choice

Write the letter that best answers the question or completes the statement on the line provided.

____ **1.** Which of the following contains a nucleus?
a. mitochondria
b. eukaryotes
c. prokaryotes
d. DNA

____ **2.** The portion of the cell outside the nucleus is called the
a. cell wall.
b. cytoplasm.
c. cell tissue.
d. lipid bilayer.

____ **3.** Which of the following is a function of the cell membrane?
a. modifies, sorts, and packages proteins
b. converts sunlight into chemical energy
c. provides a place where the assembly of ribosomes begins
d. regulates what enters and leaves the cell

____ **4.** Cell specialization is a process in which
a. cells throughout an organism develop in different ways to perform different tasks.
b. materials move through a cell membrane during exocytosis.
c. biologists identify different levels of organization.
d. water moves through a selectively permeable membrane.

____ **5.** One of the principal compounds that cells use to store and release energy is
a. CO_2.
b. O_2.
c. ATP.
d. NADPH.

____ **6.** What is the principal pigment of plants?
a. carbon dioxide
b. thylakoid
c. chlorophyll
d. adenosine triphosphate

____ **7.** The Calvin cycle uses ATP and NADPH to produce
a. high-energy sugars.
b. the energy carriers ADP and $NADP^+$.
c. alcohol and lactic acid.
d. the energy of sunlight.

____ **8.** The first set of reactions in cellular respiration is
a. the Krebs cycle.
b. electron transport.
c. the Calvin cycle.
d. glycolysis.

____ **9.** What process releases energy from food molecules by producing ATP in the absence of oxygen?
a. cellular respiration
b. photosynthesis
c. glycolysis
d. fermentation

____**10.** The electron transport chain uses high-energy electrons from the Krebs cycle to
a. break glucose in half.
b. convert ADP into ATP.
c. produce high-energy sugars.
d. produce oxygen gas.

____**11.** Cellular respiration is to mitochondria as photosynthesis is to
a. osmosis.
b. microtubules.
c. chloroplasts.
d. ribosomes.

____**12.** As a cell grows larger, it has more trouble
a. replicating its chromosomes.
b. moving enough nutrients across the cell membrane.
c. moving through the cell cycle.
d. completing cell division.

____**13.** The process by which a cell divides into two daughter cells is called
a. cellular respiration.
b. cell cyclin.
c. cell division.
d. ratio of surface area to volume.

____**14.** What is cytokinesis?
a. the division of a cell's cytoplasm
b. a pigment in the chloroplast
c. the final phase of interphase
d. the spreading of cancer cells throughout the body

Name______________________________ Class_________________ Date _____________

____15. The proteins that regulate the timing of the cell cycle in eukaryotic cells are called
a. cyclins.
b. chromatids
c. tumors.
d. centrioles.

Completion

Complete each statement on the line provided.

16. The thin, flexible barrier around a cell is called the ________________.

17. A group of similar cells that perform a particular function is called a(an) ________________.

18. Plants take in the sun's energy by absorbing ________________.

19. Cellular respiration is called an aerobic process because it requires ________________.

20. Each chromosome consists of two sister ________________.

Short Answer

In complete sentences, write the answers to the questions on the lines provided.

21. What does the cell theory state?

____

____

____

22. What is the difference between an autotroph and a heterotroph?

____

____

____

23. In words, write the overall equation for photosynthesis.

____

____

24. In words, write the overall equation for cellular respiration.

____

____

25. What are the phases of mitosis, in the order in which they occur?

____

____

Name________________________ Class________________ Date ____________

Using Science Skills

Use the drawings below to answer the following questions on the lines provided.

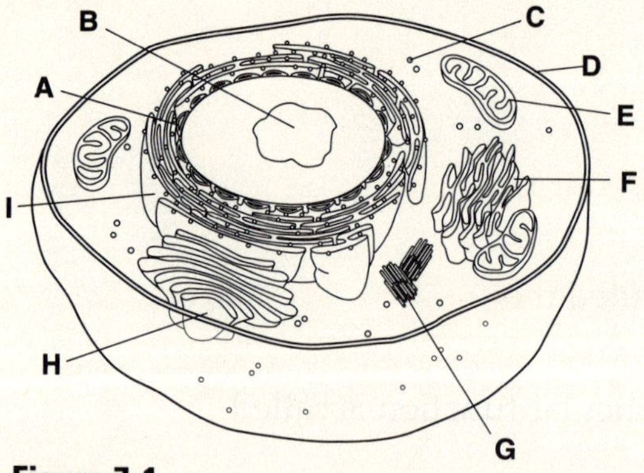

Figure 7-1

Diagram I

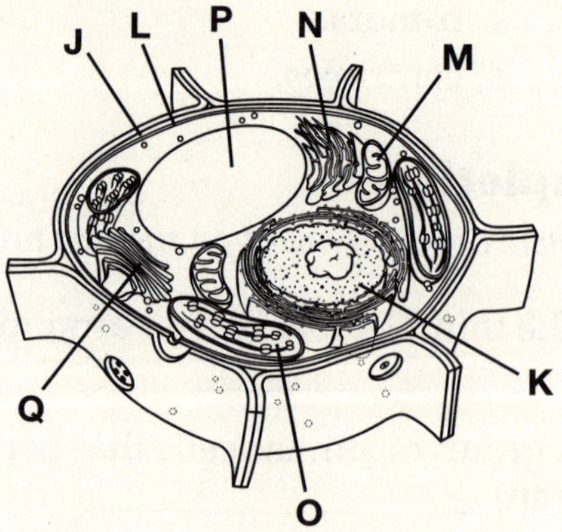

Diagram II

26. Comparing and Contrasting Look at Figure 1. Which structure letter in drawing II corresponds to structure A in drawing I? What is the name of this structure?

____

____

27. Comparing and Contrasting Look at Figure 1. Which structure letter in drawing I corresponds to structure M in drawing II? What is the name of this structure?

____

____

28. Interpreting Graphics Look at drawing II of Figure 1. What is the structure labeled J, and what is its function?

____

____

29. Comparing and Contrasting In drawing II, what is the structure labeled O? Why is there no corresponding structure in drawing I?

____

____

30. Interpreting Graphics Do the drawings in Figure 1 represent eukaryotes or prokaryotes? How can you tell?

____

____

Name________________________ Class________________ Date ____________

Multiple Choice

Write the letter that best answers the question or completes the statement on the line provided.

____ **1.** Offspring that result from crosses between true-breeding parents with different traits
   a. are true-breeding.
   b. make up the F_2 generation.
   c. make up the parental generation.
   d. are called hybrids.

____ **2.** Mendel concluded that traits are
   a. not inherited by offspring.
   b. inherited through the passing of factors from parents to offspring.
   c. determined by dominant factors only.
   d. determined by recessive factors only.

____ **3.** When Mendel crossed true-breeding tall plants with true-breeding short plants, all the offspring were tall because
   a. the allele for tall plants is recessive.
   b. the allele for short plants is dominant.
   c. the allele for tall plants is dominant.
   d. they were true-breeding like their parents.

____ **4.** In the P generation, a tall plant was crossed with a short plant. Short plants reappeared in the F_2 generation because
   a. some of the F_2 plants produced gametes that carried the allele for shortness.
   b. the allele for shortness is dominant.
   c. the allele for shortness and the allele for tallness segregated when the F_1 plants produced gametes.
   d. they inherited an allele for shortness from one parent and an allele for tallness from the other parent.

____ **5.** The principles of probability can be used to
   a. predict the traits of the offspring produced by genetic crosses.
   b. determine the actual outcomes of genetic crosses.
   c. predict the traits of the parents used in genetic crosses.
   d. decide which organisms are best to use in genetic crosses.

____ **6.** A Punnett square shows all of the following EXCEPT
   a. all possible results of a genetic cross.
   b. the genotypes of the offspring.
   c. the alleles in the gametes of each parent.
   d. the actual results of a genetic cross.

Name______________________________ Class__________________ Date ______________

____ 7. How many different allele combinations would be found in the gametes produced by a pea plant whose genotype was *RrYY?*

a. 2 c. 8

b. 4 d. 16

____ 8. A cross of a white hen with a black rooster produces erminette-color offspring. This type of inheritance is known as

a. incomplete dominance.

b. polygenic inheritance.

c. codominance.

d. multiple alleles.

____ 9. Which of the following statements about Thomas Hunt Morgan is true?

a. Morgan chose to use fruit flies in his study of genetics because fruit flies produce a large number of offspring.

b. Morgan chose to use fruit flies in his study of genetics because fruit flies take a long time to produce offspring.

c. Morgan continued Mendel's genetic study of pea plants.

d. Morgan discovered that Mendel's principles of genetics do not apply to fruit flies.

____10. If an organism's diploid number is 12, its haploid number is

a. 12. c. 24.

b. 6. d. 3.

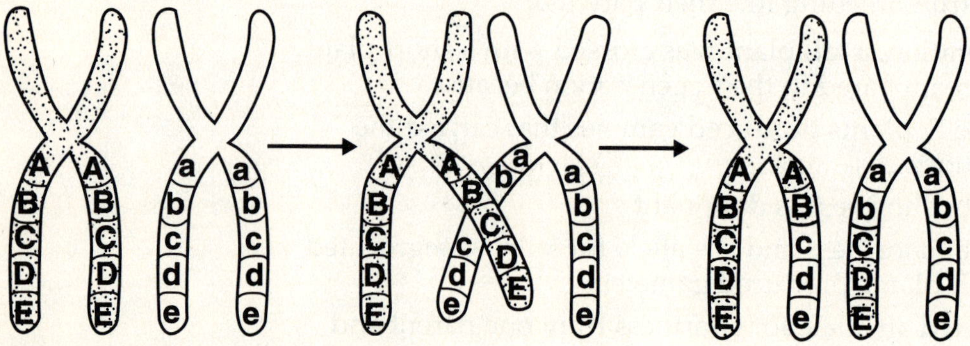

Figure 11-1

____11. What is shown in Figure 11-1?

a. independent assortment

b. anaphase I of meiosis

c. crossing-over

d. incomplete dominance

____12. Chromosomes form tetrads during

a. prophase of meiosis I.

b. metaphase of meiosis I.

c. prophase of meiosis II.

d. metaphase of meiosis II.

____13. Unlike mitosis, meiosis results in the formation of
- a. two genetically identical diploid cells.
- b. four genetically different haploid cells.
- c. four genetically identical haploid cells.
- d. two genetically different diploid cells.

____14. Linked genes
- a. are never separated.
- b. assort independently.
- c. are on the same chromosome.
- d. are always recessive.

____15. If two genes are on the same chromosome and rarely assort independently,
- a. crossing-over never occurs between the genes.
- b. crossing-over always occurs between the genes.
- c. the genes are probably located far apart from each other.
- d. the genes are probably located close to each other.

Completion

Complete each statement on the line provided.

16. The plants that Mendel crossed to produce the F_1 generation made up the ____________________ generation.

17. A pea plant that has two different alleles for the same trait is said to be ____________________.

18. Pea plants that are *TT*, ____________________, or *tt* have different genotypes.

19. An organism's gametes have ____________________ the number of chromosomes found in the organism's body cells.

20. The process shown in Figure 11-2 may occur during the stage of meiosis called ____________________.

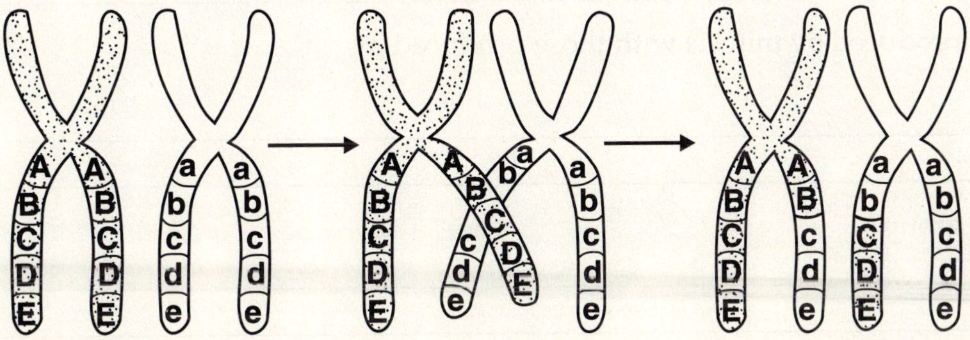

Figure 11-2

Name________________ Class________________ Date ________________

Short Answer

In complete sentences, write the answers to the questions on the lines provided.

21. How many recessive alleles for a trait must an organism inherit in order to show that trait?

____

____

RrYy (top) × **RrYy** (left)

	RY	Ry	rY	ry
RY	RRYY	RRYy	RrYY	RrYy
Ry	RRYy	RRyy	RrYy	Rryy
rY	RrYY	RrYy	rrYY	rrYy
ry	RrYy	Rryy	rrYy	rryy

Seed Shape
R – round
r – wrinkled

Seed Color
Y – yellow
y – green

Figure 11-3

22. What is the phenotype ratio of the offspring in the Punnett square shown in Figure 11-3?

____

____

23. Contrast the cells produced by mitosis with those produced by meiosis.

____

____

24. What is a linkage group?

____

____

25. What does a gene map show?

____

____

Name______________________ Class______________ Date ____________

Using Science Skills

Use the Punnett square below to answer the following questions on the lines provided.

Heterozygous male guinea pigs with black, rough hair (*BbRr*) are crossed with heterozygous female guinea pigs with black, rough hair (*BbRr*). The incomplete Punnett square in Figure 11-4 shows the expected results from the cross.

BbRr × BbRr	BR	Br	bR	br
BR	BBRR	BBRr	BbRR	BbRr
Br	BBRr	BBrr	BbRr	Bbrr
bR	BbRR	BbRr	X	bbRr
br	BbRr	Bbrr	bbRr	bbrr

Hair Color
B – black
b – white

Hair Texture
R – rough
r – smooth

Figure 11-4

26. **Using Tables and Graphs** Identify the genotype of the offspring that would be represented in the square labeled X in Figure 11-4.

____

27. **Using Tables and Graphs** Identify the phenotype of the offspring represented in the square labeled X in Figure 11-4.

____

28. **Analyzing Data** In Figure 11-4, what are the different phenotypes of the offspring?

____

____

29. **Analyzing Data** In Figure 11-4, what are the genotypes of the offspring that have black, rough hair?

____

30. **Calculating** What fraction of the offspring in Figure 11-4 is expected to have white, smooth hair?

____

Name______________________________ Class__________________ Date ______________

Essay

Write the answer to each question in the space provided.

31. Explain why short plants reappeared in Mendel's F_2 generation of pea plants.

32. You wish to determine whether a tall pea plant is homozygous or heterozygous for tallness. What cross should you perform to arrive at your answer? Explain your choice of cross.

33. Explain the difference between incomplete dominance and codominance.

34. The stages of meiosis are classified into two divisions: meiosis I and meiosis II. Compare and contrast these two divisions.

35. Define *linkage,* and explain how linkage is used to make gene maps.

Name______________________________ Class__________________ Date ______________

Multiple Choice

Write the letter that best answers the question or completes the statement on the line provided.

____ **1.** Gregor Mendel used pea plants to study
a. flowering. c. the inheritance of traits.
b. gamete formation. d. cross-pollination.

____ **2.** What are Mendel's factors called today?
a. alleles c. genes
b. traits d. characters

____ **3.** The principle of dominance states that
a. all alleles are dominant.
b. all alleles are recessive.
c. some alleles are dominant and others are recessive.
d. alleles are neither dominant nor recessive.

____ **4.** When Mendel allowed the tall F_1 pea plants (*Tt*) to self-pollinate,
a. the offspring were of medium height.
b. all of the offspring were tall.
c. all of the offspring were short.
d. some of the offspring were short.

____ **5.** Two plants with the genotypes *TT* and *Tt*
a. would have the same phenotype.
b. would have different phenotypes.
c. have all dominant alleles.
d. have all recessive alleles.

____ **6.** Organisms that have two identical alleles for a particular trait are said to be
a. hybrid. c. heterozygous.
b. homozygous. d. dominant.

____ **7.** What principle states that during gamete formation genes for different traits separate without influencing each other's inheritance?
a. principle of dominance
b. principle of independent assortment
c. principle of probabilities
d. principle of segregation

RrYy (columns); RrYy (rows)

	RY	Ry	rY	ry
RY	RRYY	RRYy	RrYY	RrYy
Ry	RRYy	RRyy	RrYy	Rryy
rY	RrYY	RrYy	rrYY	rrYy
ry	RrYy	Rryy	rrYy	rryy

Seed Shape
R – round
r – wrinkled

Seed Color
Y – yellow
y – green

Figure 11-1

____ **8.** The Punnett square in Figure 11-1 shows that the gene for pea shape and the gene for pea color
a. assort independently.
b. are linked.
c. have the same alleles.
d. are always homozygous.

____ **9.** Situations in which one allele for a gene is not completely dominant over another allele for that gene are called
a. multiple alleles.
b. incomplete dominance.
c. codominant alleles.
d. multiple genes.

____**10.** Mendel's principles of genetics apply to
a. plants only. c. pea plants only.
b. animals only. d. all organisms.

____**11.** The number of chromosomes in a gamete is represented by the symbol
a. 2N. c. N.
b. X. d. Y.

____**12.** Gametes are produced by the process of
a. mitosis. c. crossing-over.
b. meiosis. d. replication.

____**13.** Unlike mitosis, meiosis results in the formation of
a. diploid cells. c. 2N daughter cells.
b. haploid cells. d. body cells.

Name________________________________ Class____________________ Date ________________

____**14.** Traits that are produced by the interaction of several genes are said to be

a. polygenic. c. haploid.
b. codominant. d. diploid.

____**15.** Gene maps are based on

a. the frequencies of crossing-over between genes.
b. independent assortment.
c. genetic diversity.
d. the number of genes in a cell.

Completion

Complete each statement on the line provided.

16. The different forms of a gene are called ____________________ .

17. ____________________ is the likelihood that a particular event will occur.

18. In the Punnett square shown in Figure 11-2, the genotypes of the offspring are ____________________ .

19. The principle of independent assortment states that ____________________ for different traits can segregate independently during the formation of gametes.

20. Meiosis produces four genetically different ____________________ cells.

		Tt	
		T	***t***
TT	***T***	***TT***	***Tt***
	T	***TT***	***Tt***

T = *tall*
t = *short*

Figure 11-2

Short Answer

In complete sentences, write the answers to the questions on the lines provided.

21. Define *genetics.*

____

22. How many sets of chromosomes are in a diploid cell?

____

23. Define *homologous chromosomes.*

____

____

____

24. What happens to the number of chromosomes per cell during meiosis?

____

25. Why did Mendel not observe gene linkage during his experiments with pea plants?

____

____

Name______________________________ Class____________________ Date ______________

Using Science Skills

Use the diagram below to answer the following questions on the lines provided.

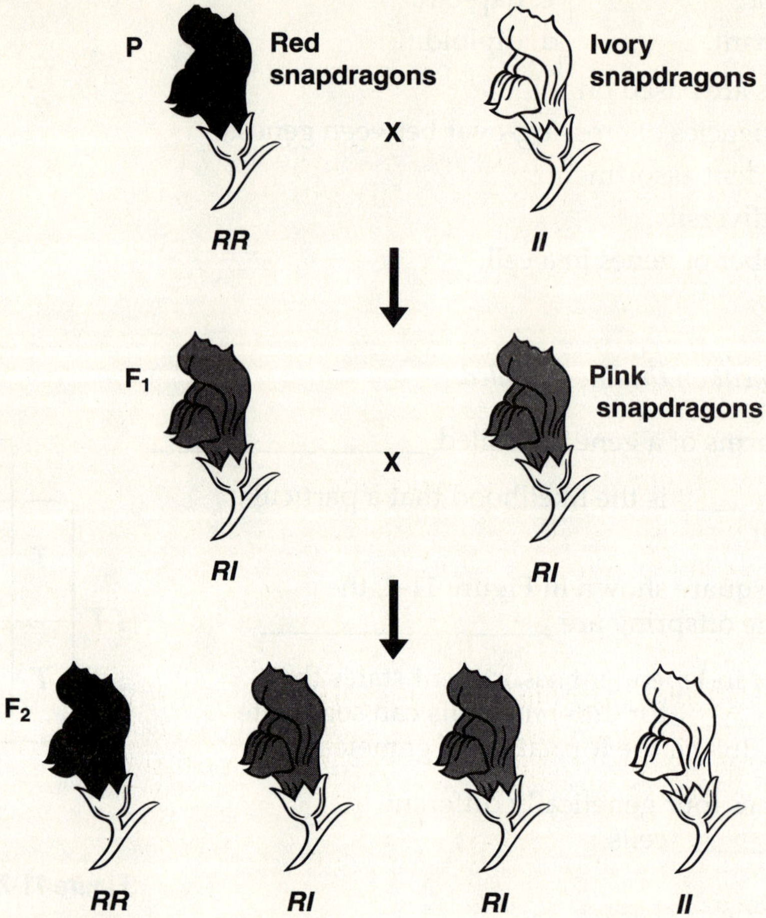

Figure 11-3

26. Inferring What do the letters *R* and *I* represent in Figure 11-3?

____

27. Interpreting Graphics In Figure 11-3, what is the genotype of the pink-flowered snapdragons?

____

28. Inferring Explain whether the alleles in Figure 11-3 show dominance, incomplete dominance, or codominance.

____

29. Inferring According to Figure 11-3, if red-flowered snapdragons and ivory-flowered snapdragons are crossed, what percent of their offspring are expected to be pink-flowered?

____

30. Inferring According to Figure 11-3, if two pink-flowered snapdragons are crossed, what percent of their offspring are expected to be pink-flowered?

____

Name______________________________ Class__________________ Date ______________

Multiple Choice

Write the letter that best answers the question or completes the statement on the line provided.

____ **1.** What did Griffith observe when he injected into mice a mixture of heat-killed disease-causing bacteria and live harmless bacteria?

a. The disease-causing bacteria changed into harmless bacteria.

b. The mice developed pneumonia.

c. The harmless bacteria died.

d. The mice were unaffected.

____ **2.** Which of the following is a nucleotide found in DNA?

a. ribose + phosphate group + thymine

b. ribose + phosphate group + uracil

c. deoxyribose + phosphate group + uracil

d. deoxyribose + phosphate group + cytosine

____ **3.** DNA replication results in two DNA molecules,

a. each with two new strands.

b. one with two new strands and the other with two original strands.

c. each with one new strand and one original strand.

d. each with two original strands.

____ **4.** During mitosis, the

a. DNA molecules unwind.

b. histones and DNA molecules separate.

c. DNA molecules become more tightly coiled.

d. nucleosomes become less tightly packed.

____ **5.** Unlike DNA, RNA contains

a. adenine. c. phosphate groups.

b. uracil. d. thymine.

____ **6.** Which type(s) of RNA is(are) involved in protein synthesis?

a. transfer RNA only

b. messenger RNA only

c. ribosomal RNA and transfer RNA only

d. messenger RNA, ribosomal RNA, and transfer RNA

____ **7.** During transcription, an RNA molecule is formed

a. that is complementary to both strands of DNA.

b. that is complementary to neither strand of DNA.

c. that is double-stranded.

d. inside the nucleus.

____ **8.** How many codons are needed to specify three amino acids?

a. 3 c. 9

b. 6 d. 12

____ **9.** Which of the following terms is LEAST closely related to the others?

a. intron c. polypeptide

b. tRNA d. anticodon

____**10.** Which type of RNA functions as a blueprint of the genetic code?

a. rRNA c. mRNA

b. tRNA d. RNA polymerase

____**11.** Which of the following is NOT a gene mutation?

a. inversion c. deletion

b. insertion d. substitution

____**12.** Which of the following statements is true?

a. A promoter determines whether a gene is expressed.

b. An expressed gene is turned off.

c. Proteins that bind to regulatory sites on DNA determine whether a gene is expressed.

d. RNA polymerase regulates gene expression.

____**13.** A *lac* repressor turns off the *lac* genes by binding to

a. the promoter.

b. tRNA.

c. the operator.

d. the *lac* genes.

____**14.** Gene regulation in eukaryotes

a. usually involves operons.

b. is simpler than in prokaryotes.

c. allows for cell specialization.

d. includes the action of DNA polymerase.

____**15.** Which of the following statements is NOT true?

a. Mutations do not occur in hox genes.

b. Hox genes that are found in different animals are very different from each other.

c. Hox genes control the normal development of an animal.

d. Hox genes occur in clusters.

Completion

Complete each statement on the line provided.

16. The structure labeled X in Figure 12-1 is a(an) ____________________.

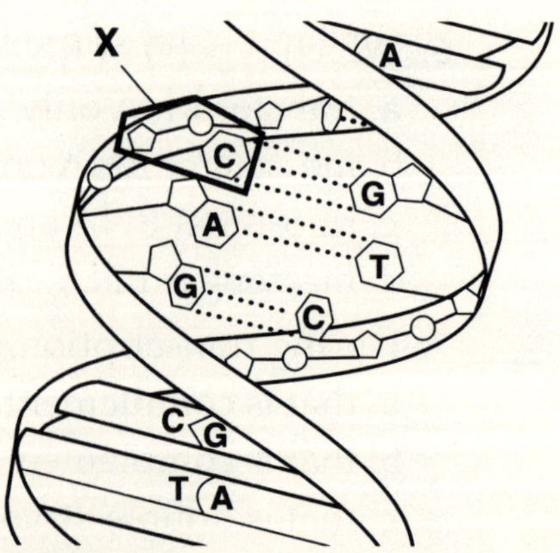

Figure 12-1

Name______________________ Class__________________ Date ______________

17. The order of nitrogenous bases in DNA determines the order of ________________ in proteins.

18. There is no ________________ that is specified by a stop codon on an mRNA molecule.

19. The *lac* repressor releases the operator in the presence of ________________ .

20. In eukaryotes, proteins that attract RNA polymerase bind to ________________ sequences in DNA.

Short Answer

In complete sentences, write the answers to the questions on the lines provided.

21. At the beginning of DNA replication, what two processes "unzip" the two strands of a DNA molecule?

____

____

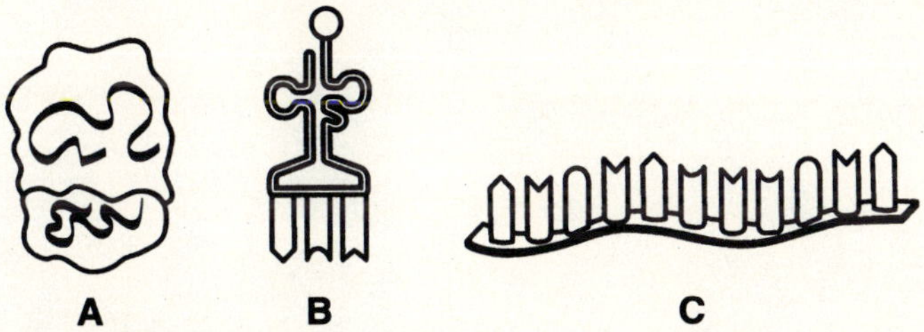

Figure 12-2

22. In Figure 12-2, which molecule is tRNA, and what is its function?

____

____

____

____

____

23. According to Figure 12-3, what codons specify the amino acid arginine?

____

____

____

____

____

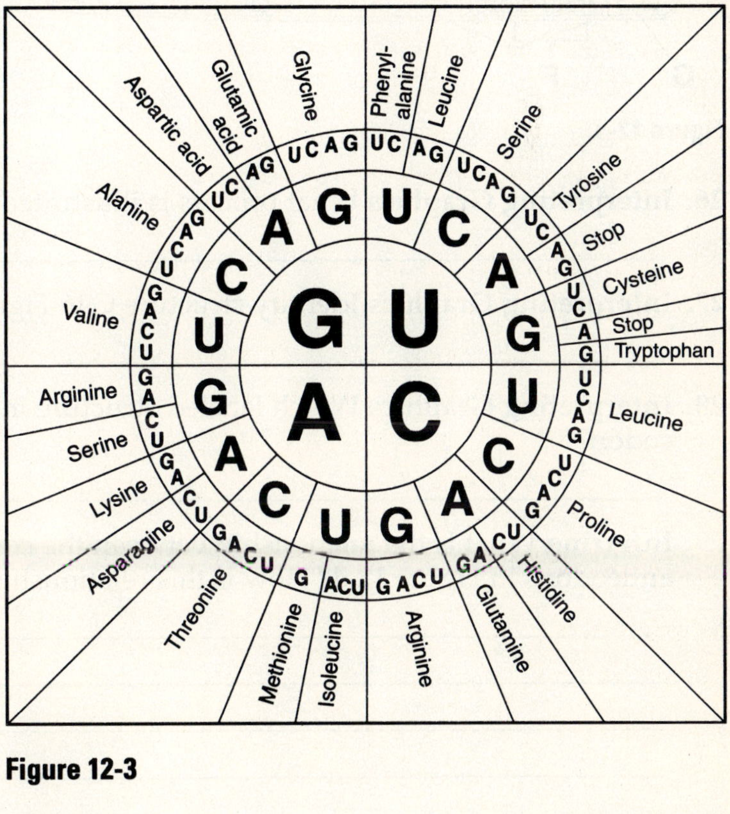

Figure 12-3

Name________________________ Class________________ Date ____________

24. What happens to *lac* repressors in *E. coli* when lactose is present?

____

____

25. Why are hox genes that are found in different animals very similar to one another?

____

____

Using Science Skills

Use the diagram below to answer the following questions on the lines provided.

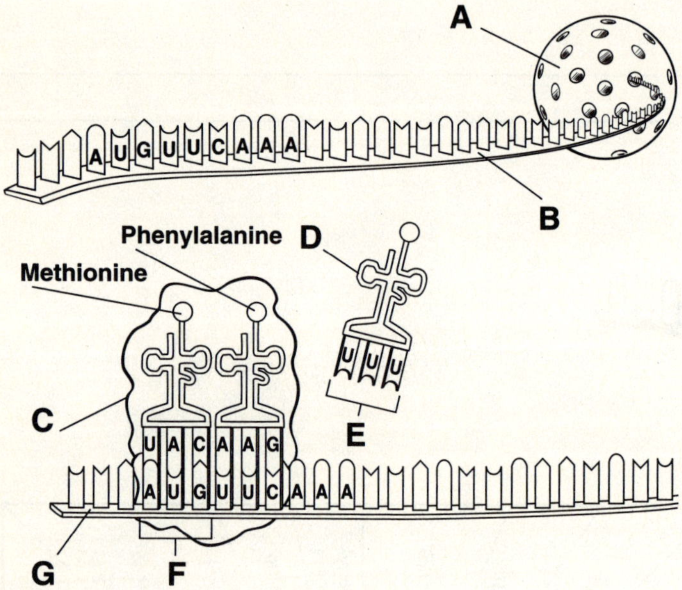

Figure 12-4

26. **Interpreting Graphics** What process is illustrated in Figure 12-4?

____

27. **Interpreting Graphics** Identify structure C in Figure 12-4.

____

28. **Interpreting Graphics** Which labeled structure in Figure 12-4 is a codon?

____

29. **Inferring** What is the relationship between the codons and anticodons in Figure 12-4? How is this relationship important?

____

____

____

____

____

Name______________________________ Class__________________ Date ______________

30. **Predicting** In Figure 12-4, what will happen after the ribosome joins the methionine and phenylalanine?

____

____

____

____

____

Essay

Write the answer to each question in the space provided.

31. Describe the Hershey-Chase experiment. Why were the results important?

32. Describe the structure of a DNA molecule.

Name_______________________ Class_______________ Date ___________

33. Contrast the functions of the three main types of RNA.

34. Mendel might have been surprised to learn that genes simply contain the instructions for assembling proteins. What do proteins have to do with the phenotype of an organism?

35. Why do some kinds of point mutations cause greater changes in proteins than others?

Name______________________________ Class_________________ Date _____________

Multiple Choice

Write the letter that best answers the question or completes the statement on the line provided.

____ **1.** Avery's experiments showed that bacteria are transformed by
- a. RNA.
- b. DNA.
- c. proteins.
- d. carbohydrates.

____ **2.** Figure 12-1 shows the structure of a(an)
- a. DNA molecule.
- b. amino acid.
- c. RNA molecule.
- d. protein.

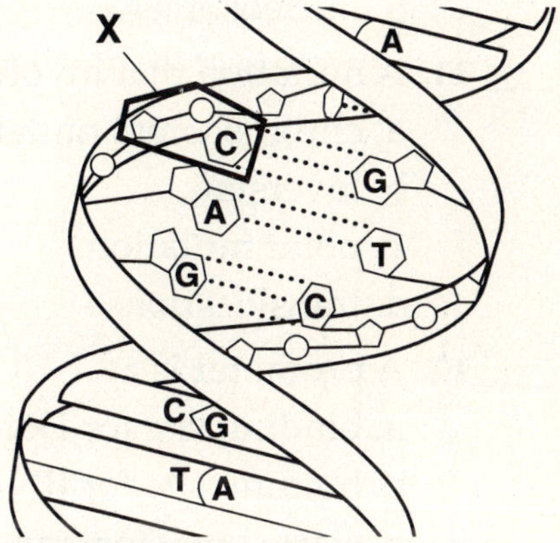

Figure 12-1

____ **3.** DNA is copied during a process called
- a. replication.
- b. translation.
- c. transcription.
- d. transformation.

____ **4.** In eukaryotes, DNA
- a. is located in the nucleus.
- b. floats freely in the cytoplasm.
- c. is located in the ribosomes.
- d. is circular.

____ **5.** RNA contains the sugar
- a. ribose.
- b. deoxyribose.
- c. glucose.
- d. lactose.

____ **6.** Which RNA molecule carries amino acids?
- a. messenger RNA
- b. transfer RNA
- c. ribosomal RNA
- d. RNA polymerase

____ **7.** What is produced during transcription?
- a. RNA molecules
- b. DNA molecules
- c. RNA polymerase
- d. proteins

____ **8.** What does Figure 12-2 show?
- a. anticodons
- b. the order in which amino acids are linked
- c. the code for splicing mRNA
- d. the genetic code

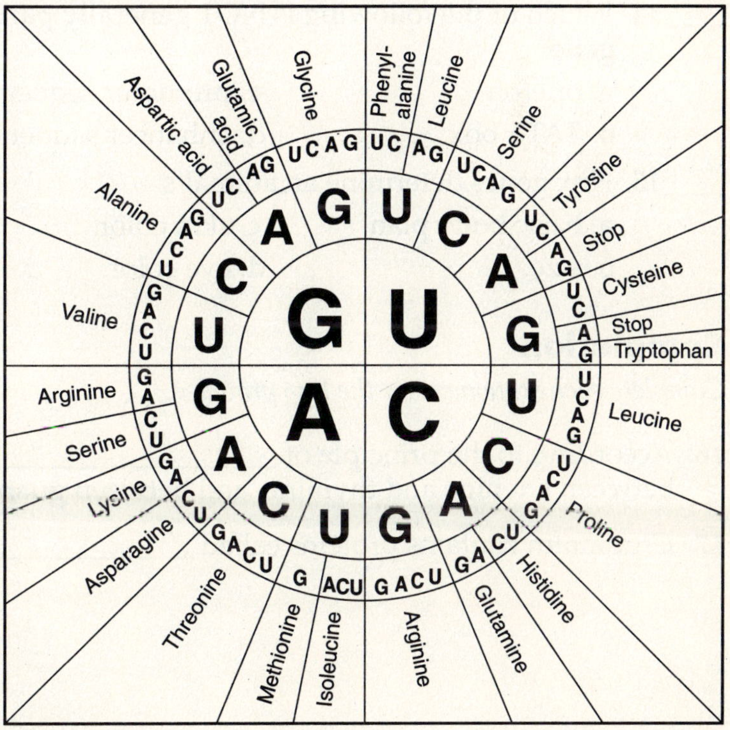

Figure 12-2

____ **9.** What happens during the process of translation?
a. Messenger RNA is made from DNA.
b. The cell uses information from messenger RNA to produce proteins.
c. Transfer RNA is made from messenger RNA.
d. Copies of DNA molecules are made.

____ **10.** Genes contain instructions for assembling
a. purines.
b. nucleosomes.
c. proteins.
d. pyrimidines.

____ **11.** A mutation that involves a single nucleotide is called a(an)
a. chromosomal mutation.
b. inversion.
c. point mutation.
d. translocation.

____ **12.** A promoter is a
a. binding site for DNA polymerase.
b. binding site for RNA polymerase.
c. start signal for transcription.
d. stop signal for transcription.

____ **13.** In *E. coli,* the *lac* operon controls the
a. breakdown of lactose.
b. production of lactose.
c. breakdown of glucose.
d. production of glucose.

____ **14.** Which of the following is NOT generally part of a eukaryotic gene?
a. operon
b. TATA box
c. promoter sequences
d. enhancer sequences

____ **15.** Hox genes determine an animal's
a. basic body plan.
b. size.
c. skin color.
d. eye color.

Completion

Complete each statement on the line provided.

16. According to the principle of ____________________, hydrogen bonds can form only between adenine and thymine, and between guanine and cytosine.

17. Chromatin contains proteins called ____________________ .

Name______________________________ Class_________________ Date _____________

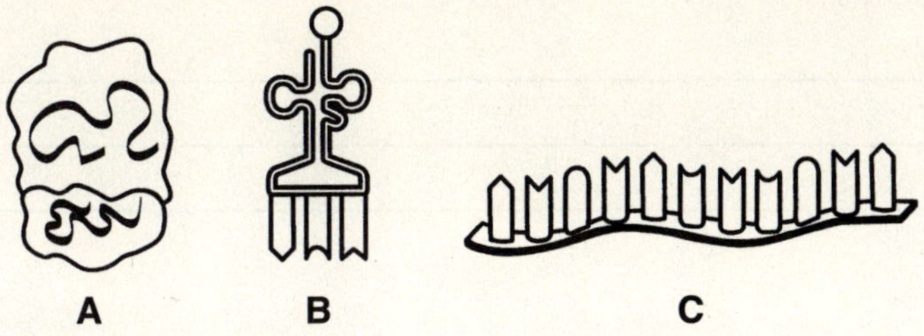

Figure 12-3

18. In Figure 12-3, A, B, and C are three types of ____________________.

19. After introns are cut out of an RNA molecule, the remaining ____________________ are spliced back together to form the final messenger RNA.

20. A mutation in a series of genes, called the ____________________, can change the organs that develop in specific parts of an embryo.

Short Answer

In complete sentences, write the answers to the questions on the lines provided.

21. What is a bacteriophage?

22. What are the three main parts of an RNA nucleotide?

23. What must happen to a DNA molecule before RNA polymerase can make RNA?

24. What causes translation to stop?

Name______________________ Class______________ Date __________

25. What is a mutation?

____

____

____

Using Science Skills

Use the diagram below to answer the following questions on the lines provided.

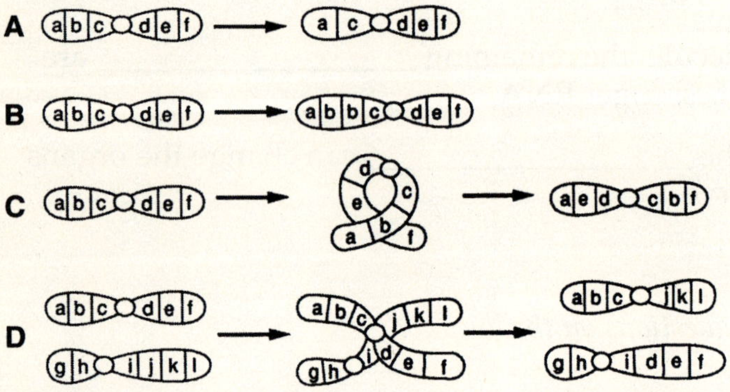

Figure 12-4

26. **Classifying** What general type of mutation results from processes A, B, C, and D in Figure 12-4?

____

____

27. **Interpreting Graphics** In Figure 12-4, which process or processes involve two chromosomes?

____

28. **Comparing and Contrasting** Contrast process A and process B in Figure 12-4.

____

____

____

29. **Interpreting Graphics** During which process in Figure 12-4 does a segment of a chromosome become oriented in the reverse direction?

____

____

30. **Interpreting Graphics** In Figure 12-4, which process produces two chromosomes with translocations?

____

Name________________________ Class__________________ Date ______________

Multiple Choice

Write the letter that best answers the question or completes the statement on the line provided.

____ **1.** Selective breeding produces
a. more offspring.
b. fewer offspring.
c. desired traits in offspring.
d. transgenic organisms.

____ **2.** The crossing of buffalo and cattle to produce beefalo is an example of
a. inbreeding.
b. hybridization.
c. genetic engineering.
d. transformation.

____ **3.** What is the ultimate source of genetic variability?
a. inbreeding
b. plasmids
c. hybridization
d. mutations

____ **4.** Breeders maintain the desired traits of an organism by
a. inbreeding.
b. using restriction enzymes.
c. hybridization.
d. inducing mutations.

____ **5.** Polyploidy instantly produces a new plant species because it
a. changes a species' number of chromosomes.
b. results from hybridization.
c. causes transgenic plants.
d. results from transformation.

____ **6.** What does Figure 13-1 show?
a. gel electrophoresis
b. DNA sequencing
c. a restriction enzyme cutting different sequences of DNA
d. polymerase chain reaction

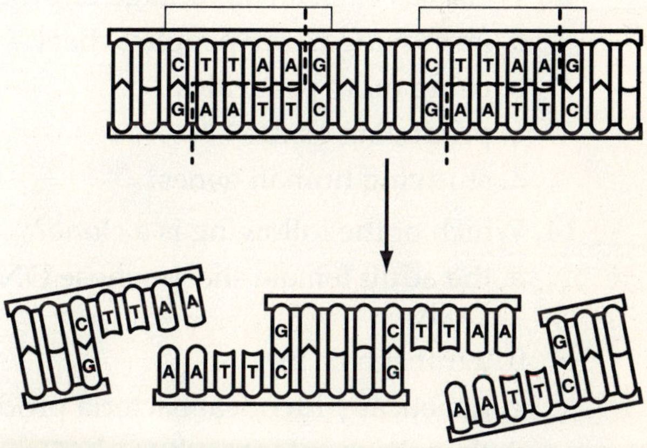

Figure 13-1

____ **7.** Analyzing DNA by gel electrophoresis allows researchers to
a. identify similarities and differences in the genomes of different kinds of organisms.
b. determine whether a particular allele of a gene is dominant or recessive.
c. compare the phenotypes of different organisms.
d. cut DNA with restriction enzymes.

Name______________________________ Class __________________ Date ______________

____ 8. Which of the following techniques do scientists use to make many copies of a gene?
a. cloning
b. transformation
c. gel electrophoresis
d. polymerase chain reaction

____ 9. Genetic engineering involves
a. reading a DNA sequence.
b. editing a DNA sequence.
c. reinserting DNA into living organisms.
d. all of the above

____10. Which of the following are NOT used to read DNA sequences?
a. nucleotides
b. gels
c. fluorescent dyes
d. double-stranded DNA molecules

____11. A recombinant plasmid gets inside a bacterial cell by
a. hybridization.
b. radiation.
c. transformation.
d. recombination.

____12. Which of the following is often used as a genetic marker in plasmids?
a. a restriction enzyme
b. a gene for antibiotic resistance
c. a DNA sequence that promotes replication
d. a nucleotide labeled with a fluorescent dye

____13. What has been an advantage of producing transgenic plants?
a. improving many people's diets
b. using more pesticides
c. producing clones
d. studying human genes

____14. Which of the following is a clone?
a. the adult female sheep whose DNA was used to produce Dolly
b. a transgenic mouse
c. genetically identical bacteria produced from a single cell
d. the tobacco plant with the luciferase gene

____15. What kind of cell or cells were used to make Dolly?
a. body cell only
b. egg cell only
c. egg cell and sperm cell
d. body cell and egg cell

Name______________________________ Class__________________ Date ______________

Completion

Complete each statement on the line provided.

16. To produce a fruit that has some characteristics of an orange and some of a grapefruit, you would use the selective breeding technique of ____________________.

17. The mutations that breeders induce in organisms are passed on to the organisms' ____________________.

18. The sequence of a DNA fragment can be determined by observing the order of the ____________________ fluorescent bands on an electrophoresis gel.

19. For a plant transformation to be successful, the recombinant DNA is integrated into one of the ____________________ of the cell.

20. Dolly is not a(an) ____________________ animal because all her genes are from the same species of organism.

Short Answer

In complete sentences, write the answers to the questions on the lines provided.

21. A mule is produced by mating a male donkey with a female horse. What can you conclude, in general, about the characteristics of a mule?

____

____

22. How are the selective breeding techniques of hybridization and inbreeding opposites?

____

____

23. Why would breeders want to increase a population's mutation rate?

____

____

____

Name________________________________ Class ____________________ Date ____________

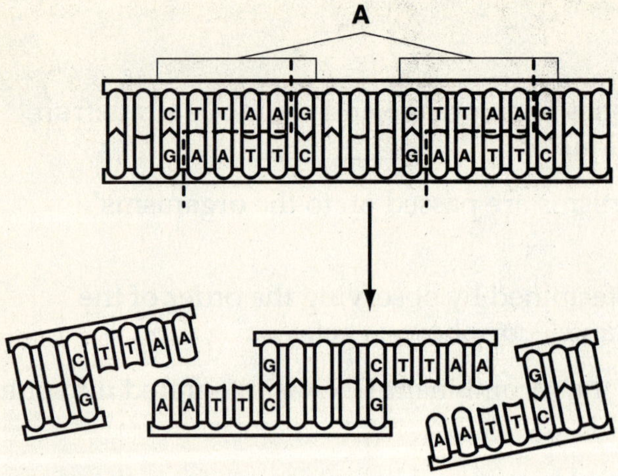

Figure 13-2

24. What is structure A in Figure 13-2, and what is its significance?

____

____

____

25. What are two ways scientists transform plant cells without using plasmids?

____

____

____

Using Science Skills

Use the diagram below to answer the questions on the following page on the lines provided.

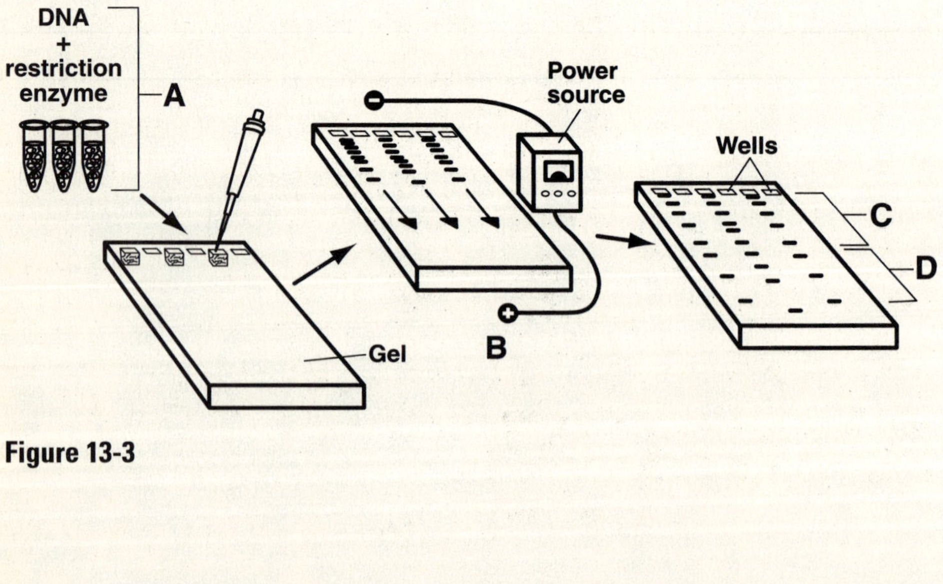

Figure 13-3

26. **Interpreting Graphics** In Figure 13-3, of what do the bands shown in B consist?

____

27. **Comparing and Contrasting** Which group of bands in Figure 13-3 moved faster, C or D? Why did they move faster?

____

28. **Inferring** What is occurring in the test tubes at A in Figure 13-3?

____

29. **Inferring** In Figure 13-3, why are the bands in B moving toward the positive end of the gel?

____

30. **Drawing Conclusions** In Figure 13-3, were the three DNA samples shown in A identical? Explain your answer.

____

____

____

Essay

Write the answer to each question in the space provided.

31. In what ways has selective breeding been useful to humans today and in the past?

32. Explain the advantage and disadvantage of inbreeding. Give an example of each.

Name________________________ Class________________ Date ____________

33. Contrast the techniques used in genetic engineering and in selective breeding to produce organisms with desired traits.

34. Explain how a plant cell might contain a recombinant plasmid but not be successfully transformed.

35. In what general ways are transgenic organisms useful to people today?

Name________________________________ Class__________________ Date ______________

Multiple Choice

Write the letter that best answers the question or completes the statement on the line provided.

____ **1.** Luther Burbank developed over 800 varieties of plants by
a. genetic engineering. c. selective breeding.
b. transformation. d. DNA sequencing.

____ **2.** Which of the following have been produced by selective breeding?
a. horse breeds c. dog breeds
b. cat breeds d. all of the above

____ **3.** Which of the following is most likely to bring together two recessive alleles for a genetic defect?
a. inbreeding c. genetic engineering
b. hybridization d. transformation

____ **4.** Two dissimilar individuals are crossed in the process of
a. inbreeding. c. transformation.
b. hybridization. d. genetic engineering.

____ **5.** Scientists produced oil-eating bacteria by
a. making bacteria polyploid.
b. inbreeding bacteria.
c. inducing mutations in bacteria.
d. hybridizing bacteria.

____ **6.** One function of gel electrophoresis is to
a. separate DNA fragments.
b. cut DNA.
c. recombine DNA.
d. make many copies of DNA.

____ **7.** The process of making changes in the DNA code of living organisms is called
a. selective breeding. c. inbreeding.
b. genetic engineering. d. hybridization.

____ **8.** A DNA molecule produced by combining DNA from different sources is known as
a. a mutant. c. a polyploid.
b. a hybrid. d. recombinant DNA.

____ 9. Knowing the sequence of an organism's DNA allows researchers to
   a. clone the organism.
   b. mutate the DNA.
   c. study specific genes.
   d. extract the DNA.

____10. During transformation,
   a. a prokaryote is changed into a eukaryote.
   b. a cell incorporates DNA from outside the cell.
   c. foreign DNA is inserted into a plasmid.
   d. a genetically identical organism is produced.

____11. Scientists can transform plant cells by
   a. using the bacterium *Agrobacterium tumefaciens.*
   b. removing the plant cell walls and then mixing the cells with DNA.
   c. injecting DNA into the plant cells.
   d. all of the above

____12. A gene that makes it possible to distinguish bacteria that carry a plasmid containing foreign DNA from those that do not is called a(an)
   a. resistance gene.   c. genetic marker.
   b. antibiotic.   d. clone.

____13. The transformation of a plant cell is successful if
   a. the plasmid that entered the cell reproduces inside the cell.
   b. the foreign DNA is integrated into one of the cell's chromosomes.
   c. the cell reproduces.
   d. a plasmid has entered the cell.

____14. Which of the following techniques do scientists use to make transgenic organisms?
   a. hybridization   c. inducing mutations
   b. inbreeding   d. genetic engineering

____15. The Scottish scientist Ian Wilmut cloned a
   a. bacterium.   c. plant.
   b. sheep.   d. cow.

Completion

Complete each statement on the line provided.

16. Humans use selective breeding to pass desired __________________ on to the next generation of organisms.

17. __________________ is the technique of selective breeding that has led to deformities in certain dog breeds.

18. Scientists use radiation and chemicals to induce ________________.

19. Plasmids are naturally found in some ________________.

20. Some plasmids have genetic markers that make them resistant to ________________.

Short Answer

In complete sentences, write the answers to the questions on the lines provided.

21. What is a polyploid organism?

____

____

____

____

22. What does PCR (polymerase chain reaction) enable scientists to make?

____

____

____

____

23. What is a plasmid?

____

____

____

____

24. What is a transgenic organism?

____

____

____

____

____

25. Compare the genes in Dolly and the sheep from which she was cloned.

____

____

____

____

____

Name________________________________ Class__________________ Date ______________

Using Science Skills

Use the diagram below to answer the following questions on the lines provided.

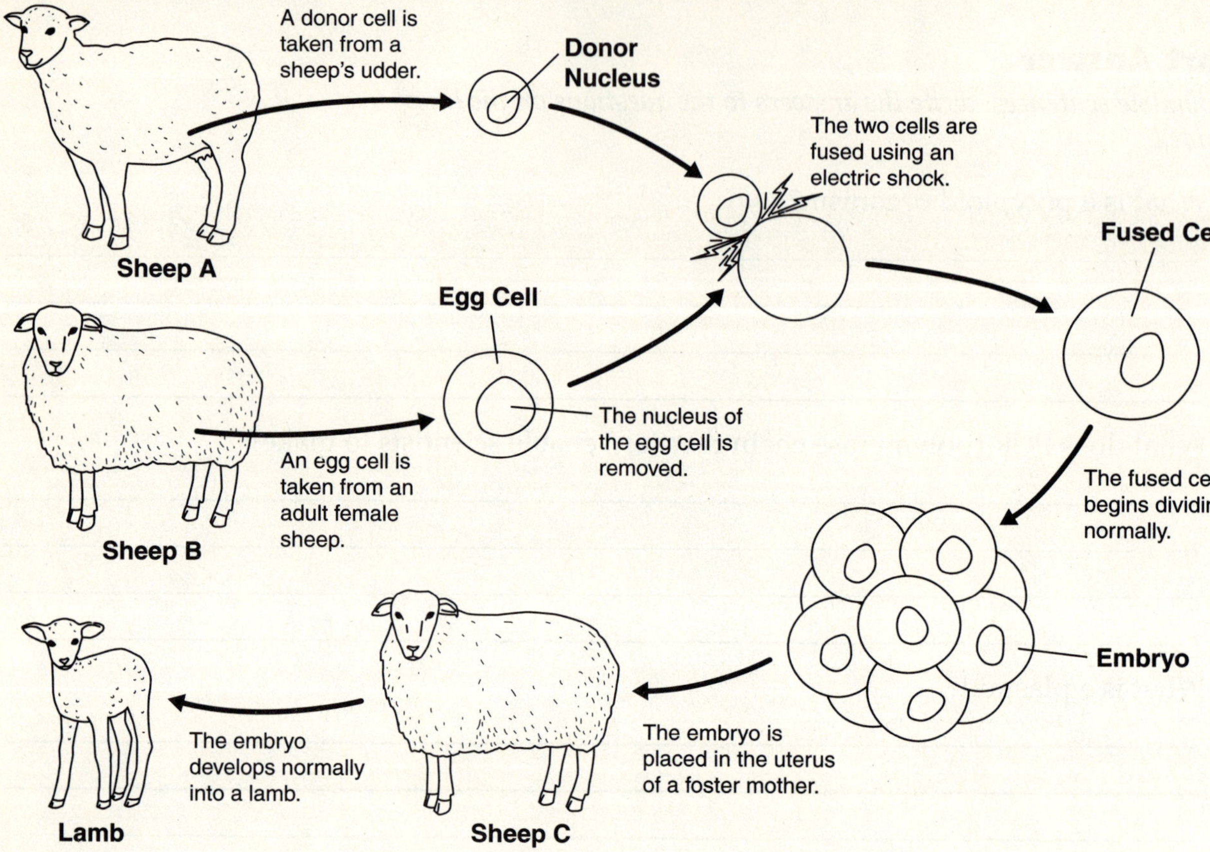

Figure 13-1

26. Interpreting Graphics In the cloning shown in Figure 13-1, which sheep is the source of the nucleus in the fused cell?

____

27. Interpreting Graphics In the cloning shown in Figure 13-1, which sheep provided an egg cell?

____

28. Inferring In Figure 13-1, why was the nucleus removed from the egg cell?

____

____

29. Interpreting Graphics Which animal in Figure 13-1 is a clone?

____

30. Inferring Which two animals in Figure 13-1 are genetically identical?

____

Name______________________________ Class________________ Date ____________

Multiple Choice

Write the letter that best answers the question or completes the statement on the line provided.

____ 1. Which of the following are shown in a karyotype?
a. homologous chromosomes
b. sex chromosomes
c. autosomes
d. all of the above

____ 2. What is the approximate probability that a human offspring will be female?
a. 10 percent c. 50 percent
b. 25 percent d. 75 percent

____ 3. What is the probability that a human sperm cell will carry an X chromosome?
a. 0 percent c. 50 percent
b. 25 percent d. 100 percent

____ 4. A pedigree CANNOT be used to
a. determine whether a trait is inherited.
b. show how a trait is passed from one generation to the next.
c. determine whether an allele is dominant or recessive.
d. none of the above

____ 5. A person who has PKU
a. inherited the allele for the trait from one parent.
b. inherited the allele for the trait from both parents.
c. is heterozygous for the trait.
d. will not pass the allele for the trait to his or her offspring.

____ 6. Which of the following genotypes result in the same phenotype?
a. I^AI^A and I^AI^B c. I^BI^B and I^AI^B
b. I^BI^B and I^Bi d. I^Bi and ii

____ 7. A change in a single gene causes the protein called CFTR to
a. become less soluble.
b. fold improperly.
c. destroy the cell membrane.
d. transport sodium ions instead of chloride ions.

____ 8. Chromosomes 21 and 22 contain long stretches of repetitive DNA, which

a. contain genes for colorblindness.

b. are able to code for proteins.

c. are unstable sites where rearrangements can occur.

d. contain genes for Lou Gehrig's disease.

____ 9. Colorblindness is more common in males than in females because

a. fathers pass the allele for colorblindness to their sons only.

b. the allele for colorblindness is located on the Y chromosome.

c. the allele for colorblindness is recessive and located on the X chromosome.

d. males who are colorblind have two copies of the allele for colorblindness.

____10. A Barr body is

a. a condensed X chromosome that is inactive.

b. a condensed Y chromosome that is inactive.

c. an activated X chromosome.

d. an activated Y chromosome.

____11. Because the X chromosome contains genes that are vital for normal development, no baby has been born

a. with one X chromosome.

b. with three X chromosomes.

c. without an X chromosome.

d. with four X chromosomes.

____12. Which of the following combinations of sex chromosomes represents a female?

a. XY   c. XXXY

b. XXY   d. XX

____13. The process of DNA fingerprinting is based on the fact that

a. the most important genes are different among most people.

b. no two people, except identical twins, have exactly the same DNA.

c. most genes are dominant.

d. most people have DNA that contains repeats.

____14. The human genome was sequenced

a. by sequencing each gene on each chromosome, one at a time.

b. using DNA fingerprinting.

c. by looking for overlapping regions between sequenced DNA fragments.

d. using open reading frames.

Name______________________________ Class__________________ Date ______________

____15. Which of the following is the first step in gene therapy?
  a. splicing the normal gene to viral DNA
  b. allowing recombinant viruses to infect human cells
  c. using restriction enzymes to cut out the normal gene from DNA
  d. identifying the faulty gene that causes the disease

Completion

Complete each statement on the line provided.

16. A(An) ____________________ can be used to determine whether a person has inherited the normal number of chromosomes.

17. The alleles I^A and ____________________ for the ABO blood group are NOT codominant.

18. People who have sickle cell disease inherited ____________________ copy(ies) of the sickle cell allele.

19. A boy who has hemophilia inherited the disorder from his ____________________ .

20. A person can be tested for the allele that causes Tay-Sachs disease because the ____________________ of that allele is different from that of the normal allele.

Short Answer

In complete sentences, write the answers to the questions on the lines provided.

21. Give two reasons why it is almost impossible to associate many human traits with single genes.

____

____

____

22. How are human chromosomes 21 and 22 similar?

____

____

____

____

23. A man who does not have hemophilia and a woman who is a carrier of the disorder have a son. What is the probability that their son has hemophilia?

____

____

24. Why is a person who has Klinefelter's syndrome (XXY) a male even though he has two X chromosomes in his cells?

____

____

____

25. Why are viruses used in gene therapy?

____

____

Using Science Skills

Use the diagram below to answer the following questions on the lines provided.

The pedigree shows the inheritance of free earlobes and attached earlobes in five generations of a family. Attached earlobes is caused by a recessive allele (f).

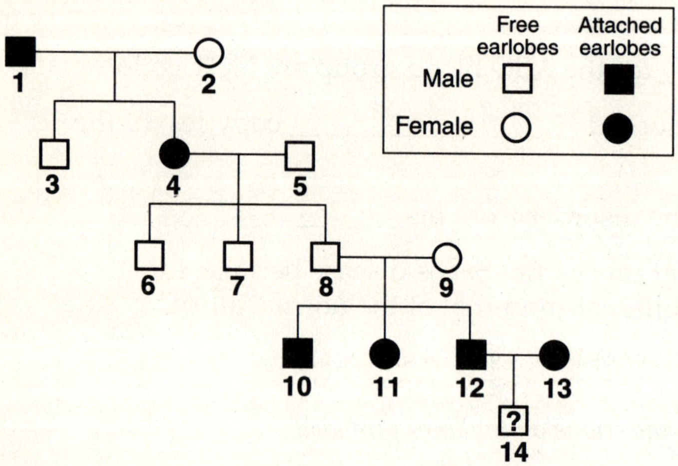

Figure 14-1

26. **Inferring** Is individual 2 in Figure 14-1 homozygous or heterozygous for free earlobes? Explain.

____

____

____

____

27. **Interpreting Graphics** In Figure 14-1, how many children of individuals 4 and 5 have attached earlobes?

____

28. **Inferring** Can you be certain of the genotype of individual 5 in Figure 14-1? Explain.

____

____

____

____

____

29. **Predicting** Predict the genotype and phenotype of individual 14 in Figure 14-1.

____

____

30. **Inferring** In Figure 14-1, are any of the descendents of individuals 1 and 2 homozygous for free earlobes? Explain your answer.

____

____

Essay

Write the answer to each question in the space provided.

31. Contrast the abnormal hemoglobin found in a person who has sickle cell disease with normal hemoglobin. How does this difference affect the person's red blood cells?

32. Explain why the father of a girl who is colorblind must also be colorblind.

Name_______________________________ Class___________________ Date _______________

33. Compare the inheritance in men and women of a trait coded for by a recessive allele on the Y chromosome.

34. What can you conclude if you observed a male cat that has both black and orange spots? Explain your answer.

35. Compare and contrast Turner's syndrome and Klinefelter's syndrome.

Name________________________ Class________________ Date ____________

Multiple Choice

Write the letter that best answers the question or completes the statement on the line provided.

____ **1.** How many chromosomes are shown in a normal human karyotype?
a. 2 c. 44
b. 23 d. 46

____ **2.** In humans, a male has
a. one X chromosome only.
b. two X chromosomes.
c. one X chromosome and one Y chromosome.
d. two Y chromosomes.

____ **3.** Human females produce egg cells that have
a. one X chromosome.
b. two X chromosomes.
c. one X or one Y chromosome.
d. one X and one Y chromosome.

____ **4.** In a pedigree, a circle represents a(an)
a. male. c. child.
b. female. d. adult.

____ **5.** Which of the following is caused by a dominant allele?
a. Huntington's disease
b. PKU
c. Tay-Sachs disease
d. none of the above

____ **6.** Which of the following is determined by multiple alleles?
a. Rh blood group c. PKU
b. ABO blood group d. Huntington's disease

____ **7.** Sickle cell disease is caused by a
a. small change in the DNA of a single gene.
b. change in the size of a chromosome.
c. change in two genes.
d. change in the number of chromosomes in a cell.

____ **8.** Compared with normal hemoglobin, the hemoglobin of a person with sickle cell disease
a. is longer.
b. is shorter.
c. has a different sequence of amino acids.
d. is sickle-shaped.

____ **9.** The sequencing of human chromosomes 21 and 22 showed that

a. some regions of chromosomes do not code for proteins.

b. all of the DNA of chromosomes codes for proteins.

c. different chromosomes have the same number of genes.

d. different chromosomes contain the same number of DNA bases.

____ **10.** Sex-linked genes are located on

a. the autosomes.

b. the X chromosome only.

c. the Y chromosome only.

d. both the X chromosome and Y chromosome.

____ **11.** Which of the following form(s) a Barr body?

a. the Y chromosome in a male cell

b. the X chromosome in a male cell

c. one of the X chromosomes in a female cell

d. both of the X chromosomes in a female cell

____ **12.** The failure of chromosomes to separate during meiosis is called

a. nondisjunction.

b. X-chromosome inactivation.

c. Turner's syndrome.

d. Down syndrome.

____ **13.** Scientists test for alleles that cause human genetic disorders by

a. making karyotypes.

b. making DNA fingerprints.

c. using DNA probes.

d. making pedigrees.

____ **14.** The Human Genome Project is an attempt to

a. make a DNA fingerprint of every person's DNA.

b. analyze the human DNA sequence.

c. cure human diseases.

d. identify alleles in human DNA that are recessive.

____ **15.** The purpose of gene therapy is to

a. cure genetic disorders.

b. determine the sequences of genes.

c. search for genes on chromosomes.

d. identify an individual using DNA.

Name______________________________ Class________________ Date ____________

Completion

Complete each statement on the line provided.

16. A(An) __________________ is a diagram that follows the inheritance of a single gene through several generations of a family.

17. A person with the alleles *ii* has blood type __________________ .

18. A person who has __________________ is unable to break down the amino acid phenylalanine.

19. The human genes associated with color vision are located on the __________________ chromosome.

20. The process of __________________ is used to identify a person based on DNA evidence.

Short Answer

In complete sentences, write the answers to the questions on the lines provided.

21. How is the DNA sequence of the allele that causes cystic fibrosis different from that of the normal allele?

____

____

____

22. Why are all X-linked alleles expressed in males, even if they are recessive?

____

____

____

23. Why are Barr bodies not found in most male cells?

____

____

____

24. Why do scientists use sections of DNA that have little or no known function to do DNA fingerprinting?

____

____

____

25. Why did scientists expect that humans would have more genes than they actually do?

____

____

____

Name____________________ Class____________________ Date____________________

Using Science Skills

Use the diagram below to answer the following questions on the lines provided.

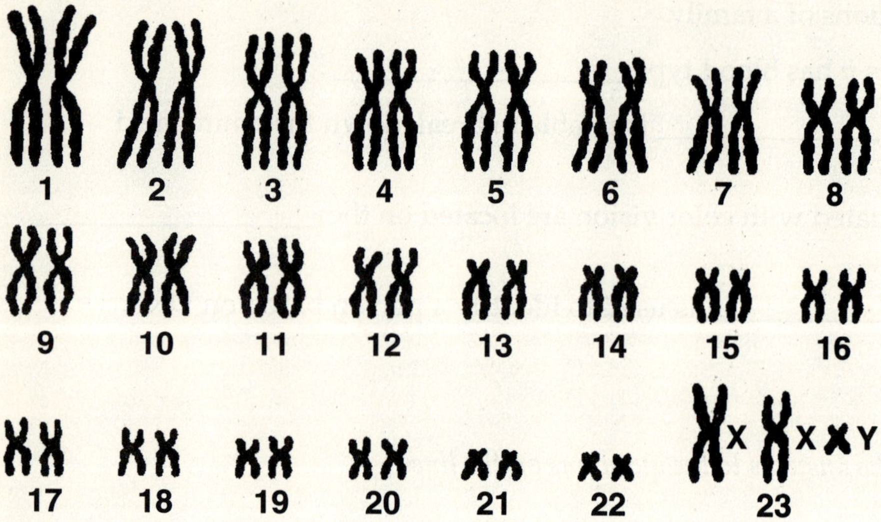

Figure 14-1

26. **Applying Concepts** In the human karyotype in Figure 14-1, what are the chromosomes in each numbered group called?

27. **Classifying** Which chromosomes in Figure 14-1 are autosomes?

28. **Interpreting Graphics** In the human karyotype in Figure 14-1, how many chromosomes are shown?

29. **Drawing Conclusions** Identify the sex chromosomes in Figure 14-1. Does the karyotype show the normal number of sex chromosomes? Explain.

30. **Inferring** Is the individual represented by the karyotype in Figure 14-1 male or female? Explain.

Name________________________________ Class__________________ Date ______________

Multiple Choice

Write the letter that best answers the question or completes the statement on the line provided.

____ **1.** When Mendel crossed true-breeding tall plants with true-breeding short plants, all the offspring were tall because
a. the allele for tall plants is recessive.
b. the allele for short plants is dominant.
c. the allele for tall plants is dominant.
d. they were true-breeding like their parents.

____ **2.** What is the probability that a heterozygous tall pea plant will produce a gamete with the allele for tallness?
a. 75 percent
b. 50 percent
c. 25 percent
d. 0 percent

____ **3.** The cross between a red flower and a white flower produces all pink flowers. This type of inheritance is known as
a. incomplete dominance.
b. polygenic inheritance.
c. codominance.
d. simple dominance.

____ **4.** Chromosomes form tetrads during
a. prophase of meiosis I.
b. metaphase of meiosis I.
c. prophase of meiosis II.
d. metaphase of meiosis II.

____ **5.** A gene map shows
a. how cross-over occurs.
b. the number of possible alleles for a gene.
c. the segregation of traits.
d. the relative locations of genes on a chromosome.

____ **6.** Hershey and Chase concluded that
a. bacteriophages do not harm bacteria.
b. the genetic material of bacteriophages is protein.
c. the genetic material of bacteriophages is DNA.
d. DNA is the transforming factor.

Name_______________ Class_______________ Date_______________

____ 7. The result of transcription is
a. a copy of the DNA molecule.
b. a strand of RNA.
c. a protein.
d. a chromosome.

____ 8. A frameshift mutation
a. changes the chromosome number.
b. includes the loss of part of a chromosome.
c. shifts the reading frame of the gene.
d. does not change the genetic material.

____ 9. The *lac* genes are turned on when
a. the repressor binds to the operator.
b. RNA polymerase binds to the repressor.
c. lactose binds to the promoter.
d. lactose binds to the repressor.

____ 10. The purpose of selective breeding is to
a. increase the genetic variation in a population.
b. produce transgenic organisms.
c. pass desired traits to the next generation.
d. produce mutations.

____ 11. Gel electrophoresis is used to
a. transform bacteria.
b. induce mutations.
c. cut DNA into fragments.
d. separate DNA fragments.

____ 12. Transgenic animals are used
a. for selective breeding.
b. to study human genes.
c. to produce plastics.
d. to increase genetic variation.

____ 13. What is the probability that a human egg cell will have an X chromosome?
a. 100 percent
b. 50 percent
c. 25 percent
d. 0 percent

Name______________________________ Class____________________ Date ______________

____**14.** Hemophilia is more common in males than in females because
a. fathers pass the allele for hemophilia to their sons only.
b. the allele for hemophilia is linked to the Y chromosome.
c. the allele for hemophilia is recessive and linked to the X chromosome.
d. males who have hemophilia have two copies of the allele for hemophilia.

____**15.** In gene therapy,
a. mutations are induced to increase genetic variation.
b. DNA is cut with restriction enzymes.
c. transgenic organisms are produced.
d. a faulty gene is replaced by a normal, working gene.

Completion

Complete each statement on the line provided.

16. The observation that wrinkled green peas result from a cross between two heterozygous smooth yellow peas is an example of Mendel's principle of ____________________.

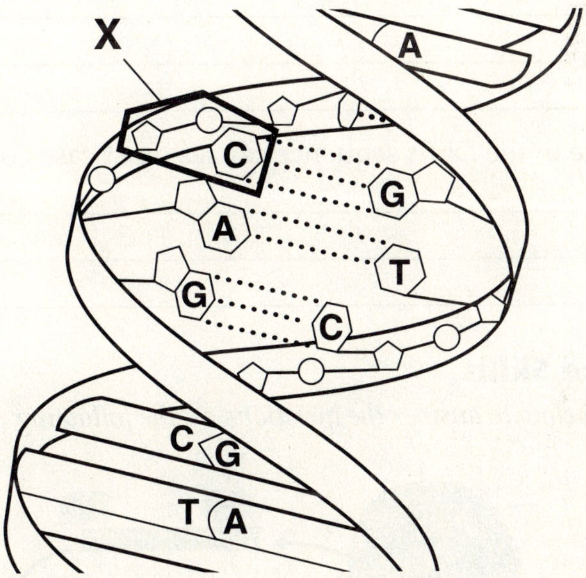

Figure 1

17. The structure labeled X in Figure 1 is a(an) ____________________.

18. In translation, the cell uses information from ____________________ to produce proteins.

19. To maintain a line of pure-breeding peas, selective breeders would use a technique called ____________________.

20. Down syndrome is caused by ____________________, resulting in three copies of chromosome 21.

Name________________________ Class__________________ Date ______________

Short Answer

In complete sentences, write the answers to the questions on the lines provided.

21. Contrast the cells produced by mitosis with those produced by meiosis.

____

____

____

22. What are two roles that DNA polymerase has in DNA replication?

____

____

____

23. Why would breeders want to increase the genetic variation in a population?

____

____

____

24. What happens during cell transformation?

____

____

____

25. What change in the DNA sequence causes most cases of cystic fibrosis?

____

____

____

Using Science Skills

Use the diagram below to answer the questions on the following page on the lines provided.

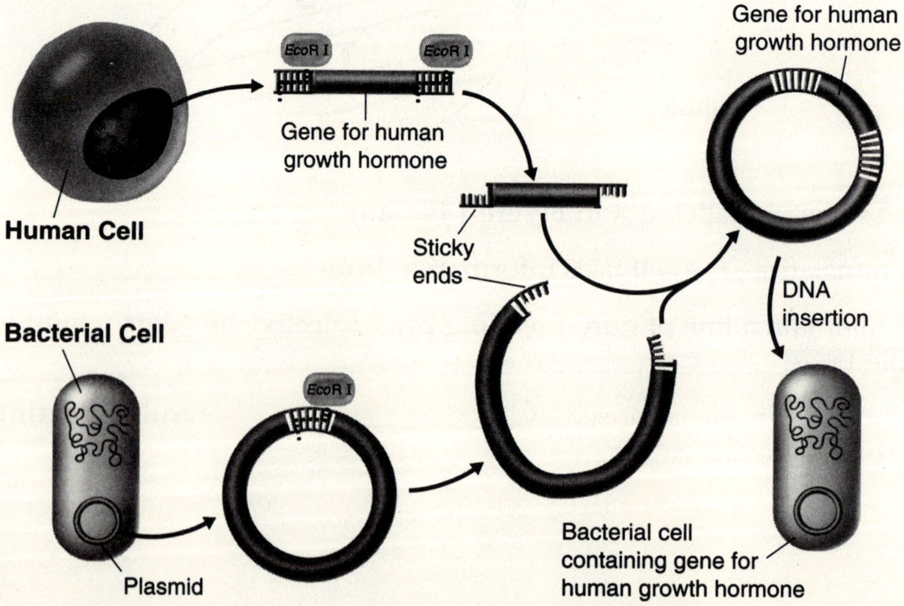

Figure 2

Name______________________________ Class________________ Date ____________

26. **Interpreting Graphics** In Figure 2, what is being transformed?

27. **Applying Concepts** Referring to Figure 2, would any stage(s) in this process require the use of gel electrophoresis? Explain.

28. **Inferring** In Figure 2, why are both the plasmid and the human gene cut with the same restriction enzyme?

29. **Applying Concepts** In Figure 2, of what is the recombinant DNA molecule made?

30. **Inferring** How do researchers identify transformed cells?

Essay

Write the answer to each question in the space provided.

31. Explain how two pea plants with round yellow peas can, when crossed, produce offspring with the recessive traits—green wrinkled peas. Use a Punnett square in your explanation.

32. Describe the structure of the DNA molecule.

Name______________________________ Class__________________ Date ______________

33. Compare and contrast the structures and functions of the three main types of RNA.

34. Why is gene regulation in eukaryotes more complex than in prokaryotes?

35. How does studying human genetic disorders help scientists understand the normal alleles of genes?

Name______________________________ Class________________ Date ____________

Multiple Choice

Write the letter that best answers the question or completes the statement on the line provided.

____ **1.** The principle of dominance states that
- a. all alleles are dominant.
- b. all alleles are recessive.
- c. some alleles are dominant and others are recessive.
- d. alleles are neither dominant nor recessive.

____ **2.** Situations in which both alleles contribute to the phenotype are called
- a. incomplete dominance.
- b. codominance.
- c. multiple alleles.
- d. polygenic alleles.

____ **3.** Unlike mitosis, meiosis results in the formation of
- a. haploid cells.
- b. identical cells.
- c. diploid cells.
- d. body cells.

____ **4.** What statement about crossing-over is true?
- a. The genetic variability of the offspring is reduced.
- b. It occurs during meiosis.
- c. Genes located far apart on a chromosome are less likely to be separated.
- d. It occurs between chromosomes that are NOT homologous.

____ **5.** Which of the following is NOT a part of the DNA molecule?
- a. deoxyribose sugar
- b. phosphate group
- c. thymine
- d. uracil

____ **6.** DNA is copied in a process called
- a. replication.
- b. transcription.
- c. translation.
- d. transformation.

____ **7.** What is the product of translation?
- a. double-stranded DNA molecules
- b. single-stranded DNA molecules
- c. single-stranded RNA molecules
- d. proteins

____ **8.** The *lac* genes are turned off when
- a. lactose is present.
- b. RNA polymerase binds to the promoter.
- c. the repressor binds to the operator.
- d. RNA polymerase binds to the repressor.

Name______________________________ Class__________________ Date ______________

____ 9. Breeders induce mutations to
a. produce hybrids.
b. produce purebred organisms.
c. decrease the genetic variation in a population.
d. increase the genetic variation in a population.

____10. One function of gel electrophoresis is to
a. separate DNA fragments.
b. cut DNA.
c. recombine DNA.
d. make many copies of DNA.

____11. During transformation,
a. a prokaryote is changed into a eukaryote.
b. a cell takes in DNA from outside the cell.
c. foreign DNA is inserted into a plasmid.
d. a clone is produced.

____12. Transgenic organisms are useful because they
a. are a source of human proteins.
b. produce crops resistant to insect damage.
c. improve the food supply.
d. all of the above

____13. In humans, a female has
a. one X chromosome only.
b. two X chromosomes.
c. one X chromosome and one Y chromosome.
d. two Y chromosomes.

____14. Nondisjunction occurs when
a. chromosomes fail to separate.
b. chromosomes become inverted.
c. the reading frame for the genetic code is shifted.
d. RNA is translated into a protein.

____15. The purpose of gene therapy is to
a. determine the sequence of genes.
b. search for genes on chromosomes.
c. cure genetic disorders.
d. determine the identify of a person.

Name______________________________ Class__________________ Date ______________

Completion

Complete each statement on the line provided.

16. The principle of ____________________ states that genes for different traits can segregate independently during the formation of gametes.

17. During ____________________, RNA polymerase uses one strand of DNA as a template to produce a strand of RNA.

18. Breeders use the process of ____________________ to bring together the best traits of two organisms.

19. Scientists use ____________________ to cut DNA molecules into smaller fragments.

20. The process of ____________________ is used to identify a person based on DNA evidence.

Short Answer

In complete sentences, write the answers to the questions on the lines provided.

Coat Color
B – black
b – white

		Bb	
		B	**b**
Bb	**B**	**BB**	**Bb**
	b	**Bb**	**bb**

Figure 1

21. In the Punnett square shown in Figure 1, what is the chance that a cross between two black guinea pigs will produce a white guinea pig?

____

____

22. What conclusion about DNA did Oswald Avery and his research group make?

____

____

23. What components make up the DNA molecule?

____

____

____

Name________________________ Class________________ Date ____________

24. What is a transgenic organism?

____

____

25. Why are all X-linked alleles expressed in males, even if they are recessive alleles?

____

____

____

Using Science Skills

Use the diagram below to answer the following questions on the lines provided.

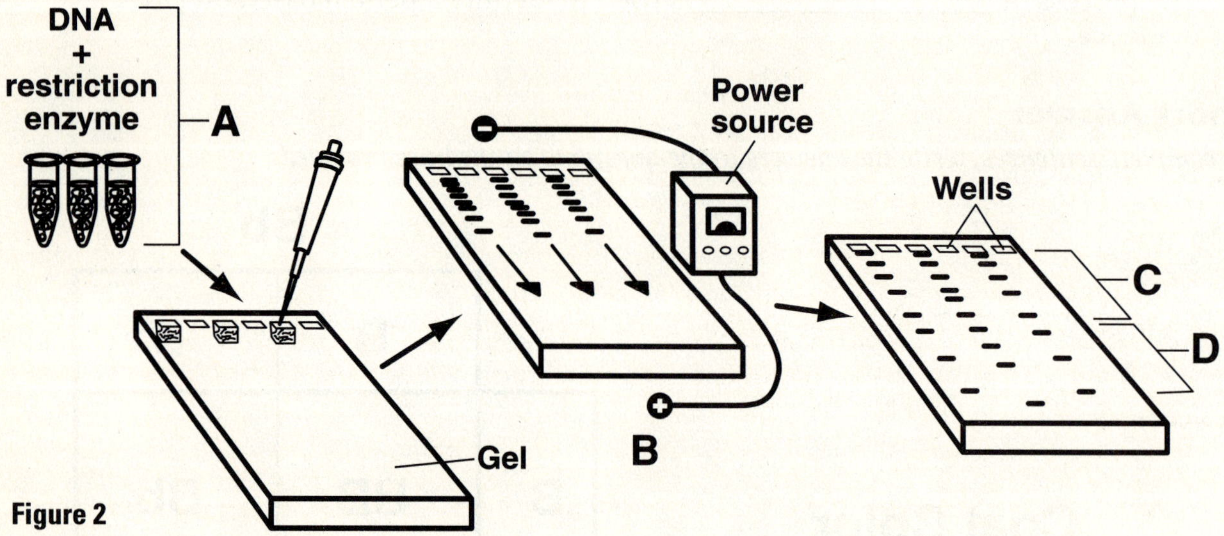

Figure 2

26. **Interpreting Graphics** What procedure is illustrated in Figure 2?

____

27. **Inferring** In Figure 2, what procedure must occur before the step at A can occur?

____

28. **Applying Concepts** In Figure 2, where would you expect to find the longer fragments, at C or D? Explain.

____

____

29. **Inferring** In Figure 2, is DNA a negatively charged molecule or a positively charged molecule? How do you know?

____

____

30. **Interpreting Graphics** In Figure 2, where is the DNA molecule being cut into fragments?

____

____

Name________________________ Class________________ Date ____________

Chapter 15 Darwin's Theory of Evolution **Chapter Test A**

Multiple Choice

Write the letter on the line provided that best answers the question or completes the statement.

____ **1.** On the Galápagos Islands, Charles Darwin observed
a. completely unrelated species on each of the islands.
b. species exactly like those found in South America.
c. somewhat similar species with traits that suited their particular environment.
d. species completely unrelated to those found in South America.

____ **2.** The species of tortoises that Darwin found on the Galápagos Islands displayed different structural adaptations. One of the adaptations that Darwin noted was the
a. similarity in the tortoises' embryos.
b. difference in shell markings of the tortoises.
c. variation in length of the tortoises' necks.
d. difference in number of eggs in each tortoise's nest.

____ **3.** Darwin began to formulate his concept of evolution by natural selection after
a. experimentation with animals.
b. observations of many species and their geographical locations.
c. reading the writings of Wallace.
d. agreeing with Lamarck about the driving force behind evolution.

____ **4.** One scientist who attempted to explain how rock layers form and change over time was
a. Thomas Malthus. c. Charles Darwin.
b. James Hutton. d. Jean-Baptiste Lamarck.

____ **5.** Lamarck proposed that organisms
a. have an innate tendency toward complexity and perfection.
b. have an innate tendency to become more simple as time passes.
c. inherit all of the adaptations they display.
d. belong to species that never change.

____ 6. Lamarck's theory of evolution includes the concept that new organs in a species appear as a result of

a. continual increases in population size.

b. the actions of organisms as they use or fail to use body structures.

c. an unchanging local environment.

d. the natural variations already present within the population of organisms.

____ 7. The idea that only famine, disease, and war could prevent the endless growth of human populations was presented by

a. Darwin. c. Malthus.

b. Lamarck. d. Lyell.

____ 8. When Darwin returned from the voyage of the *Beagle,* he

a. immediately published his ideas about evolution.

b. realized his ideas about evolution were wrong.

c. wrote about his ideas but waited many years to publish them.

d. copied the evolutionary theory of Wallace.

____ 9. When lions prey on a herd of antelope, some antelope are killed and some escape. Which part of Darwin's concept of natural selection might be used to describe this situation?

a. acquired characteristics

b. reproductive isolation

c. survival of the fittest

d. descent with modification

____10. According to Darwin's theory of natural selection, the individuals that tend to survive are those that have

a. characteristics their parents acquired by use and disuse.

b. characteristics that plant and animal breeders value.

c. the greatest number of offspring.

d. variations best suited to the environment.

____11. Darwin's concept of evolution was NOT influenced by

a. the work of Lyell.

b. knowledge of the structure of DNA.

c. his collection of specimens.

d. his trip on the H.M.S. *Beagle.*

____12. The number and location of bones of many fossil vertebrates are similar to those in living vertebrates. Most biologists would probably explain this fact on the basis of

a. the needs of the organisms.

b. a common ancestor.

c. the struggle for existence.

d. the inheritance of acquired traits.

____**13.** Darwin viewed the fossil record as

a. evidence that Earth was thousands of years old.

b. a record of evolution.

c. interesting but unrelated to the evolution of modern species.

d. evidence that traits are acquired through use or disuse.

____**14.** Darwin's theory of evolution is based on the idea(s) of

a. natural variation and natural selection.

b. use and disuse.

c. a tendency toward perfect, unchanging species.

d. the transmission of acquired characteristics.

____**15.** Which concept is NOT included in the modern theory of evolution?

a. descent with modification

b. natural selection

c. transmission of acquired characteristics

d. competition among the members of a population

Completion

Complete each statement on the line provided.

16. Hutton and Lyell held similar views about Earth's age. Both thought that Earth was ________________ of years old.

17. Although his idea was incorrect, Lamarck was one of the first people to propose a scientific explanation for ________________ .

18. Darwin concluded that in similar environments around the world, ________________ could produce similar structures in unrelated species.

19. According to Darwin's theory of evolution, all species on Earth are united by ________________ .

20. Darwin proposed that natural selection takes place as individuals best suited to the ________________ survive and reproduce.

Short Answer

In complete sentences, write the answers to the questions on the lines provided.

21. In what way did the voyage of the *Beagle* provide Darwin with an ideal opportunity for collecting and analyzing data?

____

____

____

Name______________________________ Class____________________ Date ______________

22. What was the scientific value of the specimens that Darwin brought back to England?

____

____

23. What did Darwin learn about the land tortoises of the different Galápagos Islands?

____

____

24. Were Darwin's hypotheses about natural selection and evolution similar to the ideas of most other scientists of his time? Explain.

____

____

25. Modern whales have a vestigial pelvis and femur. What does this evidence suggest about ancestors of modern whales?

____

____

Using Science Skills

Use the diagram below to answer the following questions on the lines provided.

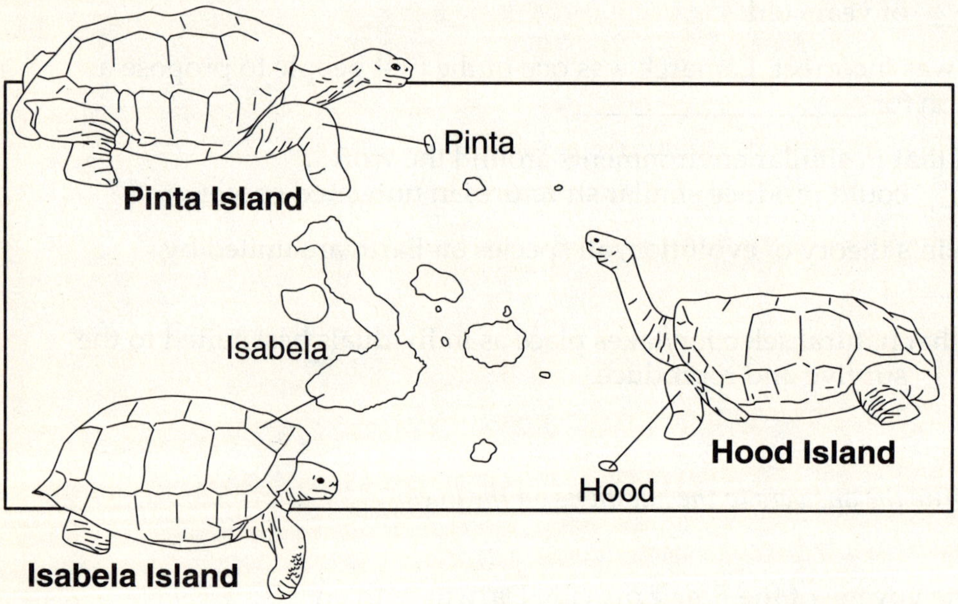

Figure 15-1

26. **Interpreting Graphics** What differences are apparent in the bodies of the three tortoise species shown in Figure 15-1?

____

____

Name____________________ Class____________ Date __________

27. **Interpreting Graphics** Which of the tortoises shown in Figure 15-1 has the longest neck?

____

28. **Applying Concepts** Can you tell from Figure 15-1 how closely the three tortoise species resemble the ancestral species? Why or why not?

____

____

____

29. **Inferring** Vegetation on Hood Island is sparse and sometimes hard to reach. How might the vegetation have affected the evolution of the Hood Island tortoise shown in Figure 15-1?

____

____

____

30. **Forming Hypotheses** Considering the body structures of the tortoises shown in Figure 15-1, which tortoises—a population from Pinta Island or a population from Isabela Island—might survive more successfully on Hood Island? Why?

____

____

____

Essay

Write the answer to each question in the space provided.

31. Explain how Darwin's observation of the tortoises on the Galápagos Islands influenced his thinking.

32. What was Darwin's greatest contribution to science, and how did he develop it?

33. Discuss Lamarck's contribution to the overall theory of evolution.

34. Suppose that selective breeding has produced a population of very similar chickens. Would that population survive if it were released into the world? Explain.

35. How is the idea of common descent supported by examples of homologous structures?

Name______________________ Class______________ Date ____________

Multiple Choice

Write the letter on the line provided that best answers the question or completes the statement.

____ **1.** During the voyage on the *Beagle,* Charles Darwin made many observations

a. in England.
c. on the Galápagos Islands.
b. in North America.
d. in Asia.

____ **2.** James Hutton's and Charles Lyell's work suggested that

a. Earth is many millions of years old.
b. Earth is several thousand years old.
c. all fossils were formed in the last one thousand years.
d. all rocks on Earth contain fossils.

____ **3.** In the 1800s, Charles Lyell emphasized that

a. the human population would outgrow the available food supply.
b. all populations evolve through natural selection.
c. Earth is a few thousand years old.
d. past geological events must be explained in terms of processes observable today.

____ **4.** What did Darwin learn from reading the work of Hutton and Lyell?

a. Earth is relatively young.
b. Earth is very old.
c. All geological change is caused by living organisms.
d. The processes that formed old rocks on Earth do not operate today.

____ **5.** In 1859, Darwin published his revolutionary scientific ideas in a work entitled

a. *Principles of Geology.*
b. *Essay on the Principle of Population.*
c. *Evolution in Malaysia.*
d. *On the Origin of Species.*

____ **6.** Darwin was prompted to publish his theory of evolution by

a. an essay by Wallace on evolution.

b. the publication of Lamarck's theory of evolution.

c. the captain of the *Beagle.*

d. the work of Hutton and Lyell.

____ **7.** When a farmer breeds only his or her best livestock, the process involved is

a. natural selection. c. artificial variation.

b. artificial selection. d. survival of the fittest.

____ **8.** An adaptation is an inherited characteristic that can be

a. physical or behavioral.

b. physical or geographical.

c. acquired during the organism's lifetime.

d. the result of artificial selection.

____ **9.** Darwin called the ability of an organism to survive and reproduce in its environment

a. diversity. c. adaptation.

b. fitness. d. evolution.

____**10.** In humans, the pelvis and the femur, or thighbone, are involved in walking. In whales, the pelvis and femur shown in Figure 15-1 are

a. examples of fossils.

b. vestigial structures.

c. acquired traits.

d. examples of natural variation.

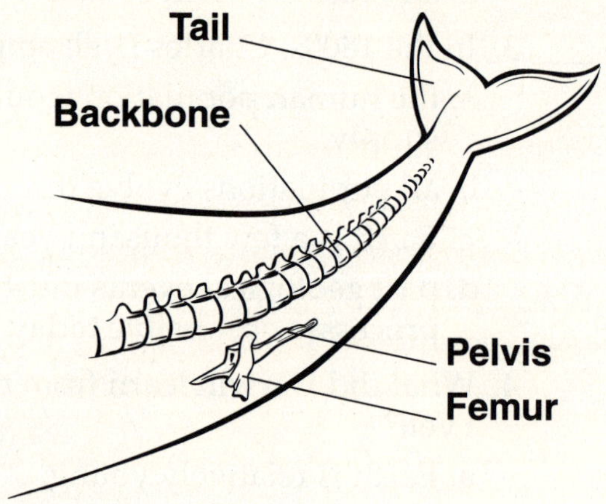

Figure 15-1

____**11.** People of Darwin's time understood that fossils

a. were preserved remains of ancient organisms.

b. were available for every organism that ever lived.

c. were unrelated to living species.

d. were evidence for the evolution of life on Earth.

____**12.** The economist Thomas Malthus suggested that

a. people die faster than babies are born.

b. there would soon be insufficient food for the growing human population.

c. in the 1700s, England needed more housing.

d. the majority of a species' offspring die.

____**13.** Darwin's theory of evolution suggests that

a. species change over time.

b. extinct species are not related to living species.

c. different species can interbreed.

d. animals that look alike are the most closely related.

Name________________________________ Class____________________ Date ______________

____**14.** Which statement is in agreement with Darwin's theory of evolution?

a. More offspring are produced than can possibly survive.

b. The organisms that are the fittest are always the largest and strongest.

c. The number of offspring is not related to fitness.

d. Acquired characteristics that are inherited are the cause of evolution.

____**15.** The hypothesis that species change over time by natural selection was proposed by

a. Hutton. c. Malthus.

b. Lamarck. d. Darwin.

Completion

Complete each statement on the line provided.

16. Hutton thought that rocks from the sea floor could be pushed up very high to form ____________________ .

17. According to Lamarck, evolution resulted from the inheritance of ____________________ traits.

18. When babies are born faster than people are dying, the size of the population ____________________ .

19. Darwin observed that instead of being perfect and unchanging, individuals in a species show____________________ .

20. According to Darwin's theory of evolution, ____________________ change over time.

Short Answer

In complete sentences, write the answers to the questions on the lines provided.

21. What evidence did Darwin collect in addition to specimens of organisms alive during his time?

____

____

22. What was Lamarck's overall belief about the way in which organisms changed over time?

____

____

23. What did Darwin discover to be true about the Galápagos birds that he thought were different kinds of wrens, warblers, and blackbirds?

____

____

Name______________________ Class______________ Date__________

24. How did the farmers observed by Darwin improve their livestock?

____

____

25. Why were fossils important to Darwin's theory of evolution?

____

Using Science Skills

Use the diagram below to answer the following questions on the lines provided.

26. **Interpreting Graphics** According to Figure 15-2, how did the body size of the horse change during its evolution?

____

27. **Observing** In Figure 15-2, how did the size of the head change as the horse evolved?

____

____

28. **Inferring** Scientists have never seen the ancient horses shown in Figure 15-2. What do you think was the main type of evidence that scientists used to prepare these diagrams?

____

____

____

____

____

29. **Comparing and Contrasting** According to Figure 15-2, how does the number of toes of *Mesohippus* compare with those of *Equus*, the modern horse?

____

____

____

30. **Inferring** Does Figure 15-2 show that all species get much larger as they evolve?

____

____

____

____

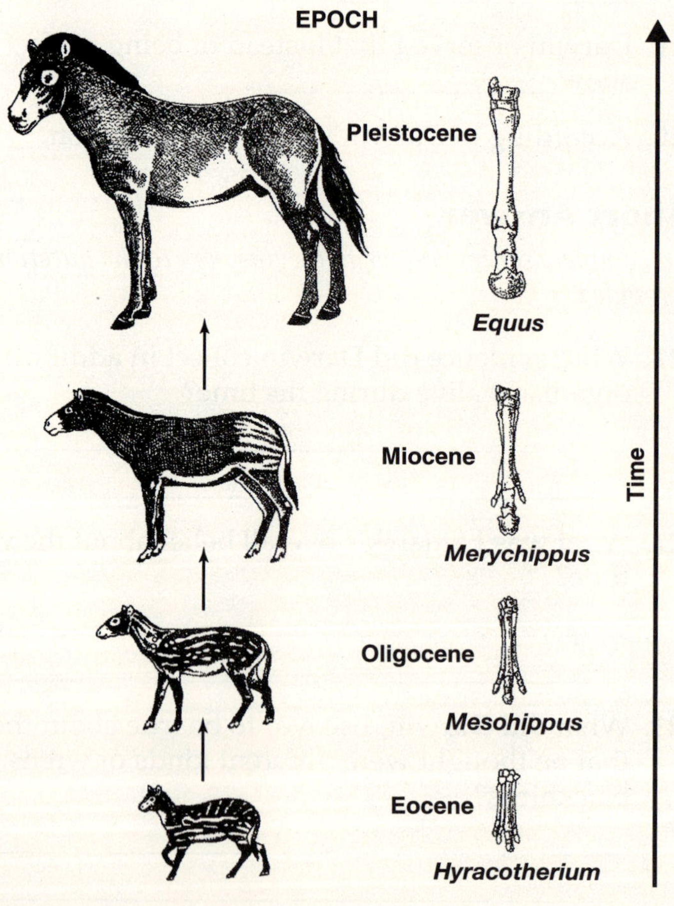

Figure 15-2

Name______________________________ Class__________________ Date ______________

Multiple Choice

Write the letter on the line provided that best answers the question or completes the statement.

____ **1.** Which of the following statements describes what all members of a population share?
a. They are temporally isolated from one another.
b. They are geographically isolated from one another.
c. They are members of the same species.
d. They have identical genes.

____ **2.** Which statement about gene pools is typically true?
a. They contain two or more alleles for each gene.
b. They contain only dominant alleles.
c. They belong to two or more interbreeding species.
d. The relative frequencies of the alleles never change.

____ **3.** If an allele makes up one fourth of a population's alleles for a given trait, its relative frequency is
a. 100 percent.
b. 75 percent.
c. 25 percent.
d. 4 percent.

____ **4.** In many kinds of organisms, most heritable differences are due to
a. mutations during gamete formation.
b. chemicals in the environment.
c. gene shuffling during gamete formation.
d. the effects of radiation.

____ **5.** Gene shuffling includes the independent movement of chromosomes during meiosis as well as
a. mutations from radiation.
b. changes in the frequencies of alleles.
c. crossing-over.
d. mutations from chemicals.

____ **6.** A single-gene trait that has two alleles and that shows a simple dominant-recessive pattern will result in
a. one phenotype.
b. two phenotypes.
c. four phenotypes.
d. millions of phenotypes.

Name________________ Class____________ Date __________

____ **7.** When individuals at only one end of a bell-shaped curve of phenotype frequencies have high fitness, the result is

a. directional selection.

b. stabilizing selection.

c. disruptive selection.

d. genetic drift.

____ **8.** When individuals with an average form of a trait have the highest fitness, the result is

a. not predictable.

b. disruptive selection.

c. directional selection.

d. stabilizing selection.

____ **9.** Genetic drift tends to occur in populations that

a. are very large.

b. are small.

c. are formed from new species.

d. have unchanging allele frequencies.

____ **10.** The type of genetic drift that follows the colonization of a new habitat by a small group of individuals is called

a. the Hardy-Weinberg principle.

b. the founder effect.

c. directional selection.

d. stabilizing selection.

____ **11.** The genetic equilibrium of a population can be disturbed by each of the following EXCEPT

a. nonrandom mating.

b. movement into and out of the population.

c. a large population size.

d. mutations.

____ **12.** The allele frequencies of a population are more likely to remain unchanged if

a. the population size is reduced.

b. frequent movement into and out of the population occurs.

c. all mating is random.

d. the mutation rate increases.

____ **13.** What situation might develop in a population having some plants whose flowers open at midday and other plants whose flowers open late in the day?

a. behavioral isolation

b. geographic isolation

c. temporal isolation

d. genetic drift

Name______________________________ Class__________________ Date ______________

____14. The Galápagos finch species are an excellent example of

a. speciation.

b. genetic equilibrium.

c. stabilizing selection.

d. selection on single-gene traits.

____15. What did Peter and Rosemary Grant learn about mate choice in the Galápagos finches?

a. Phenotype plays no role in mate choice.

b. Genotype plays no role in mate choice.

c. Finches prefer mates with smaller beaks than their own.

d. Finches prefer mates with beaks similar in size to their own.

Completion

Complete each statement on the line provided.

16. Crossing-over can occur during the meiotic divisions that produce cells called __________________ .

17. A polygenic trait can have many possible genotypes and __________________ .

18. When the phenotypes of polygenic traits are represented by a bell-shaped curve, the __________________ of individuals close together on the curve is not very different.

19. When a population is NOT evolving, it is in a situation called __________________ .

20. In the Galápagos finches that Rosemary and Peter Grant studied, a pattern of natural selection called __________________ selection favored individuals with larger, heavier beaks during a drought.

Short Answer

In complete sentences, write the answers to the questions on the lines provided.

21. Would a trait that has only two distinct phenotypes more likely be a single-gene trait or a polygenic trait? How do you know?

22. Is an allele for a trait that has no effect on a species' fitness affected by natural selection? Explain.

23. List the five conditions necessary to maintain genetic equilibrium in a population.

Name________________________ Class________________ Date ____________

24. Why might a geographic barrier such as a large river cause the formation of a new species of small rodents but not a new species of birds?

____

____

25. How does the work of Peter and Rosemary Grant in the Galápagos relate to the work of Charles Darwin?

____

Using Science Skills

Use the graph below to answer the following questions on the lines provided.

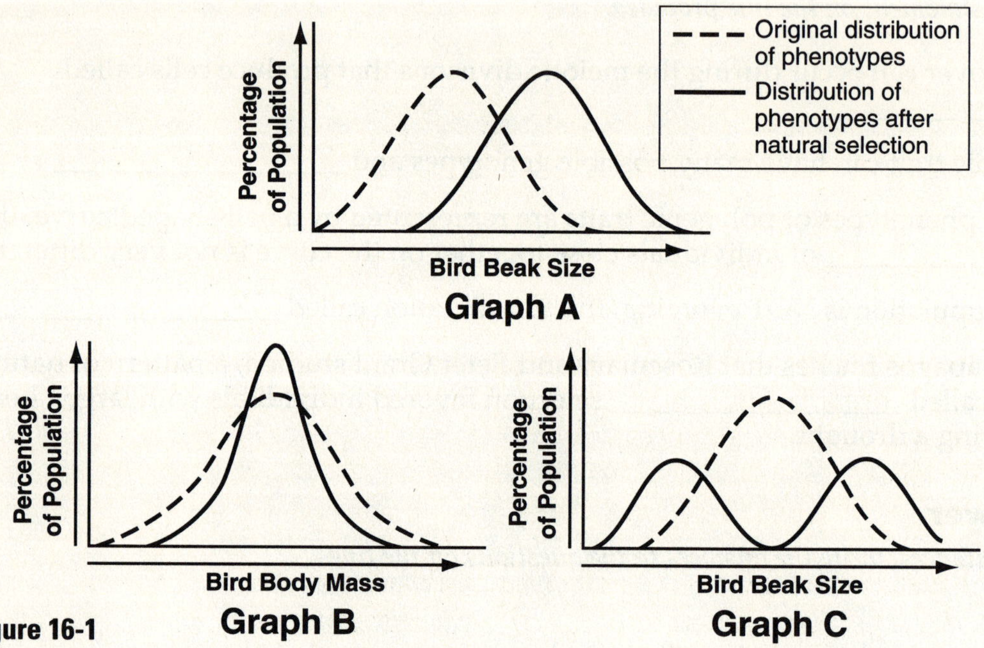

Figure 16-1

26. Interpreting Graphics According to Graph A in Figure 16-1, what has occurred?

____

27. Interpreting Graphics According to Graph B in Figure 16-1, what has occurred?

____

28. Interpreting Graphics According to Graph C in Figure 16-1, what has occurred?

____

____

29. Inferring Which of the three graphs in Figure 16-1 might show a population of birds that specialize in different types of food? Explain.

____

____

Name________________________________ Class__________________ Date ______________

30. **Inferring** What factor or condition might have led to the change shown in Graph A of Figure 16-1?

____

____

Essay

Write the answer to each question in the space provided.

31. For a trait that has many different alleles in the population, would an individual be more likely to have a particular allele if it had a low relative frequency or a high relative frequency? Why?

32. Explain how shuffling a deck of playing cards can be a good model for the effect of sexual reproduction's effect on the relative frequency of alleles in a population and the possible combinations of alleles.

Name________________________________ Class____________________ Date ______________

33. For a population, are the frequencies of phenotypes for a single-gene trait best expressed by a bar graph or a curve? Are the frequencies of phenotypes for a polygenic trait best expressed by a bar graph or a curve? Explain.

34. Describe the founder effect, and describe the conditions in which it arises.

35. The Hardy-Weinberg principle gives the conditions needed for the frequency of alleles in a population to remain unchanged. Describe those conditions.

Name______________________________ Class__________________ Date ______________

Multiple Choice

Write the letter on the line provided that best answers the question or completes the statement.

____ 1. The combined genetic information of all members of a particular population is the population's
a. relative frequency. c. genotype.
b. phenotype. d. gene pool.

____ 2. A change in a sequence of DNA is called a
a. recombination. c. single-gene trait.
b. polygenic trait. d. mutation.

____ 3. The two main sources of genetic variation are
a. genotypes and phenotypes.
b. gene shuffling and mutations.
c. single-gene traits and polygenic traits.
d. directional selection and disruptive selection.

____ 4. An example of a single-gene trait is
a. a widow's peak in humans.
b. the weight of human infants at birth.
c. height in humans.
d. beak size in the Galápagos finches.

____ 5. The number of phenotypes produced for a given trait depends upon
a. the number of genes that control the trait.
b. which form of the trait is dominant.
c. the relative frequencies of the various alleles.
d. whether or not natural selection is at work.

____ 6. The distribution of phenotypes for a typical polygenic trait can often be expressed as
a. a bar graph.
b. a bell-shaped curve.
c. Mendelian ratios.
d. allele frequencies.

____ 7. Natural selection acts directly on
a. alleles. c. phenotypes.
b. genes. d. mutations.

____ 8. Which of the following is NOT a way in which natural selection affects the distribution of phenotypes?
a. directional selection
b. stabilizing selection
c. disruptive selection
d. chance events

____ **9.** In genetic drift, allele frequencies change because of
a. mutations.
b. chance.
c. natural selection.
d. genetic equilibrium.

____ **10.** Which of the following events do biologists consider a random change?
a. directional selection
b. speciation
c. disruptive selection
d. genetic drift

____ **11.** The situation in which allele frequencies of a population remain constant is called
a. evolution.
b. genetic drift.
c. genetic equilibrium.
d. natural selection.

____ **12.** One of the conditions required to maintain genetic equilibrium is
a. natural selection.
b. mutations.
c. nonrandom mating.
d. no movement into or out of the population.

____ **13.** The separation of populations by barriers such as rivers, mountains, or bodies of water is called
a. temporal isolation.
b. geographic isolation.
c. behavioral isolation.
d. genetic equilibrium.

____ **14.** A factor that is necessary for the formation of a new species is
a. reproduction at different times.
c. different mating behaviors.
b. geographic barriers.
d. reproductive isolation.

____ **15.** Which is the first step that occurred in the speciation of the Galápagos finches?
a. establishment of genetic equilibrium
b. behavioral isolation
c. ecological competition
d. arrival of the founding population

Completion

Complete each statement on the line provided.

16. A gene pool typically contains two or more ____________________ for each gene.

17. Inheritable traits can be classified as either ____________________ traits or polygenic traits.

18. For __________________ traits, natural selection can affect the distribution of phenotypes in three ways.

19. According to the __________________ principle, allele frequencies in a population will remain constant unless one or more of five specific factors cause those frequencies to change.

20. For new species to evolve, populations must be __________________ isolated from each other.

Short Answer

Write the answers to the questions in complete sentences on the lines provided.

21. Are the members of a population necessarily the same species? Explain.

____

____

____

22. What are the two main sources of genetic variation? Which of them is more common?

____

____

____

23. Explain how stabilizing selection would affect a graph of the distribution of phenotypes for a trait.

____

____

____

24. Why might genetic drift occur if a small number of individuals colonized a new habitat?

____

____

____

25. What are three mechanisms for reproductive isolation? Which mechanism isolates two populations of similar frogs with different mating calls?

____

____

____

____

Name______________________________ Class__________________ Date ______________

Using Science Skills

Use the graph below to answer the following questions on the lines provided.

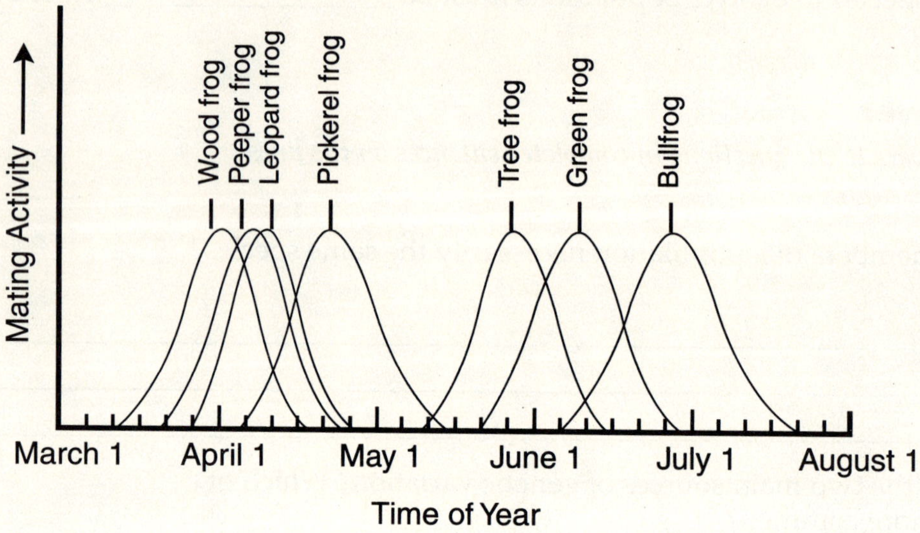

Figure 16-1

26. Interpreting Graphics Describe the information about frog species that is shown in Figure 16-1.

____

27. Interpreting Graphics According to Figure 16-1, there is a brief period during which frog mating nearly stops. When does this occur?

____

28. Inferring Based on Figure 16-1, what mechanism appears to keep bullfrogs reproductively isolated? Would that mechanism necessarily be the only isolating mechanism? Explain.

____

29. Inferring Peeper frogs and leopard frogs do not interbreed even when they share a habitat. Use the information in Figure 16-1 to determine what mechanism probably keeps the two species reproductively isolated.

____

____

30. Predicting Frog mating does not occur in cold weather. Assume that the mating times shown in Figure 16-1 are for frogs in the northern part of the United States. How might these curves change for frogs in the southern part of the United States? Explain.

____

Name________________________ Class________________ Date ____________

Multiple Choice

Write the letter that best answers the question or completes the statement on the line provided.

____ 1. Sedimentary rock is formed from
a. the soft parts of organisms.
b. the hard parts of organisms.
c. small particles of sand, silt, and clay.
d. wood, shell, and bone.

____ 2. What proportion of all species that ever lived has become extinct?
a. less than 1 percent
b. approximately one-half
c. more than 99 percent
d. 100 percent

____ 3. To be useful as an index fossil, a species must have existed for a
a. long period over a wide geographic range.
b. long period over a small geographic range.
c. short period over a wide geographic range.
d. short period over a small geographic range.

____ 4. The basic divisions of the geologic time scale from larger to smaller are
a. eras and periods.
b. periods and eras.
c. relative and absolute dates.
d. billions of years and millions of years.

____ 5. The Mesozoic Era occurred
a. before Precambrian Time.
b. during Precambrian Time.
c. after the Paleozoic Era.
d. after the Cenozoic Era.

____ 6. Which of the following was NOT characteristic of Earth before the oceans formed?
a. volcanic activity
b. bombardment by comets and asteroids
c. an atmosphere of poisonous gases
d. an atmosphere containing oxygen gas

____ 7. Two gases that probably existed in Earth's early atmosphere are
a. oxygen and hydrogen sulfide.
b. water vapor and oxygen.
c. oxygen and carbon monoxide.
d. hydrogen cyanide and carbon monoxide.

____ 8. What prevents organic molecules from forming on their own and remaining intact today?

a. Earth is too hot.

b. Atmospheric oxygen is too reactive.

c. The necessary building blocks no longer exist.

d. There is no energy source available.

____ 9. One necessary condition for the evolution of the first life on Earth was

a. the presence of DNA.

b. abundant oxygen in the atmosphere.

c. the presence of photosynthetic organisms.

d. the presence of liquid water.

____10. Which of the following functions can RNA perform under certain conditions?

a. catalyzing chemical reactions

b. processing messenger RNA after transcription

c. helping DNA replicate

d. all of the above

____11. Why did the development of sexual reproduction speed up the process of evolution?

a. Sexual reproduction occurs more rapidly than asexual reproduction.

b. The offspring of sexual reproduction are identical to their parents.

c. Sexual reproduction increases genetic variety.

d. Sexual reproduction limits genetic variety.

____12. What was the response of the various groups of early organisms that existed when oxygen levels rose in the atmosphere?

a. Some life forms became extinct.

b. Some life forms survived in only a few airless habitats.

c. Some life forms evolved metabolic pathways that used oxygen for respiration.

d. all of the above

____13. The Cambrian Explosion resulted in the evolution of the first

a. dinosaurs and mammals.

b. representatives of most animal groups.

c. bacteria.

d. land animals.

____14. The process by which two species, for example, a flower and a pollinating insect, evolve in response to each other is called

a. convergent evolution.

b. adaptive radiation.

c. coevolution.

d. punctuated equilibrium.

Name________________________ Class________________ Date ____________

____15. In the past, mass extinctions encouraged the rapid evolution of surviving species

a. by changing developmental genes.

b. by making new habitats available to them.

c. because they killed all organisms that had coevolved.

d. because they spared all organisms that had evolved convergently.

Completion

Complete each statement on the line provided.

16. In a species that has become ________________, all members have died, and the species has ceased to exist.

17. Earth's least dense elements, including hydrogen and ________________, formed the planet's first atmosphere.

18. It is possible that RNA was the first information-storing molecule, but over time, ________________ became the primary way in which genetic information is stored and transmitted.

19. ________________, which are fossilized remains of single-celled organisms similar to modern bacteria, exist in rocks that are more than 3.5 billion years old.

20. As the Paleozoic Era closed, a(an) ________________, which is the dying out of many types of living things at one time, occurred.

Short Answer

In complete sentences, write the answers to the questions on the lines provided.

21. Why is the fossil record an incomplete history of life?

____

____

____

22. Why are the divisions of geologic time NOT standard lengths, such as 100 million years?

____

____

____

____

23. What was the pattern of mammalian evolution during the Cenozoic Era?

____

____

____

____

Name________________ Class________________ Date________________

24. How does a study of modern proteinoid microspheres help scientists investigate the origin of cells?

25. What was the source of the oxygen gas that began to accumulate in the atmosphere over 2 billion years ago?

Using Science Skills

Use the graph below to answer the following questions on the lines provided.

26. **Using Tables and Graphs** Use Figure 17-1 to determine the approximate half-life of carbon-14.

27. **Using Tables and Graphs** According to Figure 17-1, how many half-lives have passed if a fossil has 1/8 of its original amount of carbon-14?

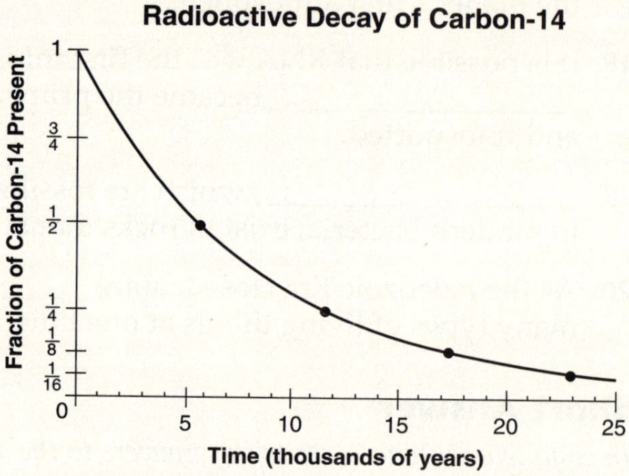

Figure 17-1

28. **Using Tables and Graphs** A fossil bone was found to contain about 1/16 of the amount of carbon-14 that was originally present. Use Figure 17-1 to determine the approximate age of the bone.

29. **Inferring** According to Figure 17-1, in which case would carbon-14 be more useful for dating: wooden beams in Native American cave dwellings, which probably are less than 7000 years old, or the fossil of an early mammal that is probably almost 100,000 years old? Explain your answer.

Name________________________ Class________________ Date ____________

30. **Predicting** As carbon-14 decays, it changes to nitrogen-14, which is released as a gas. Refer to Figure 17-1 to predict how the amount of nitrogen gas produced would change over the period of radioactive decay. Why does that change take place?

____

____

____

____

____

Essay

Write the answer to each question in the space provided.

31. Explain the principle behind radioactive dating.

32. Describe Stanley Miller and Harold Urey's famous experiment. Describe their results as well as the results of similar experiments based on more current knowledge.

Name______________________________ Class__________________ Date ____________

33. Define and explain the endosymbiotic theory.

34. Describe the milestones in the evolution of life that occurred during Precambrian Time.

35. Explain the proposed relationship between an asteroid impact and the Cretaceous mass extinction, and describe a more current view of the cause of the Cretaceous extinction.

Name______________________ Class________________ Date ____________

Multiple Choice

Write the letter that best answers the question or completes the statement on the line provided.

____ **1.** Examples of fossils include preserved
a. eggs.
b. footprints.
c. body parts.
d. all of the above

____ **2.** Most fossils form in
a. rusty water.
b. volcanic rocks.
c. sedimentary rock.
d. the sap of ancient trees.

____ **3.** The length of time required for half of the radioactive atoms in a sample to decay is its
a. half-life.
b. relative date.
c. radioactive date.
d. period.

____ **4.** To compare the relative ages of fossils, scientists sometimes use an easily recognized species called a(an)
a. carbon fossil.
b. radioactive fossil.
c. index fossil.
d. sedimentary fossil.

____ **5.** Earth's most recent era is the
a. Paleozoic.
b. Mesozoic.
c. Cenozoic.
d. Precambrian.

____ **6.** Why did oceans not exist on Earth nearly 4 billion years ago?
a. No water molecules were present.
b. Water remained a gas because Earth was very hot.
c. Water existed as ice because Earth was very cold.
d. There was no oxygen gas in the atmosphere.

____ **7.** Early in Earth's history, while the planet was in a melted state, the most dense elements formed Earth's
a. core.
b. seas.
c. crust.
d. atmosphere.

____ **8.** In addition to hydrogen, two of the gases used in Miller and Urey's experiment were
a. nitrogen and carbon monoxide.
b. hydrogen cyanide and oxygen.
c. methane and ammonia.
d. carbon dioxide and hydrogen sulfide.

____ **9.** What do proteinoid microspheres have in common with cells?
a. They can store and release energy.
b. They contain DNA.
c. They contain RNA.
d. They are communities of organisms.

Name________________________ Class __________________ Date ____________

____10. When oxygen was first released in the early seas, it combined with iron to form

a. RNA. c. proteins.
b. DNA. d. rust.

____11. The first organisms on Earth were most like today's

a. bacteria. c. multicellular organisms.
b. eukaryotes. d. DNA molecules.

____12. A very large mass extinction in which trilobites and amphibians disappeared occurred at the end of the

a. Precambrian Era. c. Paleozoic Era.
b. Cambrian Period. d. Quaternary Period.

____13. During the Jurassic and Cretaceous periods, the dominant land animals were

a. amphibians. c. grazing mammals.
b. dinosaurs. d. human ancestors.

____14. A single species that has evolved into several different forms that live in different ways has undergone

a. adaptive radiation. c. punctuated equilibrium.
b. coevolution. d. mass extinction.

____15. Sharks, dolphins, and penguins all have streamlined bodies and appendages that enable them to move through water. These similarities are the result of

a. coevolution.
b. convergent evolution.
c. asexual reproduction.
d. adaptive radiation.

Completion

Complete each statement on the line provided.

16. The ____________________ record provides evidence about the history of life on Earth.

17. After Precambrian Time, the main divisions of the geologic time scale are eras and ____________________.

18. In Miller and Urey's experiment, several ____________________, which are the building blocks of proteins, began to accumulate.

19. The ____________________ theory proposes that eukaryotic cells arose from living communities of several prokaryotic organisms.

20. ____________________ is the process by which two species evolve in response to changes in each other over time.

Name________________________ Class________________ Date__________

Short Answer

In complete sentences, write the answers to the questions on the lines provided.

21. What are two kinds of information that scientists attempt to infer from the study of fossils?

22. What information does relative dating provide?

23. Why did oceans not exist on Earth 4 billion years ago?

24. What do the membranes of microspheres and those of cells have in common?

25. What happened to the dinosaurs at the end of the Cretaceous Period?

Name______________________ Class______________ Date__________

Using Science Skills

Use the table below to answer the following questions on the lines provided.

Geologic Time Scale

Era	Period	End Date (in millions of years)
Cenozoic	Quaternary	present
	Tertiary	1.8
Mesozoic	Cretaceous	65
	Jurassic	145
	Triassic	208
Paleozoic	Permian	245
	Carboniferous	290
	Devonian	360
	Silurian	410
	Ordovician	440
	Cambrian	505
Precambrian Time	Vendian	544

Figure 17-1

26. Interpreting Graphics How many geologic periods are shown in Figure 17-1, and which is the earliest?

____

____

27. Applying Concepts The Paleozoic Era ended with the disappearance of many land and aquatic species. What is that type of event called? Use Figure 17-1 to name the period in which that event took place.

____

28. Calculating Use Figure 17-1 to determine the length in years of the Silurian Period.

____

29. Interpreting Graphics The end of the Mesozoic Era marks the extinction of dinosaurs. How long ago did the dinosaurs become extinct?

____

30. Interpreting Graphics The first marsupials, a group that includes the modern kangaroo, evolved about 100 million years ago. According to Figure 17-1, during which geologic period did marsupials evolve?

____

Name______________________________ Class__________________ Date ______________

Chapter 18 Classification Chapter Test A

Multiple Choice

Write the letter on the line provided that best answers the question or completes the statement.

____ **1.** Scientists assign each kind of organism a universally accepted name in the system known as
a. traditional classification.
b. the three domains.
c. binomial nomenclature.
d. cladistics.

____ **2.** For many species, there are often regional differences in their
a. common names.
b. scientific names.
c. taxa.
d. binomial nomenclature.

____ **3.** The second part of a scientific name is unique to each
a. order in its class.
b. family in its order.
c. genus in its family.
d. species in its genus.

____ **4.** Several different classes make up a
a. kingdom.
b. phylum.
c. family.
d. genus.

____ **5.** Which two kingdoms did Linnaeus recognize?
a. bacteria and animals
b. plants and fungi
c. plants and animals
d. protists and animals

____ **6.** Traditional classifications tended to take into account primarily
a. extinct organisms.
b. RNA similarities.
c. DNA similarities.
d. general similarities in appearance.

____ **7.** Sometimes organisms that are not closely related look similar because of
a. convergent evolution.
b. molecular clocks.
c. mutations.
d. reclassification.

____ **8.** In an evolutionary classification scheme, species within one genus should
a. be more similar to each other than they are to other species.
b. not be similar in appearance.
c. be limited to species that can interbreed.
d. have identical genes.

____ **9.** An analysis of derived characters is used to generate a
a. family tree based on external appearance.
b. family tree based on DNA structure.
c. cladogram.
d. traditional classification system.

____ **10.** What is true about dissimilar organisms such as a cow and a yeast?
a. They are not related at all.
b. Their degree of relatedness cannot be evaluated.
c. Their degree of relatedness can be determined from their genes.
d. They can interbreed and thus are the same species.

____ **11.** Scientists have found that humans and yeasts
a. have similar genes for the assembly of certain proteins.
b. share all aspects of cellular structure.
c. have nothing in common.
d. cannot be evaluated for degree of relatedness.

____ **12.** What kingdoms composed the three-kingdom classification system used by scientists in the 1800s?
a. animals, plants, fungi
b. animals, plants, bacteria
c. animals, fungi, protists
d. animals, plants, protists

____ **13.** Which of the kingdoms in the six-kingdom system of classification was once grouped with plants?
a. Animalia c. Fungi
b. Carnivores d. Protista

____ **14.** The three-domain system recognizes fundamental differences between two groups of
a. prokaryotes. c. protists.
b. eukaryotes. d. multicellular organisms.

____ **15.** Organisms in the kingdoms Eubacteria and Archaebacteria were previously grouped in a kingdom called
a. Animalia. c. Monera.
b. Fungi. d. Eukarya.

Completion

Complete each statement on the line provided.

16. In taxonomy, different classes of organisms might be grouped into the next larger category, a(an) ____________________ .

17. A(An) ____________________ is a group of closely related species.

Name________________________________ Class__________________ Date ______________

18. In cladistic analysis, a characteristic that arises as a lineage of organisms evolves over time is called a(an) ____________________ .

19. A model known as a(an) ____________________ uses DNA comparisons to estimate the length of time that species have been evolving independently.

20. The domain ____________________ is composed of the kingdom Eubacteria.

Short Answer

In complete sentences, write the answers to the questions on the lines provided.

21. Why might a particular kind of organism have more than one common name?

____

____

22. How do you know that the groups *Ursus maritimus* and *Ursus arctos* are closely related?

____

____

23. How can such different animals as fishes, amphibians, reptiles, birds, and mammals all be grouped in a single phylum?

____

____

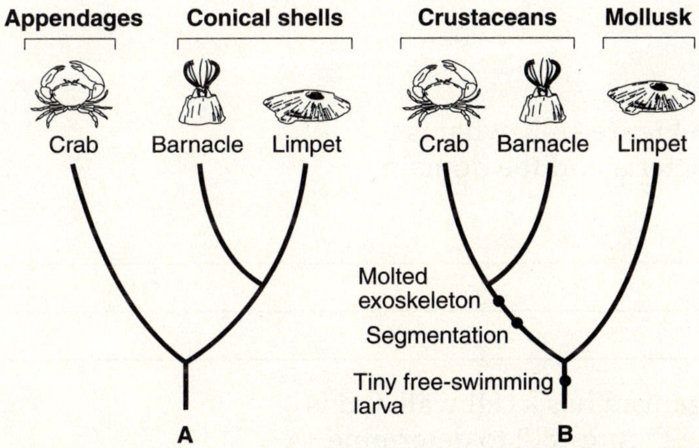

Figure 18-1

24. According to the cladogram in Figure 18-1, what two characteristics do crabs and barnacles share that limpets do not?

____

____

25. What recently developed technology allows scientists to compare the DNA of different kinds of organisms to determine classification?

____

____

Name________________________ Class________________ Date ____________

Using Science Skills

Use the table below to answer the following questions on the lines provided.

Classification of Living Things

DOMAIN	Bacteria	Archaea	Eukarya			
KINGDOM	Eubacteria	Archaebacteria	Protista	Plantae	Fungi	Animalia
CELL TYPE	Prokaryote	Prokaryote	Eukaryote	Eukaryote	Eukaryote	Eukaryote
CELL STRUCTURES	Cell walls with peptidoglycan	Cell walls without peptidoglycan	Cell walls of cellulose in some; some have chloroplasts	Cell walls of cellulose; chloroplasts	Cell walls of chitin	No cell walls or chloroplasts
NUMBER OF CELLS	Unicellular	Unicellular	Most unicellular; some colonial; some multi-cellular	Multicellular	Most multicellular; some unicellular	Multicellular
MODE OF NUTRITION	Autotroph or heterotroph	Autotroph or heterotroph	Autotroph or heterotroph	Autotroph	Heterotroph	Heterotroph
EXAMPLES	*Streptococcus, Escherichia coli*	Methanogens, halophiles	*Amoeba, Paramecium,* slime molds, giant kelp	Mosses, ferns, flowering plants	Mushrooms, yeasts	Sponges worms, insects, fishes, mammals

Figure 18-2

26. Using Tables and Graphs According to Figure 18-2, what is the main difference between the domain Bacteria and the domain Archaea?

____

____

____

27. Applying Concepts If you know an organism has a cell wall and is a multicellular autotroph, could you use Figure 18-2 to determine in which kingdom it belongs? Why or why not?

____

____

28. Using Tables and Graphs Can you determine, by examining Figure 18-2, which kingdom contains the greatest number of species? Why or why not?

____

____

____

Name____________________________ Class__________________ Date ______________

29. **Applying Concepts** If you were told only that an organism is unicellular and has chloroplasts and a nucleus, could you use Figure 18-2 to determine to which kingdom it belongs? Why or why not?

30. **Using Tables and Graphs** Considering the data presented in Figure 18-2, which characteristic seems more important in assigning an organism to a specific domain—the presence or absence of a nucleus or its mode of nutrition? Why?

Essay

Write the answer to each question in the space provided.

31. How is binomial nomenclature superior to the descriptive names used by early scientists?

32. How does traditional classification differ from evolutionary classification?

Name______________________________ Class________________ Date ____________

33. How are neutral mutations useful for estimating how long two species have been evolving independently?

34. Briefly explain the history of how microorganisms have been classified, beginning with the early systems of classification and leading to the modern six-kingdom system.

35. Identify a major source of evidence used by scientists who advocate the adoption of the three-domain system for classifying living things.

Name______________________________ Class__________________ Date ______________

Multiple Choice

Write the letter that best answers the question or completes the statement on the line provided.

____ **1.** Biologists use a classification system to group organisms in part because organisms
a. are going extinct.
b. are very numerous and diverse.
c. are too much alike.
d. share too many derived characters.

____ **2.** The study of organisms requires the use of
a. only large, general categories of organisms.
b. only small, specific categories of organisms.
c. both large and small categories of organisms.
d. no categories of organisms.

____ **3.** In binomial nomenclature, which of the two terms is capitalized?
a. the first term only
b. the second term only
c. both the first and second terms
d. neither the first nor the second term

____ **4.** In Linnaeus's system of classification, how many taxonomic categories were there?
a. one
b. three
c. five
d. seven

____ **5.** A genus is composed of a number of related
a. kingdoms.
b. phyla.
c. orders.
d. species.

____ **6.** Animals that are warm blooded, have body hair, and produce milk for their young are grouped in the class
a. Amphibia.
b. Mammalia.
c. Aves.
d. Reptilia.

____ **7.** The procedure of grouping organisms based on their evolutionary history is called
a. traditional classification.
b. binomial nomenclature.
c. derived characters.
d. evolutionary classification.

____ **8.** What do scientists consider when they perform cladistic analysis?
a. only the DNA of organisms
b. all traits of organisms
c. derived characters
d. only physical similarities

Name______________________________ Class__________________ Date ______________

____ 9. Similar genes are evidence of
a. binomial nomenclature.
b. mutations.
c. common ancestry.
d. different anatomy.

____10. What do all organisms have in common?
a. They use DNA and RNA to pass on information.
b. They are all prokaryotes.
c. They are all eukaryotes.
d. They are genetically identical.

____11. All organisms in the kingdoms Protista, Plantae, Fungi, and Animalia are
a. multicellular organisms.
b. photosynthetic organisms.
c. eukaryotes.
d. prokaryotes.

____12. Which kingdom contains heterotrophs with cell walls of chitin?
a. Protista
b. Fungi
c. Plantae
d. Animalia

____13. The domain that includes the kingdom Eubacteria is
a. Archaea.
b. Bacteria.
c. Eukarya.
d. Fungi.

____14. The domain that contains unicellular organisms that live in extreme environments is
a. Eubacteria.
b. Eukarya.
c. Archaea.
d. Bacteria.

____15. The three-domain system arose when scientists grouped organisms according to how long they have been
a. alive in their present forms.
b. going extinct.
c. evolving independently.
d. using DNA to store information.

Completion

Complete each statement on the line provided.

16. When scientists use a(an) ____________________ for an organism, they can be certain they are all discussing the same organism.

17. The use of a two-part scientific name for organisms is called ____________________ nomenclature.

18. In Linnaeus's system of classification, the two smallest categories are genus and ____________________.

Name______________________ Class______________ Date __________

19. The six kingdoms of life include eubacteria, archaeabacteria, protists, fungi, animals, and ______________ .

20. The domain ______________ contains plants, fungi, protists, and animals—which are all eukaryotes.

Short Answer

In complete sentences, write the answers to the questions on the lines provided.

21. What is evolutionary classification? How does it differ from traditional classification?

____

____

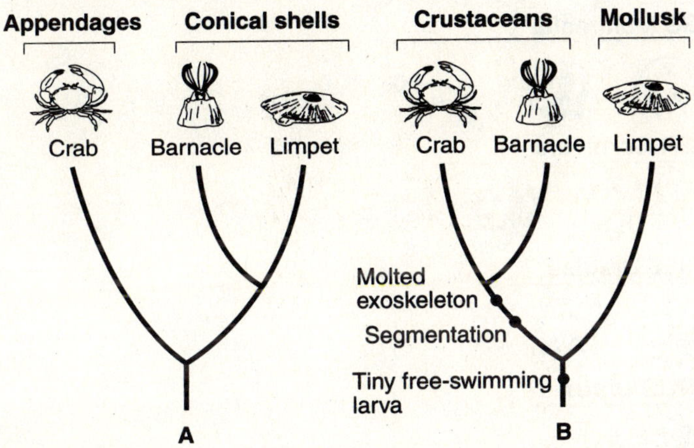

Figure 18-1

22. Which grouping in Figure 18-1, A or B, shows the older, traditional method of classifying the three animals shown? What kind of evidence was used to carry out that classification?

____

____

23. Which system of grouping in Figure 18-1, A or B, provides information about the evolution of the three animals? What is the name of the diagram used to show that information?

____

____

24. What characteristic did biologists use to reclassify some organisms from the plant or animal kingdom to the kingdom Protista?

____

____

25. What characteristic is used to place an organism in the domain Eukarya?

____

____

Name________________________ Class __________________ Date ______________

Using Science Skills

Use the diagram below to answer the following questions on the lines provided.

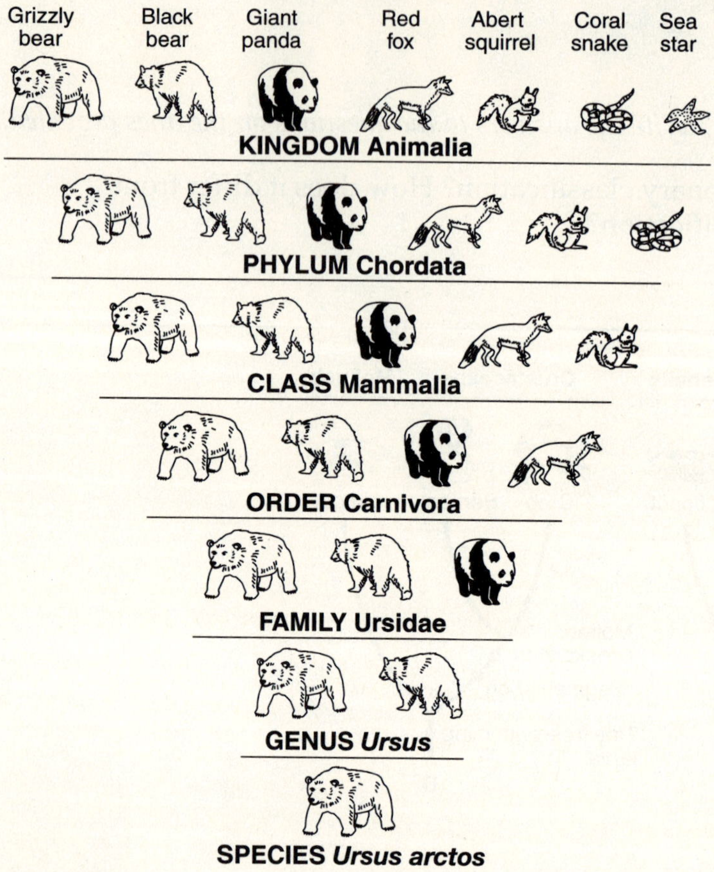

Figure 18-2

26. Interpreting Graphs Which taxonomic category shown in Figure 18-2 contains the greatest number of different organisms?

____

27. Classifying Do all organisms shown in Figure 18-2 that belong to the order Carnivora also belong to the phylum Chordata? Explain.

____

28. Classifying Do all organisms shown in Figure 18-2 that belong to the class Mammalia also belong to the genus *Ursus*? Explain.

____

29. Observing Based on the information in Figure 18-2, describe how the diversity at each level changes from species to kingdom.

____

30. Inferring Considering the information in Figure 18-2, if you were given a species name and no other information about an unfamiliar organism, what is the largest taxonomic category that you could assign it to?

____

Name________________________ Class________________ Date ____________

Multiple Choice

Write the letter that best answers the question or completes the statement on the line provided.

____ **1.** The process by which modern organisms have descended from ancient organisms is called
a. genetic drift.
b. adaptation.
c. evolution.
d. homologous descent.

____ **2.** Which of the following ideas is associated with Lamarck's evolution hypothesis?
a. disruptive selection
b. inheritance of acquired traits
c. natural selection
d. survival of the fittest

____ **3.** The struggle for existence that Darwin referred to takes place between
a. members of the same species.
b. members of different species.
c. predator and prey populations.
d. organisms and their environment.

____ **4.** The number of different genes that control a trait determine the
a. allele frequencies for the trait.
b. number of phenotypes for the trait.
c. size of the gene pool.
d. chance that natural selection will occur.

____ **5.** The weight of human infants at birth is under the influence of
a. directional selection.
b. disruptive selection.
c. stabilizing selection.
d. artificial selection.

____ **6.** Which is NOT a condition required for allele frequencies to remain the same over time?
a. natural selection
b. random mating
c. very large population
d. absence of mutation

Name________________________________ Class__________________ Date _______________

____ 7. Eastern meadowlarks do not respond to western meadowlark mating songs. This is an example of

a. founder effect.

b. behavioral isolation.

c. genetic drift.

d. temporal isolation.

____ 8. What did the research of Peter and Rosemary Grant demonstrate about finches on the Galápagos Islands?

a. The finches were no longer evolving.

b. Finch evolution was occurring slowly and gradually.

c. Differences in beak size produced differences in fitness.

d. Finches did not have enough variation to undergo natural selection.

____ 9. In relative dating, the age of a fossil is determined by its

a. radioactive isotopes.

b. half-life.

c. geographic range.

d. position in rock layers.

____10. Miller and Urey's experiments demonstrated that

a. the endosymbiotic theory is true.

b. DNA molecules can evolve from RNA molecules.

c. eukaryotes arose from living communities of prokaryotes.

d. organic compounds could form under the conditions of early Earth.

____11. Which geological era is associated with the evolution of flowering plants?

a. Mesozoic

b. Paleozoic

c. Jurassic

d. Cenozoic

____12. The disappearance of dinosaurs cleared the way for the adaptive radiation of

a. amphibians.

b. mammals.

c. birds.

d. fishes.

____13. One of Darwin's finches evolved adaptations more similar to those of woodpeckers than other finches. This is probably explained by

a. coevolution.

b. convergent evolution.

c. stabilizing selection.

d. gradualism.

Name______________________________ Class__________________ Date ______________

____14. Derived characters appear in
a. recent parts of a lineage but not its older members.
b. common ancestors of a group of species but not the species themselves.
c. organisms that have undergone convergent evolution.
d. index fossils but not other fossils.

____15. Which domain contains organisms without cell walls?
a. Animalia
b. Plantae
c. Eukarya
d. Archaea

Completion

Complete each statement on the line provided.

16. A book by ____________________ helped Darwin realize that the majority of a species' offspring die or fail to reproduce.

17. The number of times that an allele occurs in a gene pool, compared with the number of times other alleles for the same gene occur, is the ____________________ of the allele.

18. The most recent period in the geologic time scale is the ____________________ Period.

19. The endosymbiotic theory provides an explanation for the origin of ____________________.

20. A model known as a(an) ____________________ uses DNA comparisons to estimate the length of time that two species have been evolving independently.

Short Answer

In complete sentences, write the answers to the questions on the lines provided.

21. What were Lamarck's evolution hypotheses?

____

____

____

22. How can you tell when reproductive isolation has occurred?

____

____

____

Name______________________________ Class____________________ Date ______________

23. What distribution of phenotypes would be expected for a polygenic trait?

____

____

24. What does the fossil record show about evolution?

____

____

25. How do scientists know that American vultures are more closely related to storks than to African vultures?

____

____

Using Science Skills

Use the graph to answer the following questions on the lines provided.

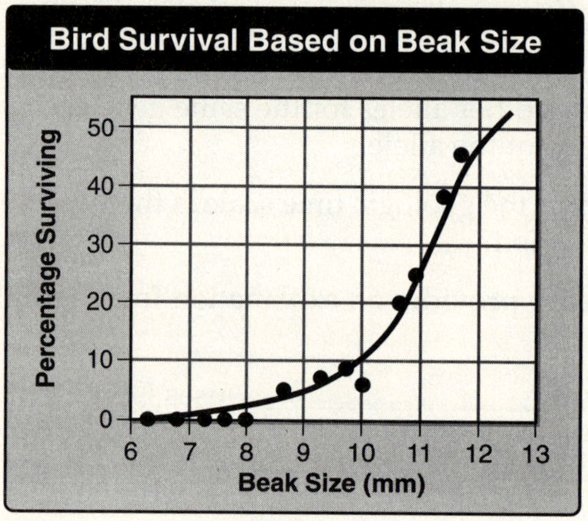

Figure 1

26. According to Figure 1, how are the two variables, bird survival and beak size, related?

____

____

27. Referring to Figure 1, if the distribution of beak size has a bell-shaped distribution, how would you expect the distribution to change through time?

____

____

28. In Figure 1, what type of natural selection appears to be occurring in the bird population?

____

____

29. According to Figure 1, below which beak size do no birds survive?

30. Referring to Figure 1, describe how the curve of survival and beak size would change if birds with smaller beaks had greater survival than birds with larger beaks.

Essay

Write the answer to each question in the space provided.

31. Contrast the three ways that natural selection can affect the distributions of phenotypes for a polygenic trait.

32. How did Hutton's and Lyell's ideas about geology influence Darwin?

Name______________________ Class ______________ Date ___________

33. Explain how most fossils form and why most organisms are not preserved as fossils.

34. How is the older six-kingdom system of classification related to the newer three-domain system of classification?

35. What different types of evidence for evolution did Darwin present in his book *On the Origin of Species*?

Multiple Choice

Write the letter that best answers the question or completes the statement on the line provided.

____ **1.** Preserved remains of ancient organisms are called
a. homologous structures.
b. fossils.
c. vestigial structures.
d. microspheres.

____ **2.** Humans select from among the naturally occurring genetic variations in a species in the process of
a. natural selection.
b. genetic equilibrium.
c. artificial selection.
d. survival of the fittest.

____ **3.** Any inherited characteristic that increases an organism's chance of survival is a(an)
a. adaptation.
b. homologous trait.
c. polygenic trait.
d. derived character.

____ **4.** Natural selection leads to an increase in a species'
a. range.
b. gene pool.
c. geographic isolation.
d. fitness.

____ **5.** Changes in the sequence of DNA are called
a. mutations.
b. gene shuffling.
c. allele frequencies.
d. genetic equilibrium.

____ **6.** Which type of selection on a polygenic trait increases variation in the trait?
a. directional selection
b. stabilizing selection
c. disruptive selection
d. polygenic selection

____ **7.** The founder effect is a type of
a. natural selection.
b. speciation.
c. isolating mechanism.
d. genetic drift.

____ **8.** A dating method that determines a fossil's age in years is
a. relative dating.
b. radioactive dating.
c. index fossil dating.
d. geological dating.

Name________________ Class____________ Date__________

____ 9. Miller and Urey's experiments produced
a. DNA and RNA.
b. proteins.
c. living cells.
d. amino acids.

____10. Which choice shows the correct order in which the evolution of traits occurred?
a. eukaryotic cells, sexual reproduction, multicellularity
b. sexual reproduction, eukaryotic cells, multicellularity
c. multicellularity, sexual reproduction, eukaryotic cells
d. eukaryotic cells, multicellularity, sexual reproduction

____11. Which geological era is called the Age of Mammals?
a. Cambrian
b. Mesozoic
c. Cenozoic
d. Tertiary

____12. Patterns of macroevolution include
a. extinction.
b. adaptive radiation.
c. convergent evolution.
d. all of the above

____13. The discipline in which scientists classify and name organisms is
a. cladistics.
b. taxonomy.
c. geology.
d. paleontology.

____14. An order is a category composed of similar
a. phyla.
b. genera.
c. families.
d. classes.

____15. All organisms that belong to the domain Archaea are
a. heterotrophs.
b. multicellular.
c. eukaryotes.
d. prokaryotes.

Name______________________________ Class__________________ Date ______________

Completion

Complete each statement on the line provided.

16. Members of a population share a common group of genes called a ____________________.

17. Structures with different mature forms that develop from the same embryonic tissues are called ____________________ structures.

18. A pattern of long, stable periods interrupted by brief periods of more rapid evolutionary change is called ____________________.

19. The term ____________________ refers to evolutionary relationships among organisms.

20. In binomial nomenclature, each ____________________ is assigned a two-part scientific name.

Short Answer

In complete sentences, write the answers to the questions on the lines provided.

21. What is the principle of descent with modification?

22. How is evolution defined in genetic terms?

23. What evolutionary event led to the accumulation of oxygen in Earth's atmosphere?

24. What is coevolution?

Name________________________ Class________________ Date ____________

25. How does a cladogram differ from a system of classification based on visible similarities?

___

___

___

Using Science Skills

Use the table to answer the following questions on the lines provided.

Effect of Color Mutations on Lizard Survival				
Color	Initial Population	Generation 10	Generation 20	Generation 30
Brown	80%	80%	70%	40%
Red	10%	0%	0%	0%
Black	10%	20%	30%	60%

Figure 1

26. Referring to Figure 1, describe how the lizard population changed through time.

___

___

27. Referring to Figure 1, explain how natural selection could cause the lizard population to change in this way.

___

___

28. Referring to Figure 1, assume that lizard color is a single-gene trait. Could the change in lizard color be an example of directional selection? Why or why not?

___

___

29. Referring to Figure 1, if a few brown lizards were separated from the rest of the lizards by a flood and formed an isolated population, the new population would have 100 percent brown lizards. What would you call this type of genetic change?

___

___

30. Assume a new mutation in generation 31 produced a green lizard that was more fit than either brown or black lizards. Using Figure 1, predict how the percentages of green, black, and brown lizards would change through time.

___

___

Name______________________________ Class __________________ Date ______________

Multiple Choice

Write the letter that best answers the question or completes the statement on the line provided.

____ 1. Prokaryotes are single-celled organisms that lack a
a. cell wall.
b. cell membrane.
c. definite shape.
d. nucleus.

____ 2. The structure labeled C in Figure 19-1 is
a. DNA.
b. an organelle.
c. the nucleus
d. a high-energy sugar.

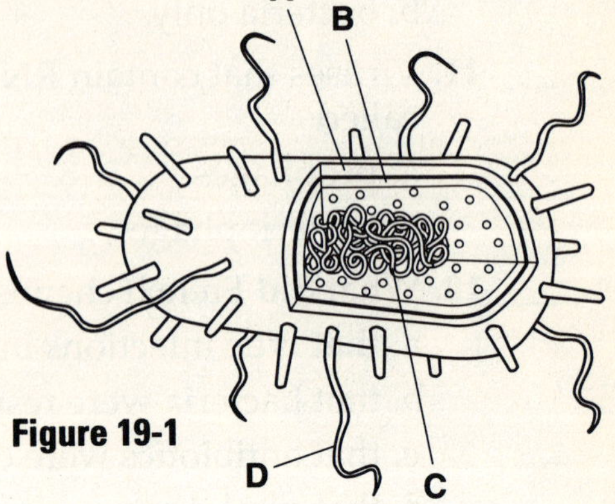

Figure 19-1

____ 3. Which of the following statements is NOT true about archaebacteria?
a. They are thought to be ancestors of eukaryotes.
b. Many live in harsh environments.
c. They lack peptidoglycan in their cell wall.
d. They make up the largest kingdom of prokaryotes.

____ 4. Spherical prokaryotes are called
a. spirilla.
b. flagella.
c. cocci.
d. bacilli.

____ 5. When treated with Gram stain, Gram-positive bacteria appear
a. violet.
b. pink.
c. yellow.
d. orange.

____ 6. Unlike photoautotrophs, chemoautotrophs obtain energy
a. directly from the sun.
b. directly from chemical reactions.
c. indirectly from carbon molecules.
d. indirectly from other organisms.

____ 7. Which of the following describes a role of bacteria in the environment?
a. carrying out photosynthesis
b. recycling nutrients
c. fixing nitrogen
d. all of the above

____ 8. Humans use bacteria in all of the following ways EXCEPT to
a. clean up small oil spills.
b. make butter and milk.
c. mine minerals from the ground.
d. synthesize drugs.

____ 9. What is the basic structure of a virus?
a. DNA or RNA surrounded by a protein coat
b. a capsid surrounded by a protein coat
c. a tail sheath surrounded by tail fibers
d. a tiny cell surrounded by a cell wall

____10. Bacteriophages infect
a. other viruses.
b. bacteria only.
c. any available host cell.
d. cells undergoing the lytic cycle.

____11. Viruses that contain RNA as their genetic information are called
a. prophages.
b. bacteriophages.
c. retroviruses.
d. capsids.

____12. What did French chemist Louis Pasteur help establish?
a. that lytic infections involved cell bursting
b. that bacteria were responsible for some diseases
c. that antibiotics were effective against bacteria
d. that viral diseases disrupt the body's equilibrium

____13. What can a vaccine do when it is injected into the body?
a. prompt the body to produce immunity to a disease
b. produce toxins that disrupt bacterial equilibrium
c. use bacterial cells for food
d. destroy new pathogens as they arise in the body

____14. Which of the following diseases is NOT caused by a virus?
a. tetanus
b. influenza
c. AIDS
d. chickenpox

____15. Which of the following is NOT true about prions?
a. They are particles composed only of protein.
b. They are single-stranded RNA molecules with no capsid.
c. They can cause disease in humans.
d. They contain no RNA or DNA.

Completion

Complete each statement on the line provided.

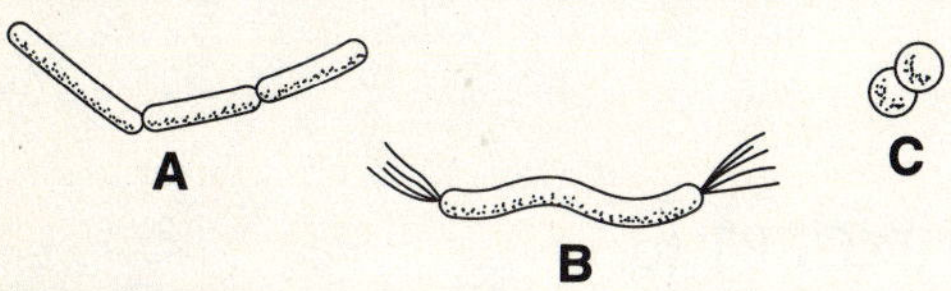

Figure 19-2

16. The organism labeled A in Figure 19-2 is an example of a(an) ____________________.

17. The knoblike root nodules of soybean plants are the sites of ____________________.

18. In a lysogenic infection, the viral DNA that is embedded in a host cell's DNA is called a(an) ___________________.

19. In Figure 19-3, the structure labeled D is a(an) ___________________.

20. The structure labeled B in Figure 19-3 is the ___________________.

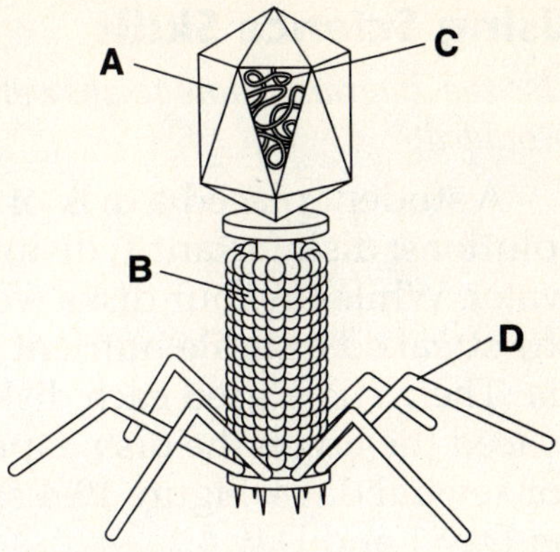

Figure 19-3

Short Answer

In complete sentences, write the answers to the questions on the lines provided.

21. List four factors that are used to identify prokaryotes.

22. Describe two roles that bacteria have in the environment.

23. What is a virus? Describe the basic structure of a virus.

24. Why are viruses not considered to be living things?

25. List three methods used to control bacterial growth.

Name________________________ Class________________ Date____________

Using Science Skills

Use the diagram below to answer the following questions on the lines provided.

A student placed a disk of filter paper in each of the following solutions: disinfectant 1, disinfectant 2, disinfectant 3, and distilled water. While the four disks were soaking in their respective solutions, she streaked a sterile nutrient agar dish with a culture of *E. coli* bacteria. Then, she placed each disk carefully onto the nutrient agar dish, placed the lid on the dish, taped it shut, and incubated the dish at 37°C for several days. Figure 19-4 shows how the nutrient agar dish looked on Day 1 and Day 4.

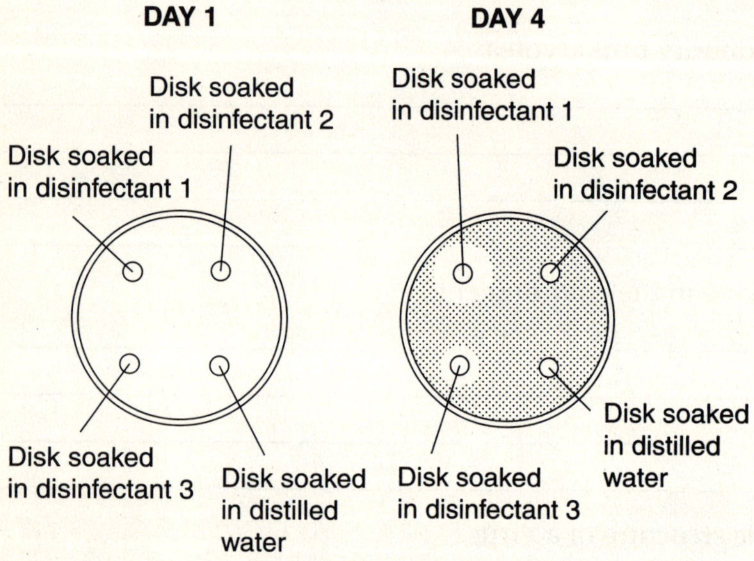

Figure 19-4

26. Designing Experiments Based on Figure 19-4, what is the student probably trying to test?

____

____

____

____

27. Controlling Variables What is the control in the experiment shown in Figure 19-4?

____

____

28. Drawing Conclusions How can the student measure the effectiveness of each disinfectant shown in Figure 19-4?

____

____

____

____

29. **Inferring** Look at the dishes in Figure 19-4. Which disinfectant was the most effective at controlling the growth of *E. coli?* How do you know?

____

____

____

30. **Drawing Conclusions** Look at the results of the experiment shown in Figure 19-4. Why do you think the different disinfectants are not equally effective against the *E. coli* bacteria?

____

____

____

Essay

Write the answer to each question in the space provided.

31. Explain how the two groups of prokaryotes differ.

32. What is the difference between binary fission and conjugation?

Name________________________________ Class____________________ Date ________________

33. Briefly describe two ways that bacteriophages infect cells.

34. Explain how bacteria cause disease. Give an example of each method.

35. Contrast a bacterial disease with a viral disease.

Name______________________________ Class__________________ Date ______________

Multiple Choice

Write the letter that best answers the question or completes the statement on the line provided.

____ **1.** The microorganisms called prokaryotes are
a. bacteria and viruses.
b. single-celled organisms that lack a nucleus.
c. heterotrophs that contain no DNA.
d. single-celled organisms that form protein clumps.

____ **2.** Which group of prokaryotes do scientists think may be ancestors of eukaryotes?
a. monera
b. eubacteria
c. archaebacteria
d. peptidoglycans

____ **3.** Which of the following are used to identify prokaryotes?
a. cell shape
b. the way prokaryotes move
c. the way prokaryotes obtain energy
d. all of the above

____ **4.** A method called Gram staining is used to tell
a. what shape a prokaryote has.
b. how a prokaryote obtains energy.
c. what kind of cell wall a prokaryote has.
d. whether a prokaryote has flagella.

____ **5.** Two groups of prokaryotes that make their own food from inorganic molecules are
a. heterotrophs and autotrophs.
b. obligate aerobes and obligate anaerobes.
c. photoautotrophs and chemoautotrophs.
d. chemoheterotrophs and photoheterotrophs.

____ **6.** Binary fission occurs when a bacterium
a. exchanges genetic information with another cell.
b. forms an endospore to survive harsh conditions.
c. forms a hollow bridge to another bacterial cell.
d. replicates its DNA and divides in half.

____ **7.** Bacteria that break down the nutrients in dead matter into simpler substances that are taken up by plant roots are called
a. endospores.
b. flagella.
c. photoautotrophs.
d. decomposers.

Name______________________________ Class____________________ Date ______________

____ 8. A typical virus is composed of a
a. cell wall made of peptidoglycan.
b. core of DNA or RNA surrounded by a protein coat.
c. bacteriophage inside a lytic infection.
d. core of protein surrounded by a coat of DNA.

____ 9. What are viruses that infect bacteria called?
a. archaebacteria c. nitrogen fixers
b. facultative anaerobes d. bacteriophages

____10. Viruses that contain RNA as their genetic information are
a. bacteriophages. c. prophages.
b. retroviruses. d. endospores.

____11. Antibiotics are compounds that can
a. disrupt a virus's normal equilibrium.
b. prompt the body to produce immunity.
c. block the growth and reproduction of bacteria.
d. control a viral infection.

____12. Which of the following is a disease caused by a virus?
a. AIDS c. tooth decay
b. strep throat d. tetanus

____13. If you wash your hands properly, ordinary soaps do a good job of removing
a. bacteria. c. viroids.
b. viruses. d. prions.

____14. Viruses have more difficulty entering plant cells than animal cells because
a. plant cells have tough cell walls.
b. nitrogen fixation harms plant cells.
c. animal cells have no cell membranes.
d. viroids only infect animals.

____15. A prion is a viruslike particle that contains only
a. DNA. c. RNA.
b. endospores. d. protein.

Completion

Complete each statement on the line provided.

16. During ____________________, a hollow bridge forms between two bacterial cells, and genes move from one cell to another.

17. The process of converting nitrogen into a form plants can use is called ____________________.

18. A virus's protein coat is called a(an) ____________________.

Name______________________________ Class____________________ Date ______________

19. A(an) ____________________ is a preparation of weakened or killed pathogens.

20. Single-stranded RNA molecules that have no surrounding capsids and infect plants are called ____________________ .

Short Answer

In complete sentences, write the answers to the questions on the lines provided.

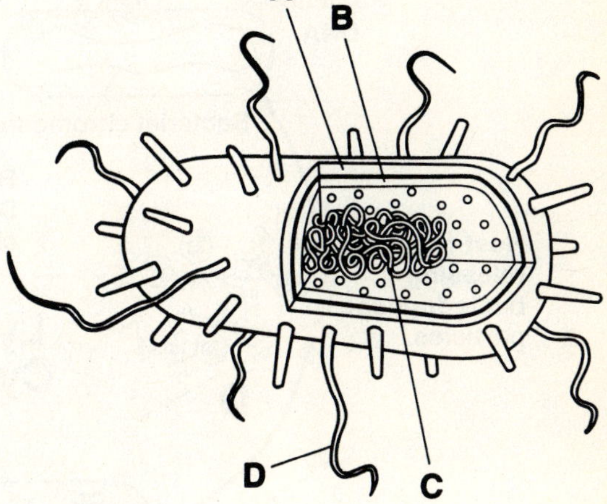

Figure 19-1

21. Identify structures A through D in Figure 19-1.

____

22. Name the two kingdoms of prokaryotes. List one way that these two groups differ from each other.

____

23. What is a pathogen?

____

24. What are two general ways that bacteria produce disease?

____

25. What are three methods used to control bacterial growth?

____

Name______________________________ Class____________________ Date ______________

Using Science Skills

Use the diagram below to answer the following questions on the lines provided.

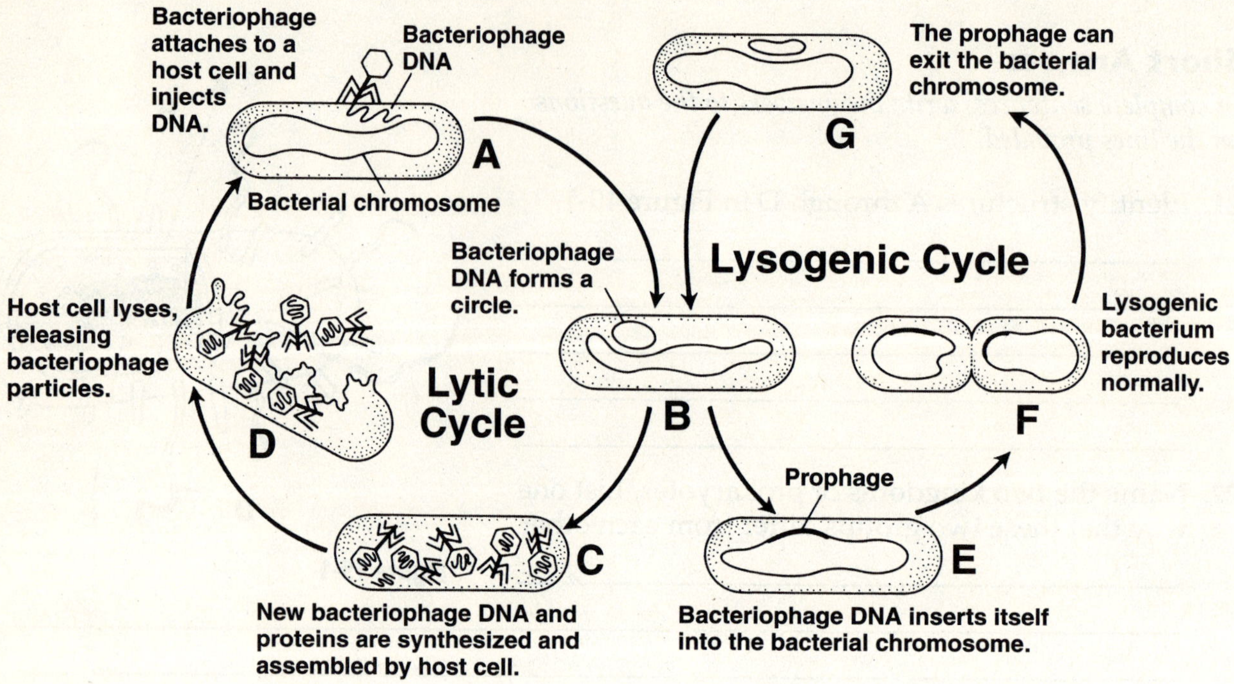

Figure 19-2

26. **Interpreting Graphics** What does Figure 19-2 represent?

____

____

27. **Comparing and Contrasting** Look at both cycles shown in Figure 19-2. During which cycle is the host cell destroyed?

____

____

28. **Interpreting Graphics** Each stage of the cycles shown in Figure 19-2 is labeled with a letter. Which letter indicates the stage at which the bacteriophage's DNA becomes a part of the host cell's DNA?

____

29. **Interpreting Graphics** Which letter in Figure 19-2 indicates the stage at which a host cell begins producing new bacteriophages?

____

30. **Interpreting Graphics** Which letter in Figure 19-2 indicates the stage at which a bacteriophage injects its DNA into a host cell?

____

Name________________________ Class________________ Date____________

Multiple Choice

Write the letter that best answers the question or completes the statement on the line provided.

____ **1.** Multicellular protists are grouped with unicellular protists because multicellular protists

a. do not resemble plants.

b. do not resemble animals.

c. do not resemble fungi.

d. are so similar to unicellular protists.

____ **2.** According to biologist Lynn Margulis, eukaryotic cells may have evolved from

a. a symbiosis of several cells.

b. mitochondria that grew very large.

c. chloroplasts that grew very large.

d. plants, animals, and fungi.

____ **3.** Many zooflagellates live in lakes and streams, where they

a. penetrate and live within cells of a host.

b. absorb food through their cell membranes.

c. feed by sweeping food particles into their gullet.

d. feed by surrounding their meal and taking it inside themselves.

____ **4.** Which of the following diseases does the animallike protist known as *Entamoeba* cause?

a. malaria

b. African sleeping sickness

c. amebic dysentery

d. potato blight

____ **5.** As sunlight passes through sea water, the sea water

a. absorbs large amounts of red and violet wavelengths.

b. reflects large amounts of red and violet wavelengths.

c. absorbs large amounts of blue wavelengths.

d. none of the above

____ **6.** What effect did the evolution of different forms of chlorophyll have in algae?

a. a decrease in the range of depths at which algae can live

b. an increase in the range of depths at which algae can live

c. no effect on the range of depths at which algae can live

d. a reduction in the photosynthetic capacity of algae

____ 7. Which of the statements is true about dinoflagellates?
a. They contain bright yellow pigments.
b. They can be both photosynthetic and heterotrophic.
c. Many species are luminescent.
d. They possess pillbox-shaped cell walls of silica.

____ 8. The population of small, photosynthetic organisms found near the surface of the ocean are called
a. chrysophytes. c. phytoplankton.
b. pyrophytes. d. diatoms.

____ 9. What characteristic of green plants is shared by green algae?
a. cell wall composition
b. photosynthetic pigments
c. multicellularity
d. all of the above

____10. Red algae lack flagella and
a. nuclei. c. accessory pigments.
b. centrioles. d. chlorophyll.

____11. In the life cycle of the green alga *Ulva*, the phase that produces male and female gametes is known as a
a. sporophyte.
b. gametophyte.
c. spore.
d. zoospore.

____12. Some products derived from algae include
a. drugs for stomach ulcers and high blood pressure.
b. thickeners for food.
c. chemicals in plastics, waxes, paints, and lubricants.
d. all of the above

____13. Funguslike protists get nutrients by
a. photosynthesis.
b. living as an animal parasite.
c. absorbing them from dead or decaying matter.
d. none of the above

____14. In oomycetes, sexual reproduction takes place in the
a. migrating colony. c. antheridium and oogonium.
b. sporangium. d. zoosporangium.

____15. The work of funguslike protists and other decomposers is important in improving the quality of
a. salt water. c. topsoil.
b. fresh water. d. potato crops.

Name________________________ Class________________ Date ____________

Completion

Complete each statement on the line provided.

16. Animallike protists that use structures called ________________ for movement and for feeding are members of the phylum Sarcodina.

17. Many members of the phylum Pyrrophyta, which are also referred to as ________________, are luminescent.

18. Red algae contain reddish accessory pigments as well as ________________, one of the three types of chlorophyll.

19. Because algae carry out ________________, they produce much of Earth's atmospheric oxygen.

20. The funguslike protists that cause several serious plant diseases, including mildews and blights of grapes and tomatoes, are the ________________.

Short Answer

In complete sentences, write the answers to the questions on the lines provided.

21. How do sporozoans differ from other animallike protists in terms of movement?

22. What is the function of chlorophyll and accessory pigments in algae?

23. What are "red tides," and why are they dangerous?

24. Compare the structure and function of the diploid sporophyte to the haploid gametophyte in the multicellular alga *Ulva*.

25. What is the primary difference between cellular slime molds and acellular slime molds?

Name______________________ Class__________________ Date ______________

Using Science Skills

Use the diagram below to answer the following questions on the lines provided.

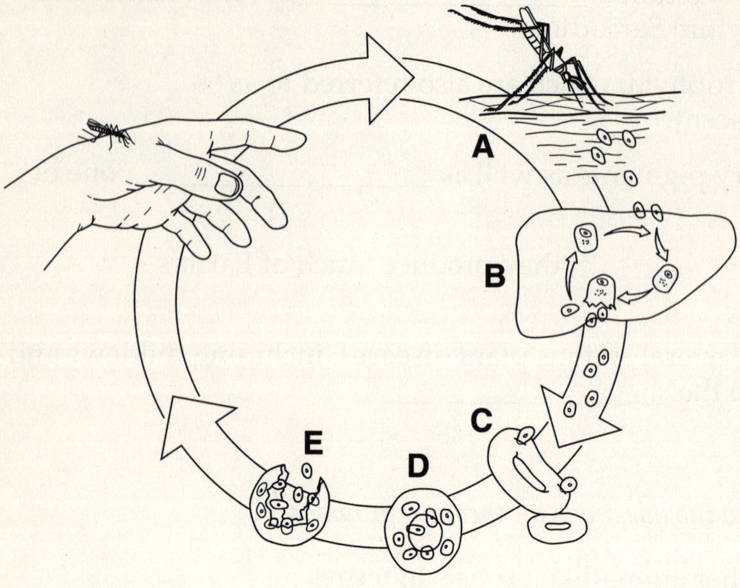

Figure 20-1

26. Interpreting Graphics What is the serious human disease whose pathway of infection is illustrated in Figure 20-1, and what organism causes the disease?

____

27. Inferring Considering the nature of the animal that carries the protist shown in Figure 20-1, in what kinds of environments would this disease be most common?

____

____

____

____

28. Interpreting Graphics According to Figure 20-1, what has occurred up to and including Step B?

____

____

____

____

29. Interpreting Graphics According to Figure 20-1, what is occurring in Steps C, D, and E?

____

____

____

Name________________________ Class__________________ Date ____________

30. **Predicting** Infection by the protist shown in Figure 20-1 produces periods of acute illness followed by periods of reduced symptoms. These periods follow each other in repeating cycles. Predict which part of the cycle of infection results in an abrupt change throughout the body, thus corresponding to the periods of acute illness and an increase in symptoms.

____

____

____

____

Essay

Write the answer to each question in the space provided.

31. Identify the cause and symptoms of African sleeping sickness, and explain how it is spread.

32. Describe the appearance of a typical euglena, and explain how it moves through water.

33. Compare the structure of the colonial alga, *Volvox*, with that of *Ulva*, a true multicellular alga.

34. Why is it difficult to classify cellular slime molds as unicellular or multicellular?

35. What might be the end result in terrestrial and aquatic ecosystems if all decomposers—including slime molds and water molds—no longer existed?

Name________________________________ Class____________________ Date ________________

Multiple Choice

Write the letter that best answers the question or completes the statement on the line provided.

____ 1. A protist is any organism that is not a plant, an animal, a fungus, or a(an)

a. eukaryote. c. eubacterium.
b. prokaryote. d. archaebacterium.

____ 2. In an amoeba, a small cavity within the cytoplasm that stores food is called a

a. gullet. c. food vacuole.
b. pseudopod. d. contractile vacuole.

____ 3. To which phylum do amoebas, foraminifers, and heliozoans belong?

a. sarcodines c. sporozoans
b. ciliates d. pseudopods

____ 4. The zooflagellate *Trypanosoma* causes the disease known as

a. malaria.
b. African sleeping sickness.
c. amebic dysentery.
d. algal bloom.

____ 5. Which substances allow algae to harvest and use the energy from sunlight?

a. cilium and fucoxanthin
b. chlorophyll and accessory pigments
c. phycobilin and flagellum
d. oogonium and antheridium

____ 6. Euglenas have an intricate, folded cell membrane called a(an)

a. cell wall.
b. trichocyst.
c. eyespot.
d. pellicle.

____ 7. An algal bloom is

a. the clouding of water by sewage.
b. an enormous mass of algae.
c. a symbiotic relationship between algae and coral.
d. none of the above

____ 8. An example of a multicellular green alga is

a. *Ulva.* c. *Volvox.*
b. *Chlamydomonas.* d. *Fucus.*

Name______________________ Class________________ Date ____________

____ 9. The switching back and forth between a diploid and haploid stage in a life cycle is called
   a. alternation of generations.
   b. fusion of opposite mating types.
   c. sexual reproduction.
   d. asexual reproduction.

____10. The green alga *Chlamydomonas* reproduces asexually by producing
   a. gametophytes.
   b. sporophytes.
   c. zygotes.
   d. zoospores.

____11. What very large type of algae lives off the coasts of North America?
   a. rockweed
   b. sea lettuce
   c. giant kelp
   d. Irish moss

____12. Which of the following groups includes only funguslike protists?
   a. cellular slime molds, brown algae, water molds
   b. cellular slime molds, acellular slime molds, water molds
   c. cellular slime molds, acellular slime molds, animallike protists
   d. cellular slime molds, fungi, water molds

____13. The funguslike protists that thrive on dead or decaying organic matter in water are
   a. water molds.
   b. acellular slime molds.
   c. cellular slime molds.
   d. true fungi.

____14. When the amoebalike cells of acellular slime molds fuse, they form structures with many nuclei called
   a. plasmodia.
   b. zoosporangia.
   c. spores.
   d. filaments.

____15. Organisms that break down organic material include the
   a. funguslike protists.
   b. animallike protists.
   c. plantlike protists.
   d. none of the above

Completion

Complete each statement on the line provided.

16. Eukaryotes that are not members of the kingdoms Plantae, Animalia, or Fungi are ________________.

17. The sporozoan *Plasmodium,* carried from host to host by mosquitoes, causes ________________, a very serious infectious disease.

18. In the human body, *Plasmodium* first infects liver cells and then ________________ cells, causing them to burst.

19. Chlorophyll and accessory pigments allow algae to harvest and use the energy of ____________________ .

20. Funguslike protists are ____________________ that absorb nutrients from dead or decaying organic matter.

Short Answer

In complete sentences, write the answers to the questions on the lines provided.

21. Explain the importance of the animallike protist *Trichonympha* to the termites in which it grows.

22. What is phytoplankton?

23. What are two human uses of algae?

24. What are slime molds?

25. What caused the destruction of much of the potato crops of 1845 and 1846 in Ireland?

Name________________________ Class________________ Date____________

Using Science Skills

Use the diagram below to answer the following questions on the lines provided.

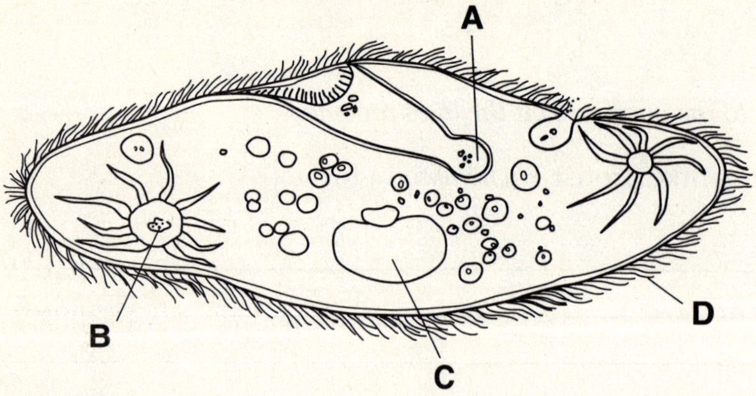

Figure 20-1

26. Interpreting Graphics Figure 20-1 shows a diagram of an organism. Is it a single-celled or a multicellular organism?

____

____

27. Inferring Structure A opens to the outside of the organism shown in Figure 20-1. How is the structure of A related to its function in the organism?

____

____

____

28. Predicting If the structure labeled B in Figure 20-1 was to malfunction, what effect would this likely have on the organism?

____

____

29. Inferring How does the structure labeled C in Figure 20-1 help the organism in its day-to-day existence?

____

____

____

____

30. Predicting If the structure labeled D in Figure 20-1 were to malfunction, what effect would this likely have on the organism?

____

____

Name_______________________________ Class___________________ Date ______________

Multiple Choice

Write the letter that best answers the question or completes the statement on the line provided.

____ **1.** Fungi do NOT
- a. carry out photosynthesis.
- b. grow on their food source.
- c. digest food outside their bodies.
- d. absorb food through their cell walls.

____ **2.** The tangled mass that makes up the body of a fungus is the
- a. hypha.
- b. rhizoid.
- c. mycelium.
- d. stolon.

____ **3.** Sporangia are found at the tops of specialized hyphae called
- a. sporangiophores.
- b. mycelia.
- c. gametangia.
- d. stolons.

____ **4.** *Rhizopus* reproduces
- a. by sexual reproduction only.
- b. by asexual reproduction only.
- c. by both asexual and sexual reproduction.
- d. all of the above

____ **5.** In the life cycle of molds, when hyphae of different mating types meet, each hypha forms a
- a. sporangium.
- b. zoospore.
- c. zygospore.
- d. gametangium.

____ **6.** The dry, powdered yeast used to bake bread actually contains
- a. zygospores.
- b. ascospores.
- c. conidia.
- d. sporangia.

____ **7.** In mushrooms, the basidia are found on the
- a. base.
- b. stalk.
- c. cap.
- d. root.

____ **8.** Over time, nutrients at the center of a large mycelium become depleted, causing new mushrooms to sprout only
- a. in a cluster at the center.
- b. in a ring at the outer edges.
- c. when the nutrients are replaced.
- d. after budding takes place.

____ **9.** *Penicillium* reproduces asexually by means of conidia, similar to reproduction in a(an)
- a. ascomycete.
- b. zygomycete.
- c. basidiomycete.
- d. lichen.

Name________________ Class____________ Date __________

____**10.** Fungi that absorb food from decaying organic matter are
- a. parasites.
- b. saprobes.
- c. mutualists.
- d. autotrophs.

____**11.** Fungi feed on
- a. only living organisms.
- b. only dead organisms.
- c. both living and dead organisms.
- d. only other fungi.

____**12.** Crop damage by fungal diseases is
- a. greatest in tropical areas.
- b. least in tropical areas.
- c. greatest in temperate areas.
- d. not affected by climate.

____**13.** The fungus that forms a mycelium within the outer layers of human skin causes
- a. ringworm and athlete's foot.
- b. thrush and reproductive-tract infections.
- c. reproductive-tract infections and ringworm.
- d. athlete's foot and thrush.

____**14.** Which statement about lichens is correct?
- a. They are not tolerant of harsh conditions.
- b. They cannot make their own food.
- c. They grow only in soil.
- d. They are composed of an alga or a cyanobacterium and a fungus living together.

____**15.** The association of plants and fungi in mycorrhizae illustrates a type of relationship called
- a. parasitism.
- b. mutualism.
- c. competition.
- d. parallelism.

Completion

Complete each statement on the line provided.

16. The cell walls of fungi are composed of the carbohydrate ____________.

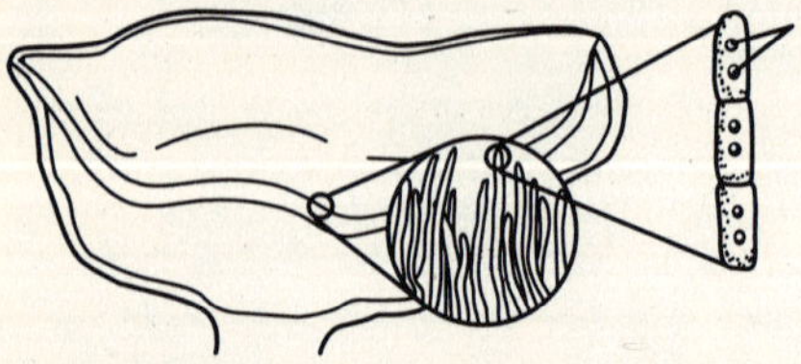

Figure 21-1

17. In the ____________ stage of the ascomycete life cycle, shown in Figure 21-1, cells contain two nuclei.

Name______________________________ Class__________________ Date ______________

18. The basidiospores of __________________ fungi are scattered by the wind.

19. Fungi break down complex organic matter by releasing digestive __________________ .

20. A serious fungal disease of __________________ needs two different plants to complete its life cycle.

Short Answer

In complete sentences, write the answers to the questions on the lines provided.

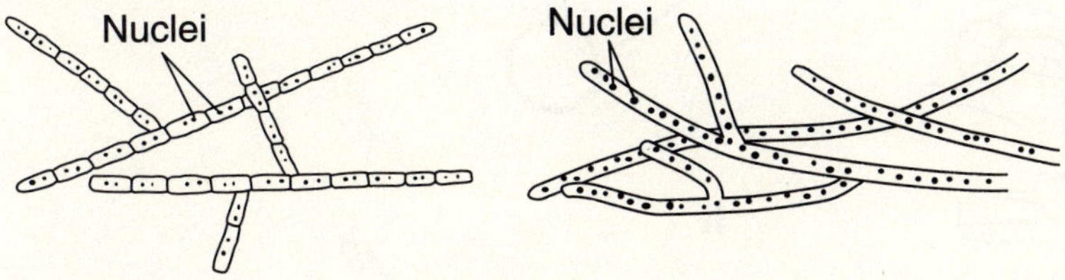

Figure 21-2

21. How do the two types of hyphae shown in Figure 21-2 differ?

____

____

____

22. What three functions do rhizoids serve in bread mold?

____

____

____

23. Suppose you like mushrooms and saw some growing in a yard. Would you pick and eat them? Why or why not?

____

____

____

24. What is the most important role of fungi in natural ecosystems, and why is this role important?

____

____

____

25. How do lichens encourage soil formation on barren rock?

____

____

____

Name________________________ Class__________________ Date ______________

Using Science Skills

Use the diagram below to answer the following questions on the lines provided.

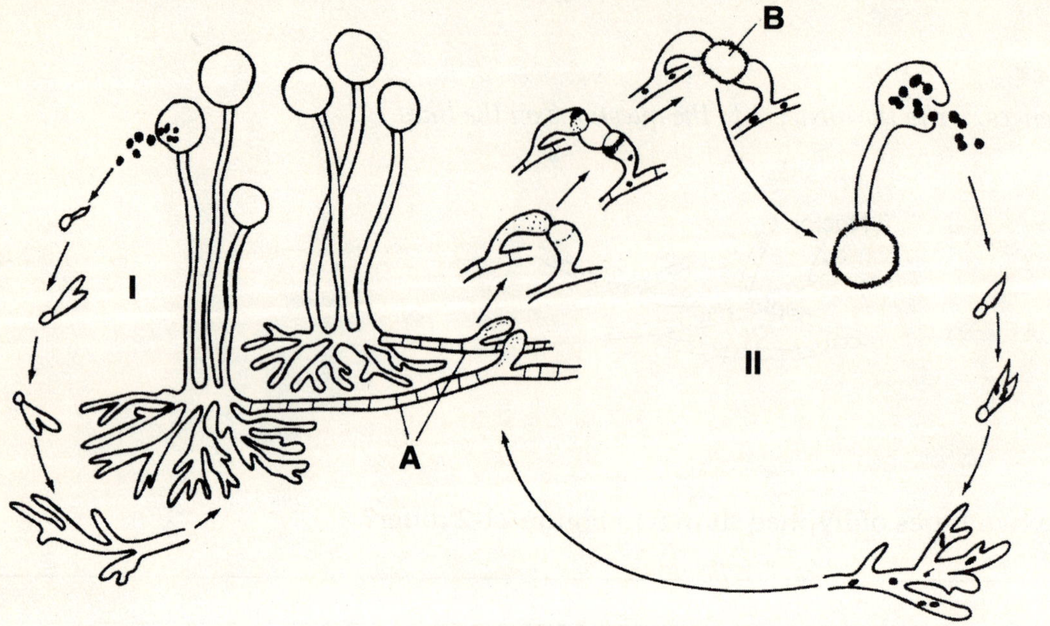

Figure 21-3

26. Inferring What mating type or types are labeled A in Figure 21-3?

____

____

27. Predicting Predict what would happen to the structure labeled B in Figure 21-3 if conditions were not favorable for reproduction.

____

____

____

28. Classifying To what phylum does the organism shown in Figure 21-3 belong?

____

29. Interpreting Graphics Which section of Figure 21-3, I or II, shows sexual reproduction?

____

30. Applying Concepts Identify the structure labeled B in Figure 21-3 and explain how it is analogous to a fertilized animal egg.

____

____

____

Name______________________ Class______________ Date ___________

Essay

Write the answer to each question in the space provided.

31. Does the haploid or diploid condition occur most in the life cycle of fungi? Explain your answer.

32. Name the four phyla of fungi. What characteristic determines the phylum into which a fungus is placed?

33. Describe the fruiting body of a mushroom. What allows a mushroom to develop so quickly that it can spring up overnight?

34. How are fungi both helpful and harmful to people?

35. Why is it not beneficial to kill all bacteria on and in the body, and what is a common result of killing naturally occurring bacteria?

Name________________________________ Class___________________ Date ______________

Multiple Choice

Write the letter that best answers the question or completes the statement on the line provided.

____ **1.** All fungi
a. make their food. c. produce mushrooms.
b. absorb their food. d. have chlorophyll.

____ **2.** A mushroom is a fungal
a. fruiting body. c. mycorrhiza.
b. lichen. d. yeast.

____ **3.** Most fungi reproduce
a. asexually only. c. both sexually and asexually.
b. sexually only. d. by budding.

____ **4.** Stinkhorns, which mimic the odor of rotting meat, have spores that are dispersed by
a. wind. c. snow.
b. birds. d. flies.

____ **5.** Dark fuzz that grows on bread is an example of
a. a toadstool. c. yeast.
b. a spore. d. mold.

____ **6.** Bread rises because fermentation by yeast produces
a. spores. c. water.
b. rhizoids. d. carbon dioxide.

____ **7.** Mushrooms are classified as
a. common molds. c. club fungi.
b. sac fungi. d. imperfect fungi.

____ **8.** Which structure is NOT found in a mushroom?
a. ascus c. cap
b. gills d. stalk

____ **9.** Which statement about *Penicillium* is correct?
a. It produces mushrooms.
b. It causes bread to rise.
c. It is the source of an antibiotic.
d. It causes athlete's foot.

____ **10.** An important role of fungi in an ecosystem is
a. photosynthesis.
b. breaking down dead organisms.
c. making alcohol.
d. killing bacteria.

Name________________________________ Class____________________ Date ______________

____**11.** Fungi might have been important to the evolution of

a. worms. c. bacteria.

b. grasshoppers. d. plants.

____**12.** The human disease ringworm is caused by

a. worms. c. a fungus.

b. bacteria. d. yeasts.

____**13.** The growth of yeasts in moist regions of the body is kept in check by competition from

a. antibiotics. c. rusts.

b. bacteria. d. mildews.

____**14.** The normal balance between bacteria and yeasts in the body can be upset by

a. eating yeast-leavened bread.

b. eating edible mushrooms.

c. using antibiotics.

d. being exposed to mushroom spores.

____**15.** Which of the following is NOT a single organism?

a. rust c. yeast

b. smut d. lichen

Completion

Complete each statement on the line provided.

16. ____________________ reproduction takes place when cells or hyphae break off a fungus and begin to grow.

17. Spore-bearing structures called basidia are found in the ____________________ of mushrooms.

18. Imperfect fungi do not have a(an) ____________________ phase in their life cycles.

19. The symbiotic associations of plant roots and fungi are called ____________________.

Name______________________ Class______________ Date ____________

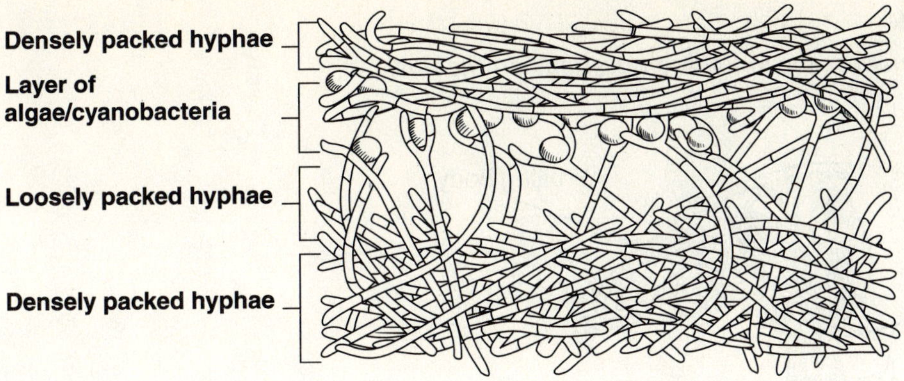

Figure 21-1

20. The cyanobacterium or algal layer of the ______________ shown in Figure 21-1 performs photosynthesis.

Short Answer

In complete sentences, write the answers to the questions on the lines provided.

21. What are the individual threads of a fungus called, and what structure do they form when they appear as a tangled mass?

____

____

22. In the ascomycete fruiting body shown in Figure 21-2, how many nuclei are found in each cell of the hyphae, and what separates each cell of the hyphae?

____

____

____

____

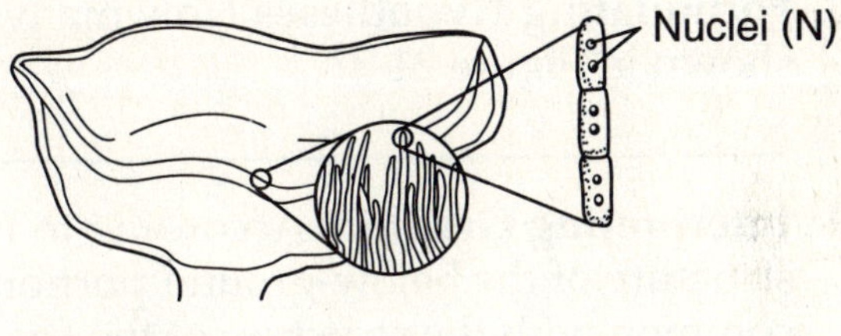

Figure 21-2

23. Name two commercial uses of yeast.

____

____

24. Why are basidiomycetes called club fungi?

____

____

25. What is an example of a competition between humans and fungi for nutrients?

____

____

Name________________________ Class__________________ Date ______________

Using Science Skills

Use the diagram below to answer the following questions on the lines provided.

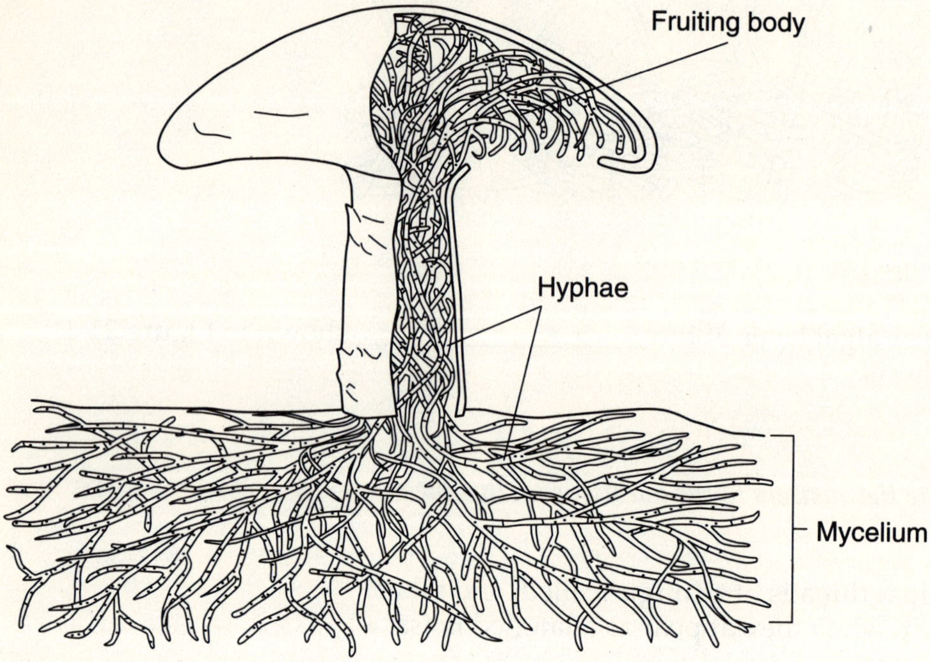

Figure 21-3

26. **Formulating Hypotheses** How many fungal fruiting bodies are shown in Figure 21-3?

____

27. **Interpreting Graphics** According to Figure 21-3, what does the structure of the below-ground portion of the mycelium have in common with the structure of the fruiting body?

____

28. **Predicting** Mushrooms sometimes grow from the trunks of trees. Which structure shown in Figure 21-3 would likely be embedded in the tree's bark?

____

____

29. **Inferring** Look at the mycelium shown in Figure 21-3. Would it be possible to easily detect the presence of a mycelium underground if no fruiting bodies were present? Explain your answer.

____

30. **Interpreting Graphics** What kind of a structure is the fruiting body shown in Figure 21-3?

____

____

Name____________________ Class________________ Date ____________

Multiple Choice

Write the letter that best answers the question or completes the statement on the line provided.

____ 1. Which of the following is NOT a characteristic used to identify prokaryotes?

a. type of cell wall c. shape
b. habitat d. movement

____ 2. Which of the following describes binary fission in prokaryotes?

a. Spores form and then germinate when conditions are good.
b. Genetic information is exchanged through a hollow bridge.
c. Gametes from opposite mating types fuse.
d. A cell replicates its DNA and divides in half.

____ 3. A typical virus is composed of a

a. prophage embedded within RNA or DNA.
b. core of RNA or DNA surrounded by a protein coat.
c. viroid surrounded by a double helix of DNA.
d. clump of peptidoglycan within a coat of chitin.

____ 4. The process in which the host cell makes copies of a virus indefinitely is called a(an)

a. lysogenic infection.
b. algal bloom.
c. lytic infection.
d. alternation of generations.

____ 5. Which of the following is a disease caused by a bacterium?

a. tuberculosis c. hepatitis C
b. common cold d. warts

____ 6. Which of the following is NOT true about protists?

a. They are eukaryotes.
b. They are not plants, animals, or fungi.
c. They lack a nucleus in their cells.
d. Most are unicellular.

____ 7. Accessory pigments allow many algae to absorb

a. nutrients from decaying organic matter.
b. chlorophyll *a* during photosynthesis.
c. light at different wavelengths other than chlorophyll.
d. delicate cell walls rich in silicon.

____ 8. Euglenophytes are like zooflagellates in that members of both groups
a. are green and contain chloroplasts.
b. move and feed using cilia.
c. are autotrophs that carry out photosynthesis.
d. swim using flagella.

____ 9. The largest known alga is giant kelp, which is an example of
a. red algae.
b. green algae.
c. brown algae.
d. yellow algae.

____10. The plasmodium of an acellular slime mold is a
a. single-celled phase of the organism's life cycle.
b. single structure with many nuclei.
c. colony of filaments in which cells are stacked.
d. thick mass of hyphae tangled together.

____11. All fungi are eukaryotic
a. autotrophs.
b. chemoautotrophs.
c. parasites.
d. heterotrophs.

____12. Molds seem to spring up in many different locations because
a. most fungi reproduce sexually.
b. fungal spores are found in almost every environment.
c. fungi lack membrane-bound organelles.
d. most molds are obligate anaerobes.

____13. The process of asexual reproduction in yeasts is called
a. budding.
b. basidia.
c. conjugation.
d. alternation of generations.

____14. Organisms that obtain food from decaying organic matter are
a. autotrophs. c. parasites.
b. sporangia. d. saprobes.

____15. The green alga or cyanobacterium in a lichen provides the fungus with
a. water and minerals collected from the environment.
b. protection from water molds and other pathogens.
c. a source of energy through photosynthesis.
d. shade and protection from intense sunlight.

Name______________________________ Class __________________ Date ______________

Completion

Complete each statement on the line provided.

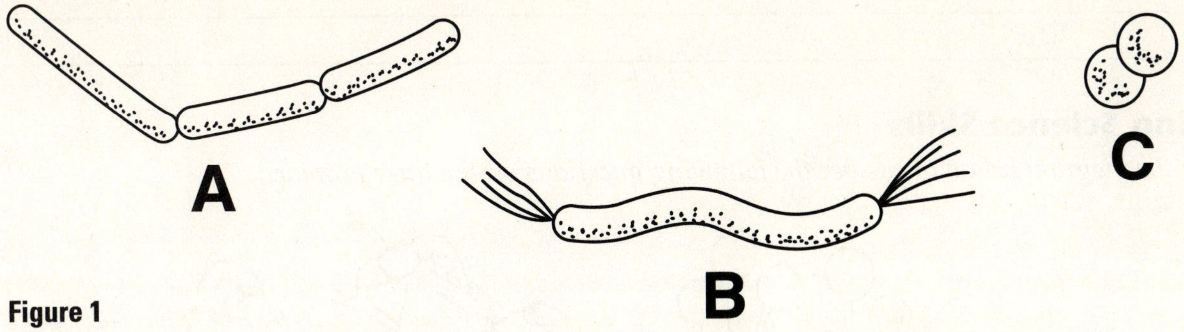

Figure 1

16. The bacteria labeled C in Figure 1 are examples of ____________________.

17. Viruses that contain RNA as their genetic information are called ____________________.

18. A(an) ____________________ is a preparation of weakened or killed pathogens used to prompt the body to produce immunity to a disease.

19. The population of small, photosynthetic organisms found near the surface of oceans is called ____________________.

20. *Penicillium notatum,* which has never been observed to reproduce sexually, is classified as a(an) ____________________ fungus.

Short Answer

In complete sentences, write the answers to the questions on the lines provided.

21. In what ways are archaebacteria different from eubacteria?

____

____

22. Why can't viral diseases be treated with antibiotics?

____

____

____

23. What characteristics do plants and green algae share?

____

____

____

24. What is alternation of generations?

____

____

____

Name________________ Class________________ Date________________

25. How do fungi help maintain equilibrium in ecosystems?

____

____

____

Using Science Skills

Use the diagram below to answer the following questions on the lines provided.

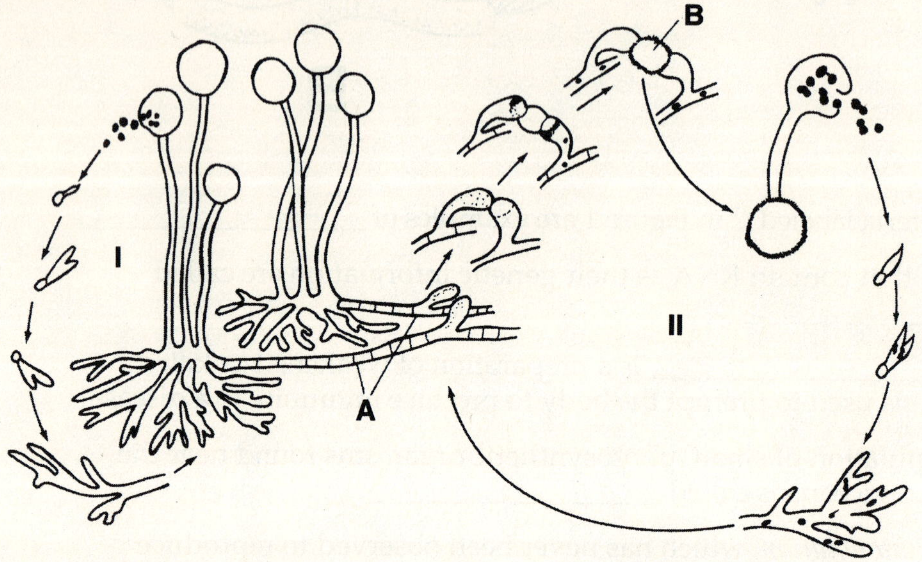

Figure 2

26. **Classifying** Identify the organism whose life cycle is illustrated in Figure 2. Name the kingdom, phylum, and specific organism.

____

____

27. **Interpreting Graphics** Explain in general terms what Figure 2 depicts, including labels for section I and section II.

____

____

28. **Interpreting Graphics** What are the two structures labeled A in Figure 2, and how are they involved in reproduction?

____

____

29. **Interpreting Graphics** What is the structure labeled B in Figure 2? Explain how that structure became either haploid or diploid.

____

____

30. **Applying Concepts** What follows the formation of structure B in the life cycle of this organism, and what is that step's significance for the organism's survival?

____

____

Name________________________________ Class__________________ Date ______________

Essay

Write the answer to each question in the space provided.

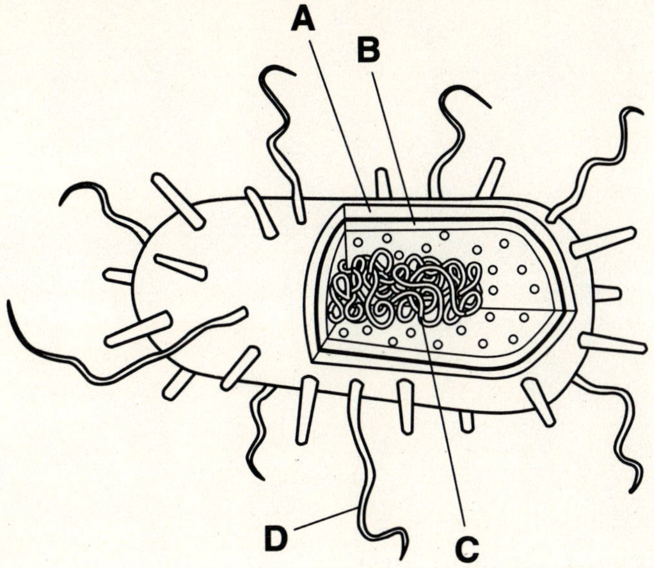

Figure 3

31. Describe the structure of a typical eubacterium. Use Figure 3 for reference. Explain the function of each structure identified.

32. Describe what occurs in a lytic infection.

Name________________________________ Class____________________ Date ______________

33. Identify the cause and symptoms of malaria, and explain how it is spread.

34. Describe the structure of a typical multicellular fungus.

35. What are mycorrhizae, and why are they essential for the growth of many plants?

Name______________________________ Class__________________ Date ______________

Multiple Choice

Write the letter that best answers the question or completes the statement on the line provided.

____ **1.** What are the two kingdoms of prokaryotes?
a. monera and bacteria
b. eubacteria and archaebacteria
c. archaebacteria and chemoheterotrophs
d. eubacteria and monera

____ **2.** In the process of binary fission, a bacterium
a. forms an endospore and waits for favorable conditions.
b. replicates its DNA and divides in half.
c. forms a hollow bridge to another bacterium.
d. embeds itself into a host's DNA.

____ **3.** A virus's protein coat is called its
a. prophage.
b. rhizoid.
c. holdfast.
d. capsid.

____ **4.** What is a pathogen?
a. a virus that infects bacteria
b. a bacterium undergoing conjugation
c. a disease-causing agent
d. a plant infected with a virus

____ **5.** Which of the following is a disease caused by a bacterium?
a. AIDS
b. measles
c. strep throat
d. common cold

____ **6.** Eukaryotes that are not plants, animals, or fungi are classified as
a. protists.
b. heterotrophs.
c. prokaryotes.
d. saprobes.

____ **7.** The sporozoan *Plasmodium* causes the disease called
a. tuberculosis.
b. athlete's foot.
c. African sleeping sickness.
d. malaria.

____ 8. What allows algae to harvest and use the energy from sunlight?
   a. chlorophyll and accessory pigments
   b. cell walls rich in silicon
   c. DNA surrounded by a protein coat
   d. alternation of generations

____ 9. A haploid reproductive cell is called a(an)
   a. gamete.
   b. trichocyst.
   c. spore.
   d. hypha.

____10. What is an algal bloom?
   a. a plasmodium of an acellular slime mold
   b. an enormous mass of algae
   c. a clump of accessory pigments
   d. a loss of phytoplankton

____11. Eukaryotic heterotrophs that have cell walls are known as
   a. methanogens.
   b. water molds.
   c. fungi.
   d. prions.

____12. What are rhizoids?
   a. the accessory pigments found in red algae
   b. the rootlike hyphae of bread molds that penetrate the bread's surface
   c. hairlike projections similar to flagella
   d. bacteria that convert nitrogen gas into a form plants can use

____13. The unicellular fungi used for baking and brewing are the
   a. imperfect fungi.
   b. water molds.
   c. chrysophytes.
   d. common yeasts.

____14. The mushroom seen above ground is the organism's
   a. mycelium.
   b. fruiting body.
   c. plasmodium.
   d. sporophyte.

____15. Mycorrhizae are mutualistic relationships between
   a. bacteria and viroids.
   b. fungi and algae.
   c. plant roots and fungi.
   d. eukaryotes and eubacteria.

Name________________________ Class________________ Date ____________

Completion

Complete each statement on the line provided.

16. Scientists think that ________________ may be ancestors of eukaryotes.

17. Compounds that block the growth and reproduction of bacteria are called ________________.

18. Sarcodines move via temporary cytoplasmic projections known as ________________.

19. Green algae share many characteristics with ________________, including their photosynthetic pigments and cell wall composition.

20. Multicellular fungi are composed of thin filaments called ________________.

Short Answer

In complete sentences, write the answers to the questions on the lines provided.

21. What are three methods used to control bacterial growth?

22. Describe a typical virus.

23. In your textbook, protists are classified into three groups: animallike protists, plantlike protists, and funguslike protists. What is the basis for that classification?

24. What is the process of alternation of generations in algae?

25. What are the four main groups of fungi?

Name______________________________ Class__________________ Date ______________

Using Science Skills

Use the table below to answer the following questions on the lines provided.

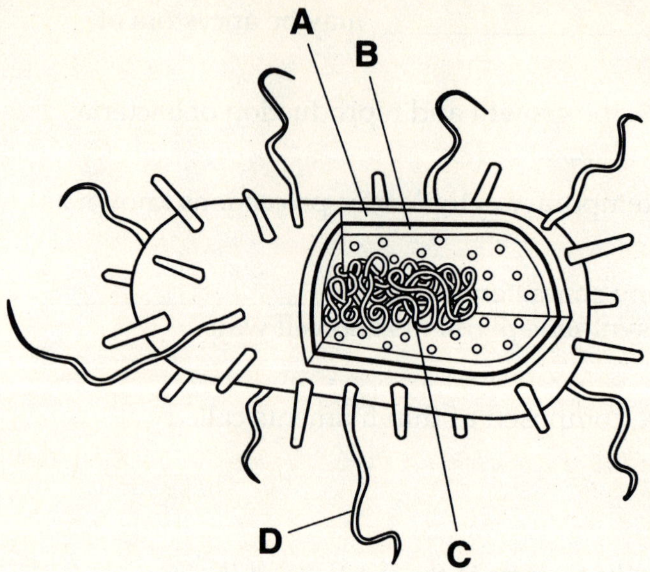

Figure 1

26. Interpreting Graphics Is the organism illustrated in Figure 1 unicellular or multicellular?

____

27. Inferring Structure B in Figure 1 is a cell membrane. What does structure A represent?

____

28. Interpreting Graphics Structure C in Figure 1 is DNA. How is this genetic material found in this organism?

____

____

29. Interpreting Graphics Structure D in Figure 1 is a flagellum. What is the function of that structure?

____

____

30. Classifying Identify the kind of organism shown in Figure 1, and describe its main characteristics.

____

____

____

Name________________________________ Class__________________ Date ______________

Multiple Choice

Write the letter that best answers the question or completes the statement on the line provided.

____ **1.** Which of the following is NOT a characteristic of all plants?
a. are eukaryotic c. produce seeds
b. have cell walls d. are multicellular

____ **2.** Without gas exchange, a plant would be unable to
a. make food. c. make minerals.
b. absorb sunlight. d. absorb water from the soil.

____ **3.** Living on land required that plants
a. evolve photosynthetic pigments.
b. conserve water.
c. exchange gases.
d. have cell walls.

____ **4.** Which of the following statements is true about bryophytes?
a. They have specialized tissues that conduct water.
b. They draw up water by osmosis.
c. They are not dependent on water.
d. They are a group of plants made up of algae and mosses.

____ **5.** Which of the following structures has functions similar to that of roots?
a. gemma c. capsule
b. rhizoid d. stalk

____ **6.** Which of the following is LEAST related to the others?
a. protonema c. gametophyte
b. sporophyte d. rhizoid

____ **7.** Xylem and phloem are NOT
a. transport subsystems. c. present in bryophytes.
b. vascular tissues. d. present in ferns.

____ **8.** Horsetails do NOT
a. have xylem tissue. c. have roots.
b. produce seeds. d. have phloem tissue.

____ **9.** Which of the following structures in ferns is diploid?
a. gametophyte c. egg
b. sporangium d. spore

Name____________________ Class__________________ Date ______________

____**10.** Seed-bearing plants differ from all other plants in that
- a. they have vascular tissue.
- b. they do not have a gametophyte generation.
- c. their gametes do not require water for fertilization to occur.
- d. all of the above

____**11.** The most ancient surviving seed plants are the
- a. mosses.
- b. liverworts.
- c. ferns.
- d. gymnosperms.

____**12.** Which group of gymnosperms produces seeds that are exposed?
- a. conifers only
- b. conifers and cycads only
- c. conifers and ginkgoes only
- d. conifers, cycads, gnetophytes, and ginkgoes

____**13.** Which term below is LEAST closely related to the others?
- a. fruit
- b. seed
- c. ovary
- d. cone

____**14.** The number of seed leaves distinguishes
- a. club mosses from mosses.
- b. angiosperms from gymnosperms.
- c. two classes of angiosperms.
- d. seed plants from seedless plants.

____**15.** Which type of plant lives the longest?
- a. annual
- b. biennial
- c. perennial
- d. none of the above

Name______________________ Class______________ Date__________

Completion

Complete each statement on the line provided.

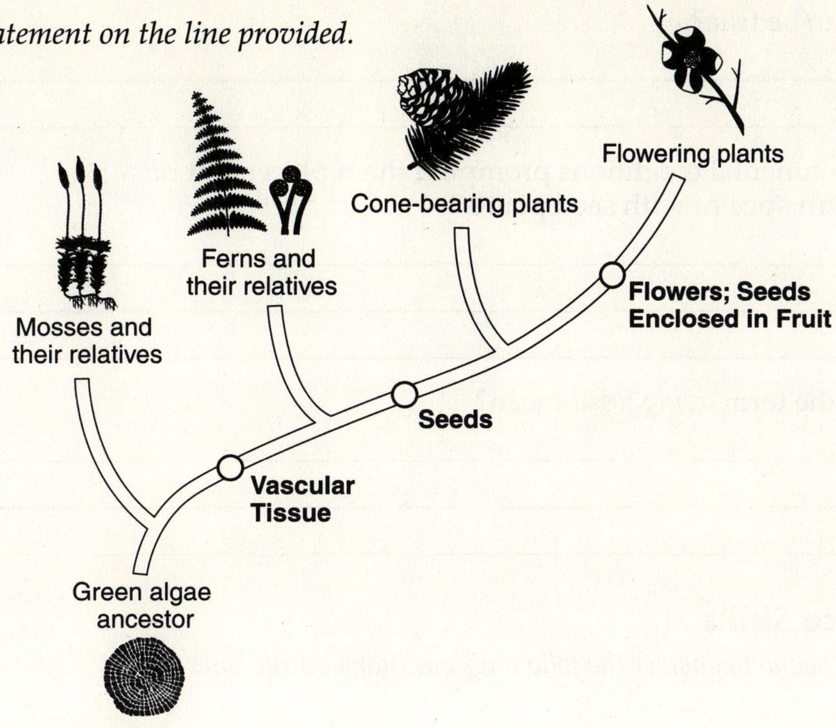

Figure 22-1

16. According to Figure 22-1, cone-bearing plants evolved before ______________ .

17. Figure 22-1 shows that plants evolved ______________ before they evolved seeds.

18. If a sorus on a fern contains 1000 spores, it will produce about 1000 ______________ .

19. Most seeds can survive long periods of drought because they have a(an) ______________ .

20. A seed that has two cotyledons is classified as a(an) ______________ .

Short Answer

In complete sentences, write the answers to the questions on the lines provided.

21. Why are bryophytes considered to be successful land plants even though they are not fully terrestrial?

____

____

____

22. How are the rhizoids of mosses similar to roots? How are they different?

____

____

____

Name______________________________ Class__________________ Date ______________

23. Why are the roots, stems, and leaves of ferns and their relatives considered to be true?

____

____

24. What environmental conditions prompted the replacement of moss and fern species with seed plants?

____

____

____

25. What does the term *living fossil* mean?

____

____

____

Using Science Skills

Use the diagram below to answer the following questions on the lines provided.

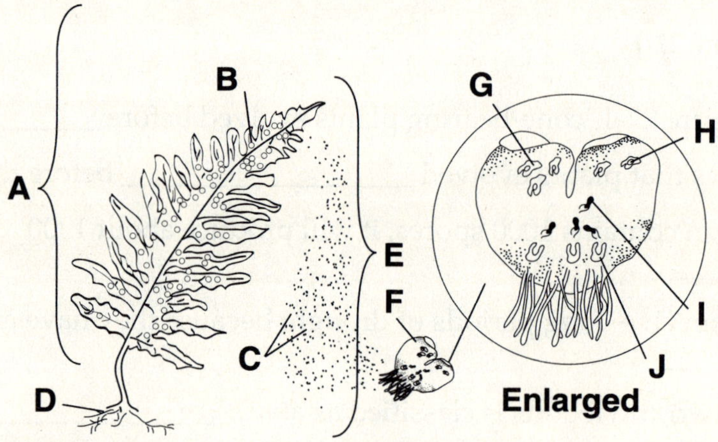

Figure 22-2

26. **Classifying** In Figure 22-2, which labeled structures are part of the gametophyte stage of a fern?

____

____

27. **Interpreting Graphics** From which labeled structure in Figure 22-2 are structures C released? Name the structures.

____

____

28. **Comparing and Contrasting** How do structures G and J in Figure 22-2 differ? Explain.

____

____

Name______________________________ Class__________________ Date ______________

29. **Predicting** In Figure 22-2, what will the structures labeled C become?

____

30. **Inferring** In Figure 22-2, what is the structure labeled E? Name the structure from which E grows.

____

____

Essay

Write the answer to each question in the space provided.

31. Describe similarities between multicellular green algae and plants. What do these similarities suggest?

32. Describe how a sporophyte is formed during the life cycle of a moss plant.

Name______________________________ Class__________________ Date ______________

33. Explain the importance of vascular tissue in plants.

34. How do animals aid in the reproduction of seed plants?

35. Compare and contrast angiosperms and gymnosperms.

Name______________________ Class________________ Date ____________

Multiple Choice

Write the letter that best answers the question or completes the statement on the line provided.

____ **1.** A plant is a
a. unicellular prokaryote.
b. multicellular prokaryote.
c. unicellular eukaryote.
d. multicellular eukaryote.

____ **2.** Plants use the energy of sunlight to
a. exchange gases with the atmosphere.
b. take in water from the soil.
c. carry out cellular respiration.
d. carry out photosynthesis.

____ **3.** The first plants evolved from
a. mosses.
b. an organism similar to multicellular green algae.
c. a protist that lived on land.
d. photosynthetic prokaryotes.

____ **4.** Bryophytes need water to
a. reproduce.
b. draw up water by osmosis.
c. undergo photosynthesis.
d. grow tall.

____ **5.** In liverworts, the structures that produce eggs and sperm look like
a. fronds. c. liver.
b. horns. d. tiny, green umbrellas.

____ **6.** What is the sperm-producing structure of a bryophyte?
a. gemma c. rhizoid
b. archegonium d. antheridium

____ **7.** Which of the following includes all the others?
a. xylem c. phloem
b. vascular tissue d. tracheids

____ **8.** Club mosses are
a. nonvascular plants. c. seedless vascular plants.
b. seed plants. d. none of the above

____ 9. Fern spores are
a. produced by the gametophyte.
b. produced in the rhizomes.
c. called sori.
d. produced in sporangia.

____10. Which of the following includes a plant embryo, a food supply, and a protective covering?
a. pollen grain
b. spore
c. seed
d. gametophyte

____11. When land environments became drier millions of years ago,
a. many moss and fern species became extinct.
b. mosses evolved vascular tissue.
c. mosses and ferns no longer required water for reproduction.
d. many angiosperm species became extinct.

____12. The four groups of gymnosperms are conifers, cycads, ginkgoes, and
a. bryophytes.
b. horsetails.
c. liverworts.
d. gnetophytes.

____13. Angiosperms produce seeds inside protective structures called
a. pollen grains.
b. cones.
c. ovaries.
d. petals.

____14. A monocot is an angiosperm that has
a. a taproot.
b. branched veins.
c. one seed leaf.
d. two seed leaves.

____15. Flowering plants that complete a life cycle within a single growing season are called
a. annuals.
b. dicots.
c. perennials.
d. monocots.

Completion

Complete each statement on the line provided.

16. The __________________ of all plants are made of cellulose.

17. Plants need to exchange __________________ with the atmosphere in order to carry out the processes of photosynthesis and respiration.

18. Bryophytes are low-growing because they lack __________________.

19. The capsule of a moss produces __________________ .

20. Solutions of nutrients and carbohydrates produced by photosynthesis are transported through __________________ tissue.

Name______________________________ Class_________________ Date ______________

Short Answer

In complete sentences, write the answers to the questions on the lines provided.

21. Describe four basic needs of plants.

____

____

22. Describe the structure of a fern gametophyte.

____

____

____

23. What is a pollen grain?

____

____

24. Define *fruit*.

____

____

25. Contrast the number of cotyledons in a monocot and dicot.

____

____

Using Science Skills

Use the diagram and table below to answer the following questions on the lines provided.

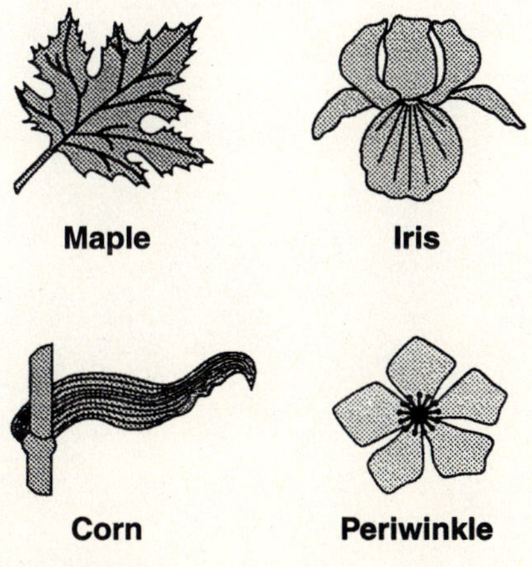

	Monocots	Dicots
Leaves	Parallel veins	Branching veins
Flowers	Parts in multiples of three	Parts in multiples of four or five
Vascular bundles in stems	Scattered throughout stem	Arranged in a ring
Roots	Fibrous	Taproot
Seeds	One seed leaf	Two seed leaves

Figure 22-1

26. Using Tables and Graphs A corn seed is a monocot. According to the table in Figure 22-1, does a corn seed have one or two seed leaves?

____

____

Name________________________ Class________________ Date____________

27. **Classifying** Is the maple leaf in Figure 22-1 a monocot or a dicot? How do you know?

____

____

____

28. **Using Tables and Graphs** How are the vascular bundles in the stem of the corn plant in Figure 22-1 arranged?

____

____

____

29. **Comparing and Contrasting** Which flower in Figure 22-1 is a monocot? How do you know?

____

____

____

30. **Predicting** A bean plant is a dicot. According to Figure 22-1, what type of root will grow from its seed?

____

____

____

Name______________________________ Class__________________ Date ______________

Multiple Choice

Write the letter that best answers the question or completes the statement on the line provided.

____ **1.** Ground tissue is found in plant
- a. stems only.
- b. stems and leaves only.
- c. roots and stems only.
- d. roots, stems, and leaves.

____ **2.** Which of the following should a student examine under a compound microscope to observe cell division?
- a. epidermis of a leaf
- b. tip of a shoot
- c. xylem from a tree trunk
- d. phloem from the leaf of a plant

____ **3.** If some of the xylem of a young oak tree was destroyed, it would most likely interfere with the tree's ability to
- a. conduct sugars to the roots.
- b. absorb sunlight.
- c. absorb carbon dioxide from the air.
- d. conduct water to the leaves.

____ **4.** Which of the following are found mainly in monocots?
- a. taproots
- b. long, thick primary roots
- c. extensive root systems
- d. small secondary roots

____ **5.** The vascular cylinder of a root consists of
- a. xylem only.
- b. phloem only.
- c. xylem and phloem.
- d. xylem, phloem, and ground tissue.

____ **6.** There is a one-way movement of water and minerals from the cortex into the vascular cylinder of a root because
- a. water molecules are in high concentration in the cortex.
- b. nutrients are in low concentration in the cortex.
- c. the root pressure is low.
- d. the Casparian strip is waterproof.

Name________________________ Class________________ Date ____________

____ **7.** Root pressure

a. causes a plant's roots to absorb water.

b. forces the water in xylem downward.

c. is produced in the cortex of the root.

d. is produced in the vascular cylinder by active transport.

____ **8.** The vascular tissue in a plant's stem

a. has buds.

b. is continuous from the roots.

c. carries nutrients up the stem but not down.

d. consists of nodes.

____ **9.** Vascular bundles are

a. surrounded by pith in monocot stems.

b. found only in dicot roots.

c. scattered throughout dicot stems.

d. surrounded by parenchyma in both monocot and dicot stems.

____**10.** In dicot plants, secondary growth

a. changes primary xylem and phloem to secondary xylem and phloem.

b. makes the roots longer.

c. results from an increase in the primary xylem and phloem.

d. produces wood.

____**11.** Most of the photosynthetic activity in plants takes place in the

a. mesophyll. c. guard cells.

b. stomata. d. xylem.

____**12.** The stomata of leaves are usually open in

a. light if a plant has enough water.

b. light if a plant has too little water.

c. darkness if a plant has enough water.

d. darkness if a plant has too little water.

____**13.** Which of the following terms is LEAST related to the other terms?

a. oxygen c. carbon dioxide

b. stomata d. companion cells

____**14.** Water will move higher in a narrow glass tube than in a wide glass tube because of

a. adhesion only. c. pressure.

b. capillary action. d. cohesion only.

Name______________________________ Class__________________ Date ______________

____15. When a plant moves sugars from its leaves to its stems, the sink is the
- a. leaves.
- b. roots.
- c. stems.
- d. same as the source.

Completion

Complete each statement on the line provided.

16. When sugars move through phloem cells called __________________, they pass through small holes between the cells.

17. Root hairs take in water from the soil through the process of __________________.

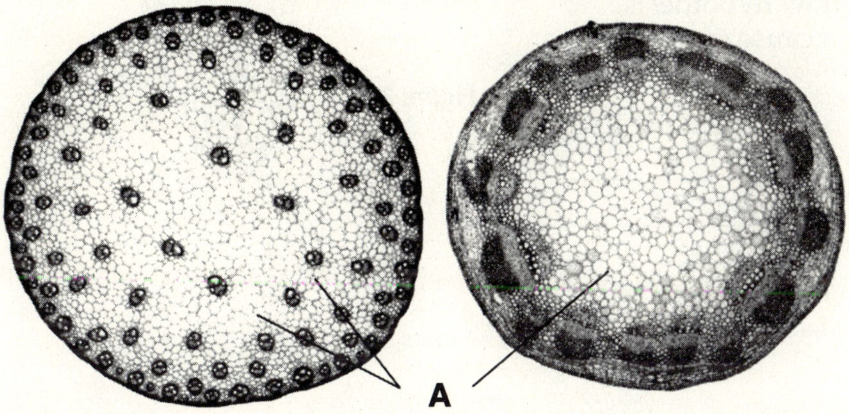

Figure 23-1

18. In Figure 23-1, A is pointing to __________________ tissue.

19. Stomata open into the __________________ layer of a leaf.

20. Water rises to the top of a giant redwood tree by __________________.

Short Answer

In complete sentences, write the answers to the questions on the lines provided.

21. Contrast the flow of materials in xylem and phloem.

____

____

____

____

22. Contrast the growth of fibrous roots and taproots in soil.

____

____

____

____

Name________________ Class________________ Date________________

23. In Figure 23-2, what is structure A, and what tissues is it made up of?

24. How is the tissue makeup of a stem similar to that of a root?

25. According to the pressure-flow hypothesis, how does water from xylem cause sugars to flow through phloem?

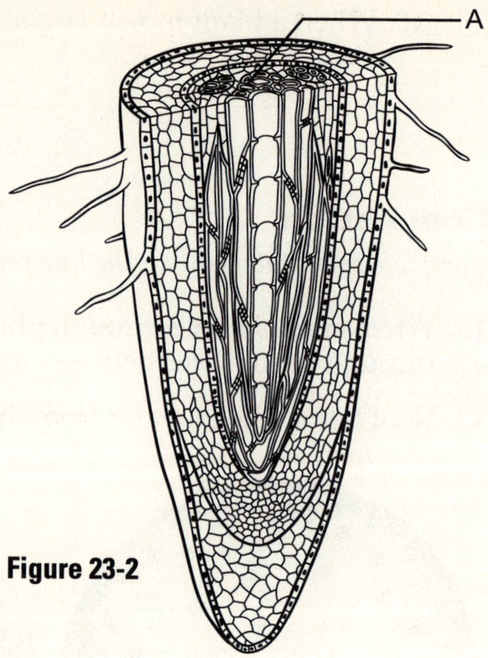

Figure 23-2

Using Science Skills

Use the diagram below to answer the following questions on the lines provided.

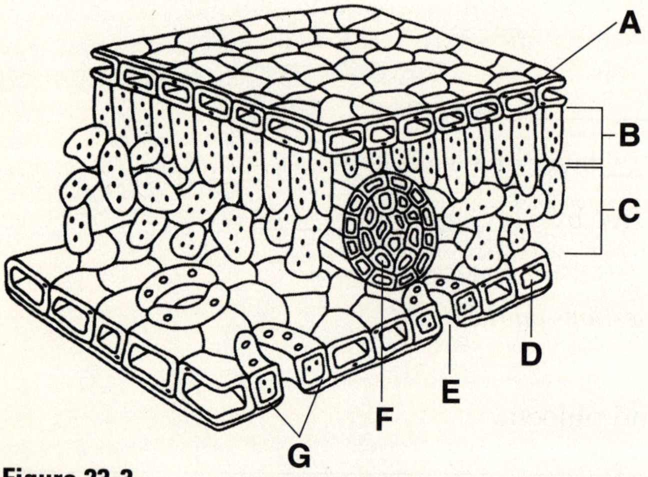

Figure 23-3

26. **Interpreting Graphics** Are the stomata in Figure 23-3 open or closed? Identify the letter of the stomata.

27. **Drawing Conclusions** What is the importance of the spaces between the cells labeled C in Figure 23-3?

28. Interpreting Graphics In Figure 23-3, what is structure F? What two types of tissues make up this structure?

____

____

29. Using Models Which labeled structures in Figure 23-3 form a waterproof barrier that protects tissues and limits water loss? Identify the structures.

____

____

30. Interpreting Graphics In Figure 23-3, which cells do NOT contain chlorophyll?

____

____

Essay

Write the answer to each question in the space provided.

31. In what ways do the leaves of a plant depend on the plant's roots and stem?

32. How do the functions of the three kinds of cells that form ground tissue differ?

Name______________________________ Class__________________ Date ______________

33. Explain the role of active transport in the movement of water and dissolved nutrients from the soil to the root of a plant.

34. Describe three main functions of a plant leaf.

35. Explain why the stomata of a plant open after the plant has been watered.

Multiple Choice

Write the letter that best answers the question or completes the statement on the line provided.

____ **1.** A seed plant is anchored in the ground by its
a. stems. c. leaves.
b. roots. d. trichomes.

____ **2.** Which of the following is the only tissue that produces new plant cells?
a. meristematic tissue c. ground tissue
b. phloem d. xylem

____ **3.** Vascular tissue in plants consists of
a. meristem.
b. xylem and phloem.
c. parenchyma and collenchyma cells.
d. epidermal cells.

____ **4.** In angiosperms, xylem consists of tracheids and
a. sieve tube elements.
b. companion cells.
c. vessel elements.
d. parenchyma cells.

____ **5.** A carrot is a(an)
a. taproot. c. monocot.
b. fibrous root. d. extensive root system.

____ **6.** The layer of cells that encloses the vascular tissue in the central region of a root is the
a. endodermis. c. epidermis.
b. cortex. d. mesophyll.

____ **7.** Minerals from the soil move into roots by
a. diffusion. c. active transport.
b. transpiration. d. root pressure.

____ **8.** One of the main functions of stems is to
a. carry out photosynthesis.
b. transport substances between roots and leaves.
c. store carbohydrates.
d. store water.

Name______________________________ Class____________________ Date ______________

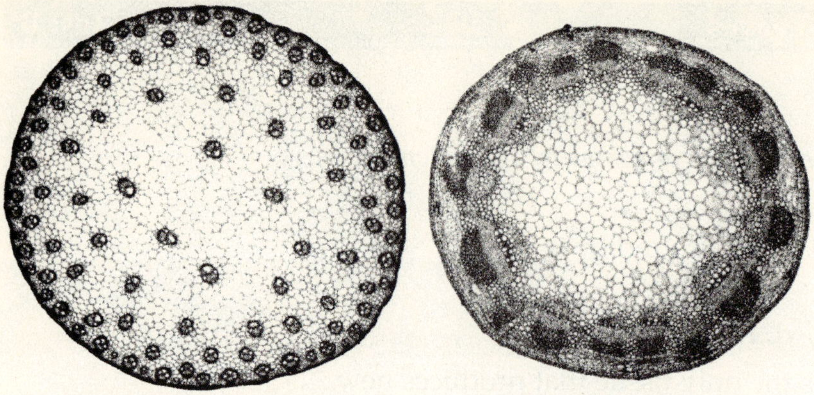

Figure 23-1

____ **9.** Figure 23-1 shows cross sections of monocot and dicot
a. roots. c. root hairs.
b. leaf veins. d. stems.

____**10.** The primary growth of a stem
a. is produced by cell division in the cambium.
b. is produced by cell division in the xylem and phloem.
c. increases the length of the stem.
d. increases the width of the stem.

____**11.** The movement of gases in and out of the leaf is regulated by
a. palisade mesophyll. c. phloem.
b. guard cells. d. stomata.

____**12.** In Figure 23-2, the X points to a
a. guard cell. c. vein.
b. tracheid. d. stoma.

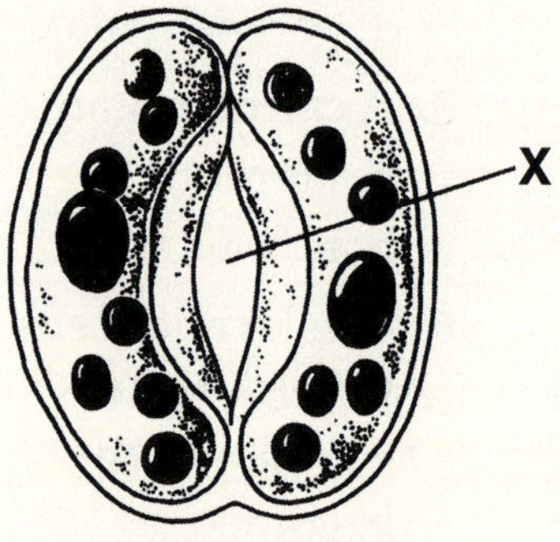

Figure 23-2

____**13.** The attraction of water molecules to other molecules is called
a. adhesion. c. capillary action.
b. cohesion. d. transpiration pull.

____**14.** Through which plant cells does water move by capillary action?
a. phloem cells c. mesophyll cells
b. guard cells d. xylem cells

____**15.** The movement of sugars in a plant can be explained by
a. capillary action. c. the pressure-flow hypothesis.
b. transpiration pull. d. root pressure.

Name________________________ Class________________ Date ____________

Completion

Complete each statement on the line provided.

16. The three main organs of seed plants are roots, leaves, and ________________.

17. The roots of grasses are a type of root called a(an) ________________ root.

18. Structure A in Figure 23-3 is called the ________________, which is made up of xylem and phloem.

19. The ________________ on a stem contain undeveloped tissue that can produce new stems and leaves.

20. Stomata open into the ________________ layer of a leaf.

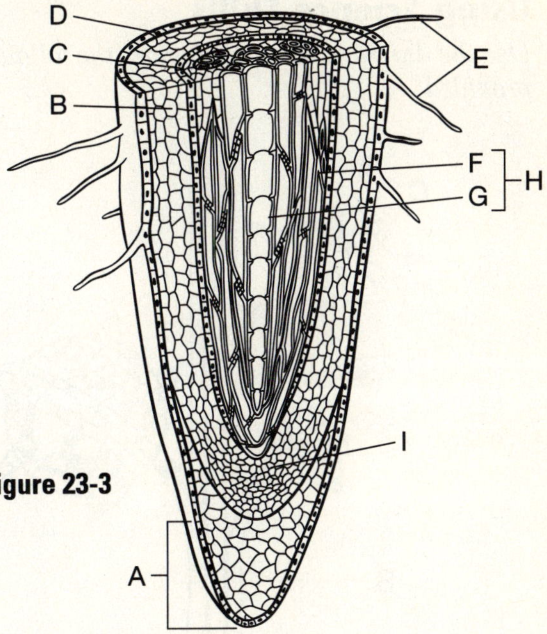

Figure 23-3

Short Answer

In complete sentences, write the answers to the questions on the lines provided.

21. What three kinds of tissues does meristematic tissue develop into?

22. What do roots absorb from the soil?

23. Contrast the arrangement of vascular bundles in monocot stems and dicot stems.

24. By what three processes does water rise from the roots to the top of a tree?

25. In the pressure-flow hypothesis, what does the term *sink* refer to?

Name________________________ Class________________ Date ____________

Using Science Skills

Use the diagram below to answer the following questions on the lines provided.

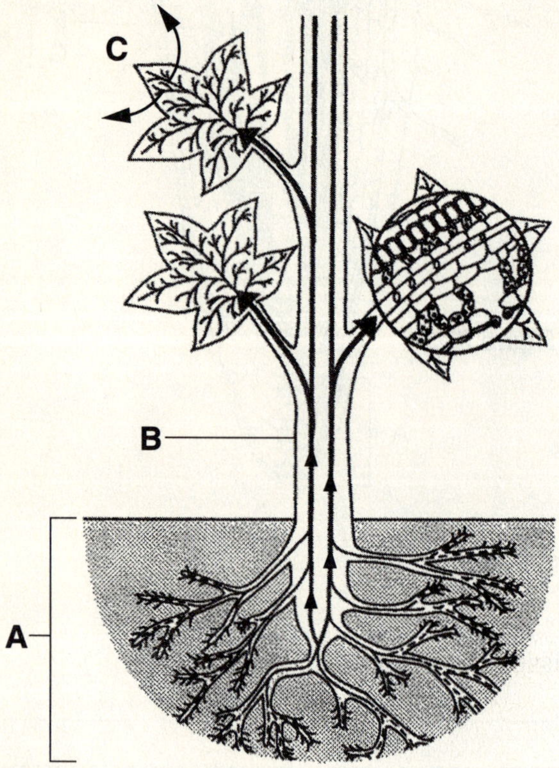

Figure 23-4

26. **Interpreting Graphics** What do the arrows in Figure 23-4 represent?

____

27. **Inferring** What kind of vascular tissue is involved in the processes represented in Figure 23-4?

____

28. **Inferring** In Figure 23-4, what process causes water to move upward in part B of the plant?

____

29. **Inferring** In part C of Figure 23-4, what process is helping to bring water to the top of the plant?

____

30. **Inferring** What process is helping to move water upward through part A of Figure 23-4?

____

Name______________________________ Class__________________ Date ______________

Multiple Choice

Write the letter that best answers the question or completes the statement on the line provided.

____ **1.** In gymnosperms, ovules are located near the base of each
a. scale. c. carpel.
b. filament. d. cone.

____ **2.** The sterile leaves of a flower are the
a. carpel and stamens.
b. filaments and anthers.
c. stigma and style.
d. sepals and petals.

____ **3.** A pollen grain landing near an ovule produces a
a. gametophyte. c. flower.
b. pollen tube. d. stamen.

____ **4.** In angiosperms, the nucleus of each pollen grain undergoes
a. one meiotic division.
b. two meiotic divisions.
c. one mitotic division.
d. two mitotic divisions.

____ **5.** The endosperm of a corn seed develops through the process of
a. pollination.
b. fertilization.
c. double fertilization.
d. germination.

____ **6.** Which of the following is NOT true of all fruits?
a. All fruits contain seeds.
b. All fruits are sweet.
c. All fruits help disperse seeds.
d. All fruits are formed from the ovary wall.

____ **7.** A seed that is dispersed to an area far away from the parent plant might face less
a. alternation. c. germination.
b. pollination. d. competition.

____ **8.** Seeds dispersed by animals typically are contained in
a. fleshy, nutritious fruits.
b. cones.
c. buoyant structures.
d. lightweight structures.

Name_______________________________ Class___________________ Date _______________

____ 9. A period during which the embryo of a seed is alive but not growing is
a. fruit production.
b. seed production.
c. germination.
d. dormancy.

____10. What is true of the seed coat during germination?
a. It must crack open.
b. It cannot be damaged if germination is to occur.
c. It dissolves in water.
d. It is absorbed by the growing plant.

____11. When a plant reproduces vegetatively, its offspring
a. are genetically different.
b. are genetically identical.
c. remain dormant until the spring.
d. grow from seeds.

____12. The horizontal stems of the strawberry plant shown in Figure 24-1 are called
a. scions.
b. stolons.
c. buds.
d. plantlets.

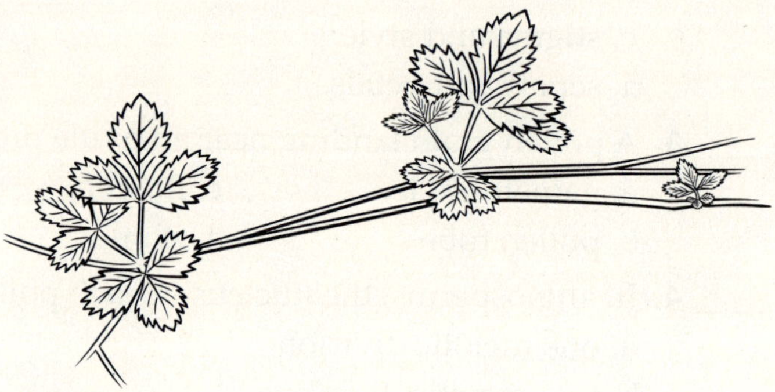

Figure 24-1

____13. What would be the best method of propagating most woody plants?
a. using cuttings
b. grafting or budding
c. planting seeds
d. hand pollination

____14. Which two methods of plant propagation are most similar?
a. planting seeds and rooting cuttings
b. rooting cuttings and grafting
c. rooting cuttings and budding
d. grafting and budding

____15. Corn, sugar beets, cauliflower, and cabbage were all developed by
a. plant propagation.
b. germination.
c. pollination.
d. selective breeding.

Completion

Complete each statement on the line provided.

16. In gymnosperms, ____________________ produce the male gametophytes.

17. In the gymnosperm seed, the ____________________ is part of the old sporophyte generation.

Name________________________ Class________________ Date ____________

18. In the flower diagram in Figure 24-2, the structure labeled C is called the ________________.

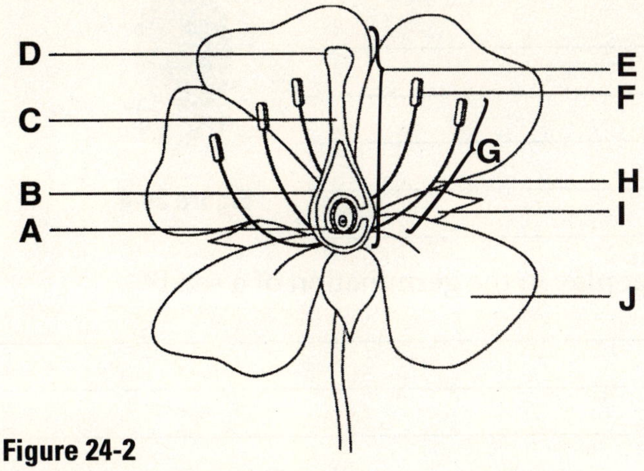

Figure 24-2

19. A coconut is a very large ________________ that contains a milky endosperm layer.

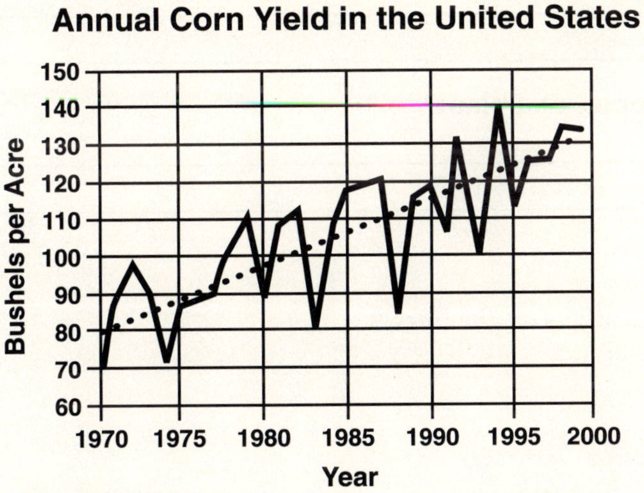

Figure 24-3

20. Figure 24-3 shows that, from 1970 to 2000, the annual yield of corn in the United States fluctuated up and down. The overall trend however, is that corn yield has ________________.

Short Answer

In complete sentences, write the answers to the questions on the lines provided.

21. What is the function of a pollination drop?

Name______________________ Class______________ Date ____________

22. Of the seeds shown in Figure 24-4, which is more typically dispersed by wind? How can you tell?

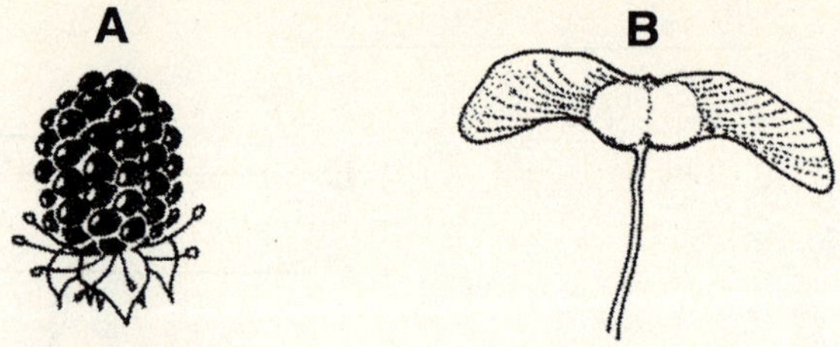

Figure 24-4

____

____

____

____

____

23. What role does water play in the germination of a seed?

____

____

____

24. Name three ways that new plants are produced by vegetative reproduction.

____

____

25. How might a horticulturist make exact copies of a plant?

____

____

Using Science Skills

Use the diagram below to answer the following questions on the lines provided.

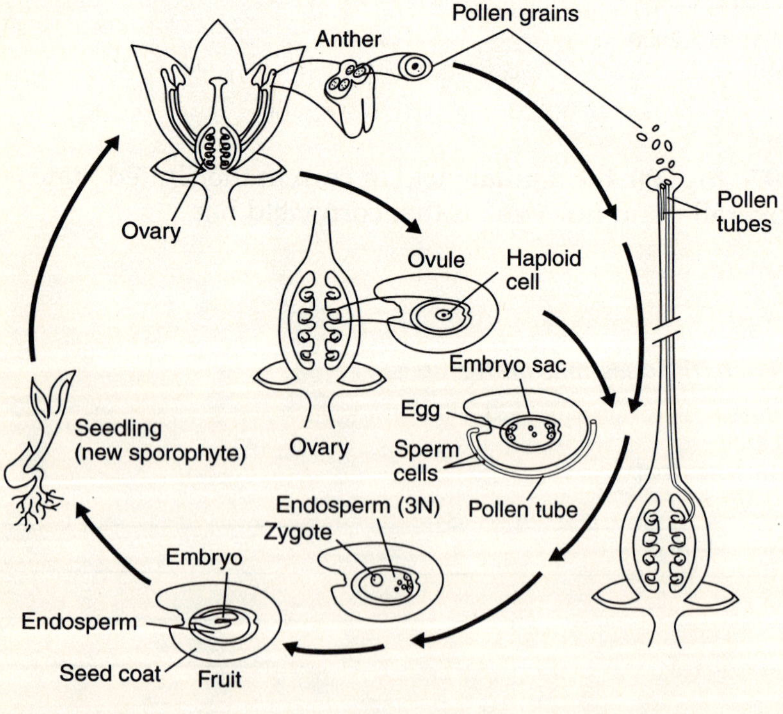

Figure 24-5

26. **Interpreting Graphics** Which structure shown in Figure 24-5 allows pollen to enter the ovary?

____

27. **Applying Concepts** How many nuclei are in the embryo sac shown in Figure 24-5? From what type of cell (haploid or diploid) and through what type of process did these nuclei arise?

____

____

28. **Applying Concepts** Which of the following structures in Figure 24-5 are diploid: anther, pollen grains, ovary, embryo sac, endosperm, or seedling?

____

29. **Interpreting Graphics** According to Figure 24-5, what does the zygote develop into?

____

30. **Predicting** What is the most likely way that the pollen in Figure 24-5 will disperse? Explain your answer.

____

____

____

Essay

Write the answer to each question in the space provided.

31. What does the term *alternation of generations* refer to? Where are these two generations in seed plants?

Name________________________________ Class____________________ Date ________________

32. Describe the formation of a fruit in an angiosperm.

33. Explain how a forest fire can affect the germination of certain pine seeds, and what effect this has on the recovery of the forest from a fire.

34. Describe grafting and budding, and explain why these methods of propagation are used.

35. Describe two important ways in which agricultural efficiency has been improved.

Name________________________________ Class__________________ Date ____________

Multiple Choice

Write the letter that best answers the question or completes the statement on the line provided.

____ 1. Pollen grains are produced by
a. male reproductive structures.
b. female reproductive structures.
c. ovules.
d. flowers.

____ 2. In angiosperms, reproduction takes place in
a. leaves. c. cones.
b. flowers. d. pollen.

____ 3. A sticky secretion on the scales of seed cones traps
a. sporophytes. c. pollen grains.
b. pollen cones. d. egg cells.

____ 4. In an angiosperm, pollen grains are produced in the
a. stigma. c. carpel.
b. filament. d. anther.

____ 5. The triploid cell that grows into a food-rich tissue in the seed is called the
a. endosperm. c. embryo sac.
b. fruit. d. cotyledon.

____ 6. A ripened ovary that contains seeds is called a(an)
a. embryo. c. fruit.
b. ovule. d. vegetable.

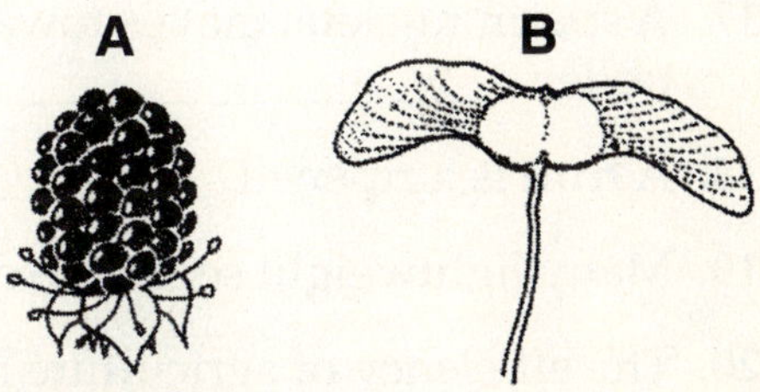

Figure 24-1

____ 7. The seed type shown in Figure 24-1 that is generally dispersed by animals is(are)
a. A c. both A and B
b. B d. neither A nor B

____ 8. Seeds that are dispersed by wind and water typically are
a. lightweight. c. nutritious.
b. large. d. sweet and fleshy.

____ 9. Seeds of temperate plants tend to germinate in
a. winter. c. summer.
b. spring. d. fall.

____10. The early growth stage of a plant embryo is called
a. fertilization. c. germination.
b. dormancy. d. pollination.

____**11.** Some plants reproduce vegetatively by producing
a. plantlets. c. flowers.
b. seeds. d. pollen.

____**12.** A parent plant and its plantlets
a. are not similar.
b. are genetically identical.
c. develop inside protective structures.
d. must always remain attached.

____**13.** Grafting usually works best when plants are
a. growing. c. dormant.
b. germinating. d. pollinating.

____**14.** In a grafting, the cut stem that is attached to the parent plant is called the
a. vascular cambium. c. scion.
b. stock. d. cutting.

____**15.** Most people in the world depend on food crops such as
a. sugar beets, cabbage, and broccoli.
b. strawberries, chilies, and avocadoes.
c. wheat, rice, and corn.
d. apples, grapes, and strawberries.

Completion

Complete each statement on the line provided.

16. In a flower's stamen, the filament is topped by a(an) ___________________ .

17. A sperm nucleus moves toward a flower's ovary through a long passageway called a pollen ___________________ .

18. A fruit is a ripened ___________________ that contains angiosperm seeds.

19. Many lightweight seeds are dispersed by wind or ___________________ .

20. The efficiency of agriculture has been improved through the ___________________ of crop plants and improvements in farming techniques.

Short Answer

In complete sentences, write the answers to the questions on the lines provided.

21. Name the four types of specialized leaves in a flower.

____

____

____

____

Name ______________________ Class ______________ Date ____________

22. Name two environmental factors that can end a seed's dormancy.

23. How does a plantlet become a new plant?

24. Why do horticulturists use plant propagation?

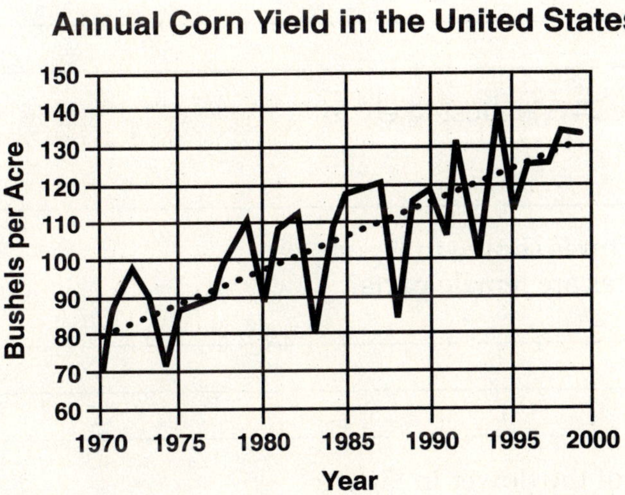

Figure 24-2

25. According to Figure 24-2, what has happened to overall yield of corn in the United States during the past 30 years?

Name________________________ Class________________ Date ____________

Using Science Skills

Use the diagram below to answer the following questions on the lines provided.

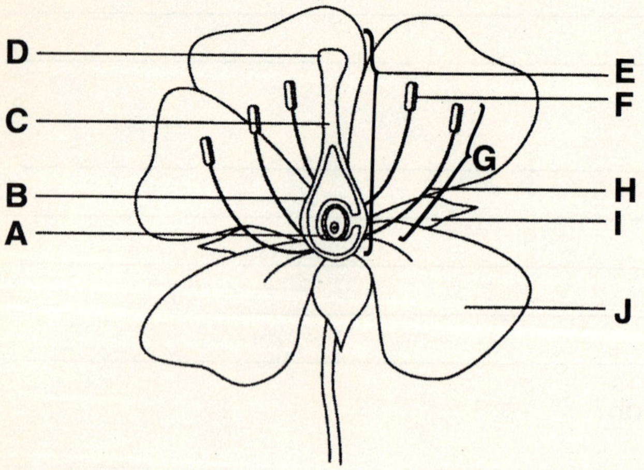

Figure 24-3

26. **Interpreting Graphics** What is the structure labeled C in Figure 24-3 called?

____

____

____

27. **Inferring** Which labeled structure in Figure 24-3 is most likely to be brightly colored?

____

28. **Interpreting Graphics** Which labeled structures shown in Figure 24-3 are male parts? Which labeled structures are female parts?

____

____

____

29. **Interpreting Graphics** Which labeled part of the flower in Figure 24-3 produces pollen grains?

____

30. **Interpreting Graphics** Which labeled structure shown in Figure 24-3 receives pollen during pollination?

____

Multiple Choice

Write the letter that best answers the question or completes the statement on the line provided.

____ **1.** The regions of tissue in a plant that produce cells that later become specialized tissues are the
a. roots.
b. stems.
c. leaves.
d. meristems.

____ **2.** Most plant hormones are produced in rapidly dividing tissues in
a. the buds of flowers.
b. flowers in full bloom.
c. the underside of leaves.
d. the tips of roots and stems.

____ **3.** What is the source of ethylene gas in a plant?
a. leaf buds
b. root hairs
c. fruit tissues
d. lateral meristems

____ **4.** Powerful synthetic auxins are sometimes used as
a. fertilizers.
b. herbicides.
c. insecticides.
d. fruit ripeners.

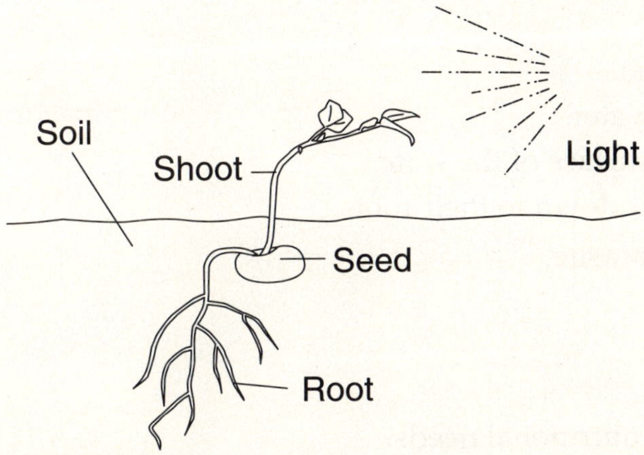

Figure 25-1

____ **5.** What is the phenomenon that is causing the bean seedling to bend as shown in Figure 25-1?
a. gravitropism
b. phototropism
c. rapid response
d. thigmotropism

____ **6.** Gravitropism ensures that the plant in Figure 25-1 will
a. grow roots into the soil.
b. wrap tightly around a trellis.
c. grow toward a light source.
d. have shoots that grow toward the soil.

_____ 7. The growth patterns of plants such as ivy and pole beans are regulated by

a. gravitropism. c. thigmotropism.
b. phototropism. d. all of the above

_____ 8. Phytochrome is a pigment that is responsible for

a. bud scaling. c. photosynthesis.
b. gravitropism. d. photoperiodism.

_____ 9. What hormones influence the shedding of leaves?

a. auxin and ethylene
b. cytokinin and ethylene
c. gibberellin and auxin
d. ethylene and gibberellin

_____10. Which of the following seals a leaf off from the rest of the plant?

a. bud c. waxy scales
b. petiole d. abscission layer

_____11. Bald cypress trees have "knees" that function to

a. support the tree in the mud.
b. bring oxygen down to the roots.
c. produce seeds that float in water.
d. photosynthesize in the deep shade.

_____12. Aquatic plants that grow in mud that has little or no oxygen have adaptations that

a. allow them to survive without oxygen.
b. keep them from taking in too much water.
c. let them photosynthesize below the surface of the water.
d. help them diffuse oxygen from the air down to their roots.

_____13. The dodder plant *Cuscuta,* which is a parasite,

a. helps the plant on which it grows.
b. gathers its own water and nutrients.
c. harms the plant on which it grows.
d. captures small organisms to meet its nutritional needs.

_____14. Epiphytes differ from parasites in that epiphytes

a. harm their host plants.
b. grow directly on the bodies of other plants.
c. extract water from their host plants.
d. produce their own food.

_____15. The tobacco plant produces a chemical that

a. is probably a natural insecticide.
b. protects the plant from insects.
c. affects the human nervous system.
d. all of the above

Name______________________ Class______________ Date__________

Completion

Complete each statement on the line provided.

16. A target cell must contain a hormone ______________ to which the hormone can bind.

17. Very high concentrations of auxins inhibit the ______________ of cells.

18. As cold weather approaches, ______________ plants turn off photosynthetic pathways.

19. Mangroves survive in oxygenless mud by growing specialized ______________ that have air spaces in them.

20. A desert plant that is transplanted to an aquatic environment likely would not survive because it is not ______________ to that environment.

Short Answer

In complete sentences, write the answers to the questions on the lines provided.

21. Explain what is meant by the phrase "forever young" with regard to most plants.

22. What three major changes do plants undergo in the fall in preparation for winter dormancy?

23. How is the cactus in Figure 25-2 adapted to prevent water loss through transpiration?

24. How are epiphytes similar to and different from parasites?

Figure 25-2

25. How are some of the chemical defenses of plants beneficial to humans?

Name____________________ Class______________ Date__________

Using Science Skills

Use the diagram below to answer the following questions on the lines provided.

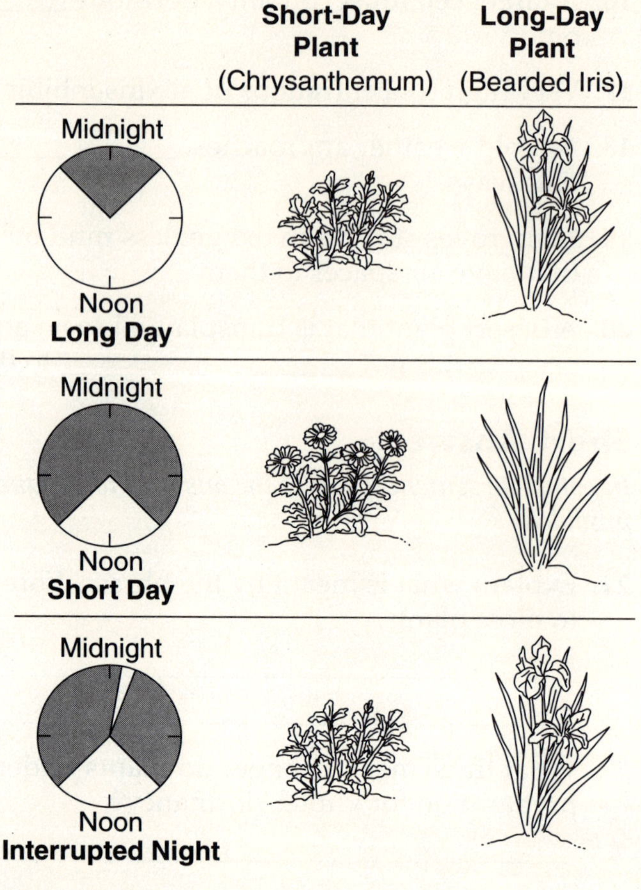

Figure 25-3

26. Predicting Which of the plants in Figure 25-3 would you expect to have the highest level of cytokinin in the fall? Explain your answer.

27. Comparing and Contrasting Describe how the plants in Figure 25-3 respond differently to phototropism and photoperiodism.

28. Interpreting Graphics According to Figure 25-3, what will happen to each of the plants if they receive more than twelve hours of darkness each day? Why?

29. Inferring Describe the adaptation that prevents the chrysanthemum in Figure 25-3 from blooming when exposed to light in the middle of the night.

Name______________________________ Class__________________ Date ______________

30. **Applying Concepts** Describe three effects that auxins have on the chrysanthemum in Figure 25-3.

____

____

____

____

Essay

Write the answer to each question in the space provided.

31. Explain why sealing fruit in a bag might cause the fruit to ripen quickly.

32. Briefly describe how plants respond to light and gravity.

Name________________________ Class________________ Date ____________

33. What happens to a plant when a gardener snips off the growing tip?

34. How are water lilies adapted to growing in mud without oxygen?

35. Describe the similarities and differences that pitcher plants and parasitic plants have in obtaining nutrients.

Name________________________ Class________________ Date______________

Multiple Choice

Write the letter that best answers the question or completes the statement on the line provided.

____ 1. Plants that are thousands of years old have
a. no meristems.
b. a rigid body organization.
c. the ability to continue growing.
d. predetermined growth patterns.

____ 2. Plants can respond to changing environmental conditions by the action of which substances?
a. sugars
b. enzymes
c. target cells
d. hormones

____ 3. Which part of a plant is affected by a particular hormone?
a. pigments
b. apical meristems
c. hormone-producing cells
d. target cells

____ 4. Hormones that stimulate cell elongation and are produced in the rapidly growing region near the tip of the plant's root or stem are called
a. auxins.
b. ethylenes.
c. cytokinins.
d. gibberellins.

____ 5. What type of hormones cause increased plant growth?
a. auxins
b. cytokinins
c. gibberellins
d. ethylenes

____ 6. The responses of plants to environmental stimuli are called
a. tropisms.
b. hormones.
c. receptors.
d. adaptations.

____ 7. Roots growing with gravity and stems growing against gravity are examples of
a. gravitropism.
b. phototropism.
c. thigmotropism.
d. photoperiodism.

____ 8. The response of a plant to changes in the length of day or night is called
a. abscission.
b. thigmotropism.
c. dormancy.
d. photoperiodism.

____ 9. A period of decreased activity in a plant is called
a. dormancy. c. germination.
b. abscission. d. tropism.

____10. What is a pigment that regulates changes in activity that prepare plants for winter?
a. cytokinin c. ethylene
b. meristem d. phytochrome

____11. The prickly-pear cactus shown in Figure 25-1 has a leaf adaptation that
a. pumps out salt.
b. reduces water loss.
c. produces thick, green stems.
d. produces seeds when it rains.

Figure 25-1

____12. The closing of the leaves of a Venus' flytrap is called
a. a stimulus. c. a rapid response.
b. phototropism. d. tropism.

____13. A carnivorous plant obtains nitrogen
a. directly from the air.
b. through its roots from the soil.
c. by trapping and digesting insects.
d. by extracting it from a host plant.

____14. Plants that get their water and nutrients directly from a host plant are
a. epiphytes. c. xerophytes.
b. carnivorous plants. d. parasites.

____15. A drug in tobacco that might be a natural insecticide protecting the tobacco plants is
a. nicotine. c. aspirin.
b. codeine. d. ethylene.

Completion

Complete each statement on the line provided.

16. High concentrations of auxins are used as ____________________ that are used to kill weeds.

17. Fruit ripening can be stimulated by ____________________.

Name______________________________ Class__________________ Date ______________

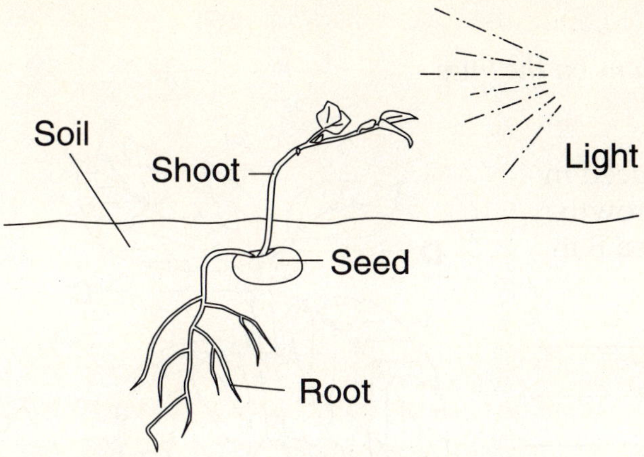

Figure 25-2

18. The ability of the young seedling in Figure 25-2 to turn toward the light in just a few hours is called ____________________ .

19. As a grapevine grows, it curls and twists around objects because of its response to touch called ____________________.

20. A plant that grows in fresh water would not survive in salt water because its leaves do not have specialized ____________________ to pump out salt the way salt-tolerant plants do.

Short Answer

In complete sentences, write the answers to the questions on the lines provided.

21. List two environmental factors to which plants respond.

22. What is a plant hormone?

23. What did Charles and Francis Darwin demonstrate in their experiment with oat seedlings?

24. Why do hollyhocks that are planted outside bloom only in the summer?

25. What is an abscission layer?

Name________________________________ Class____________________ Date ______________

Using Science Skills

Use the diagram below to answer the following questions on the lines provided.

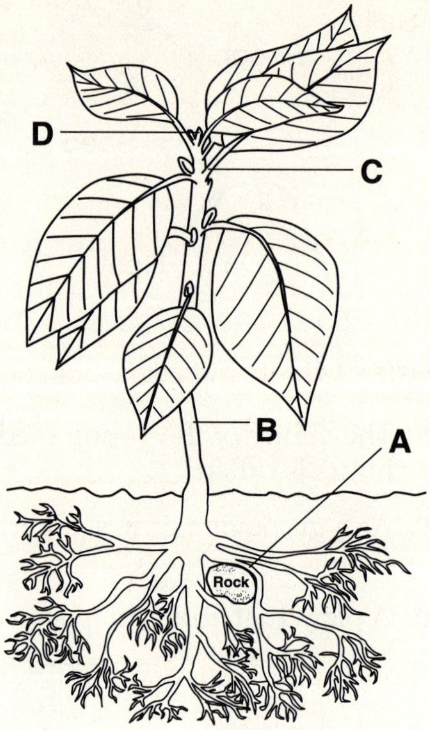

Figure 25-3

26. Predicting How will a hormone that is produced in area D of the plant in Figure 25-3 affect the growth of the plant in area B? Identify area D and area B in your answer.

____

____

____

____

____

27. Applying Concepts In Figure 25-3, how does the root in area A "know" how to grow around an obstacle and then down again instead of up or out?

____

____

____

____

____

28. Inferring In Figure 25-3, what might the structure in area D produce in the fall if this plant is a long-day flowering plant?

____

____

____

29. Applying Concepts If the plant in Figure 25-3 is a deciduous plant, what occurs at area C as ethylene production increases in preparation for winter?

____

____

____

30. Inferring Do you think the plant in Figure 25-3 has any adaptations for growing in its natural habitat? Why or why not?

____

____

____

Multiple Choice

Write the letter that best answers the question or completes the statement on the line provided.

____ **1.** Which of the following is NOT a characteristic of plants?
- a. unicellular
- b. contain cell walls made of cellulose
- c. make their own food
- d. eukaryotes

____ **2.** Plants that lack vascular tissue and depend on water for reproduction are classified as
- a. bryophytes.
- b. ferns.
- c. gymnosperms.
- d. angiosperms.

____ **3.** In the life cycles of the following plant groups, which have the diploid sporophyte as the dominant, recognizable stage?
- a. bryophytes only
- b. ferns only
- c. gymnosperms and angiosperms
- d. ferns, gymnosperms, and angiosperms

____ **4.** Both gymnosperms and angiosperms
- a. require water for reproduction.
- b. reproduce with flowers.
- c. produce seeds.
- d. reproduce with cones.

____ **5.** All annuals
- a. are short-day plants.
- b. complete a life cycle in one growing season.
- c. reproduce with cones.
- d. require water for reproduction.

____ **6.** Which tissue systems make up roots, stems, and leaves?
- a. xylem and phloem
- b. parenchyma, collenchyma, and sclerenchyma
- c. dermal, vascular, and ground
- d. nodes, internodes, and buds

____ 7. Unlike primary growth in stems, secondary growth
a. causes an increase in length.
b. causes an increase in width.
c. is produced by the apical meristem.
d. occurs in all seed plants.

____ 8. On a hot and sunny day, a wilting plant will
a. release pollen.
b. have open stomata.
c. have closed stomata.
d. become dormant.

____ 9. Water moves from the roots to the leaves
a. by root pressure, capillary action, and transpiration.
b. as described by the pressure-flow hypothesis.
c. by the force of adhesion.
d. by root pressure alone.

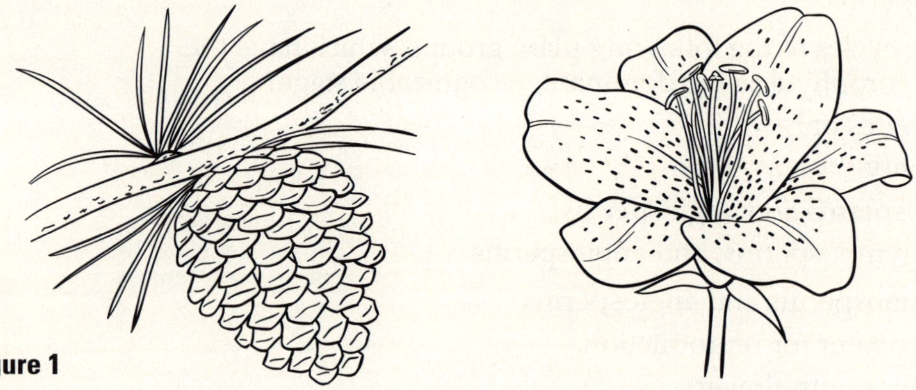
Figure 1

A B

____10. Which of the reproductive structures in Figure 1 is pollinated by wind?
a. A only
b. B only
c. both A and B
d. neither A nor B

____11. Which of the following are fruits?
a. peaches
b. beans
c. tomatoes
d. all of the above

____**12.** In plant propagation, horticulturists produce plants that are
a. genetically different from the parent.
b. genetically identical to the parent.
c. grown from seeds.
d. grown from stolons.

____**13.** The plant hormone that stimulates the growth of lateral buds is
a. auxin.
b. cytokinin.
c. gibberellin.
d. ethylene.

____**14.** In preparation for winter, deciduous plants
a. flower during short days.
b. increase their rate of photosynthesis.
c. produce less ethylene and more auxin.
d. form waxy scales around new leaf buds.

____**15.** Like parasites, carnivorous plants have
a. specialized cells to pump out salt.
b. extensive root systems.
c. specialized features to obtain nutrients.
d. air-filled spaces in their tissues.

Completion

Complete each statement on the line provided.

16. The plant life cycle alternates between the haploid _______________ stage and the diploid _______________stage.

17. The first plants were limited in height because they lacked _______________.

18. The functions of _______________ include absorbing water and dissolved nutrients from the soil.

19. As angiosperm seeds mature, the _______________ thicken to form a fruit that encloses the developing seeds.

20. Plant _______________ demonstrate the ability of plants to respond to external stimuli, such as gravity, light, and touch.

Name________________ Class__________ Date________

Short Answer

In complete sentences, write the answers to the questions on the lines provided.

21. What must plants have to survive?

22. List five ways in which monocots differ from dicots.

23. What is the function of meristematic tissue?

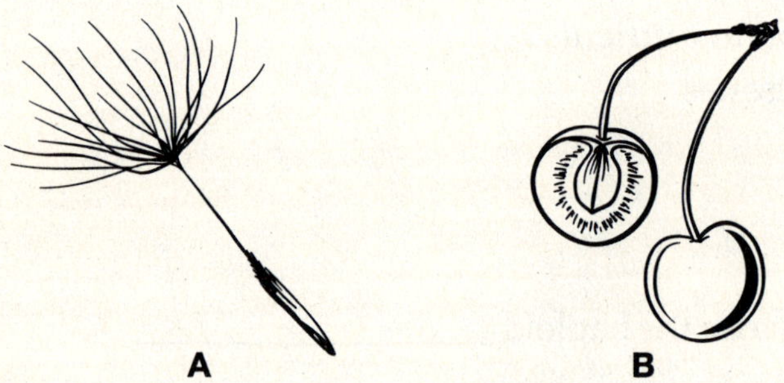

Figure 2

24. How are the seeds in Figure 2 dispersed?

25. How do plants defend themselves from insects?

Name________________________ Class________________ Date ____________

Using Science Skills

Use the diagram below to answer the following questions on the lines provided.

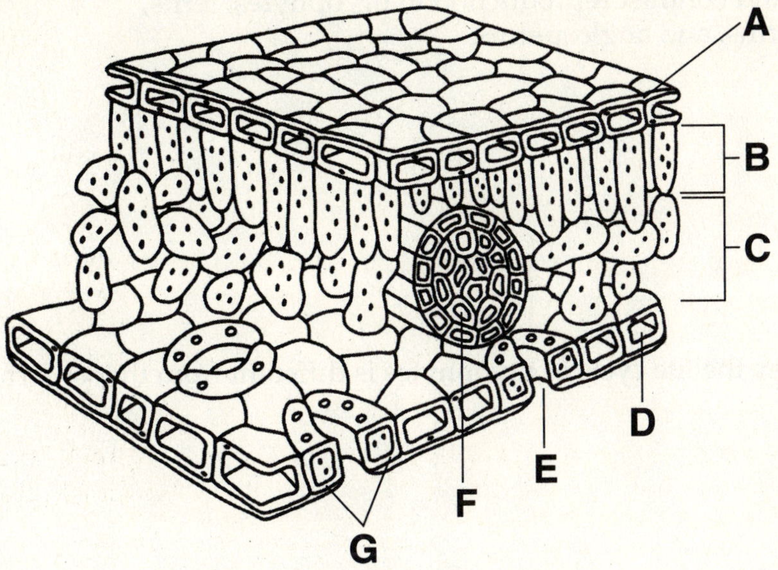

Figure 3

26. **Applying Concepts** What two functions does the structure in Figure 3 have?

____

____

27. **Interpreting Graphics** Would the structure in Figure 3 have come from a wilting plant? Explain.

____

____

____

28. **Interpreting Graphics** What labeled structure in Figure 3 transports water and the products of photosynthesis?

____

____

____

29. **Inferring** What happens in the structures labeled B and C in Figure 3 when cold weather approaches?

____

____

30. **Applying Concepts** If the structure in Figure 3 were part of a salt-tolerant plant, what adaptation would it have?

____

____

____

Name______________________________ Class__________________ Date ______________

Essay

Write the answer to each question in the space provided.

31. Compare and contrast reproduction in bryophytes, ferns, gymnosperms, and angiosperms.

32. Explain how the life cycle of bryophytes is different from that of ferns.

33. Summarize the organization of plant organs, tissues, and cells.

34. Although plants are stationary organisms, they are able to get everything they need. Explain how angiosperms "use" animals to meet their needs.

35. Explain the effect of auxins on the growth, structure, and responses of flowering plants.

Name________________________ Class________________ Date __________

Multiple Choice

Write the letter that best answers the question or completes the statement on the line provided.

____ **1.** In order to survive, plants need
a. sunlight.
b. gas exchange.
c. movement of water and nutrients.
d. all of the above

____ **2.** All of the following are bryophytes EXCEPT
a. mosses.
b. club mosses.
c. hornworts.
d. liverworts.

____ **3.** One characteristic of ferns is that they
a. produce seeds.
b. have rhizoids.
c. have vascular tissue.
d. do not require water for reproduction.

____ **4.** Which of the following is a gymnosperm?
a. monocot
b. gnetophyte
c. horsetail
d. hornwort

____ **5.** Flowering plants that live for more than two years are called
a. annuals.
b. biennials.
c. perennials.
d. long-day plants.

____ **6.** The three main plant organs are
a. roots, stems, and leaves.
b. parenchyma, sclerenchyma, and collenchyma.
c. dermal, vascular, and ground.
d. spores, flowers, and cones.

____ **7.** Which of the following is a function of roots?
a. anchor a plant into the ground
b. absorb light
c. produce leaves
d. produce seeds

____ **8.** Secondary growth takes place in the
a. apical meristems.
b. vascular cambium.
c. spongy mesophyll.
d. cortex.

____ **9.** Which of the following does NOT help move water through a plant?
a. root pressure
b. capillary action
c. transpiration pull
d. pressure flow

Name________________________________ Class___________________ Date ______________

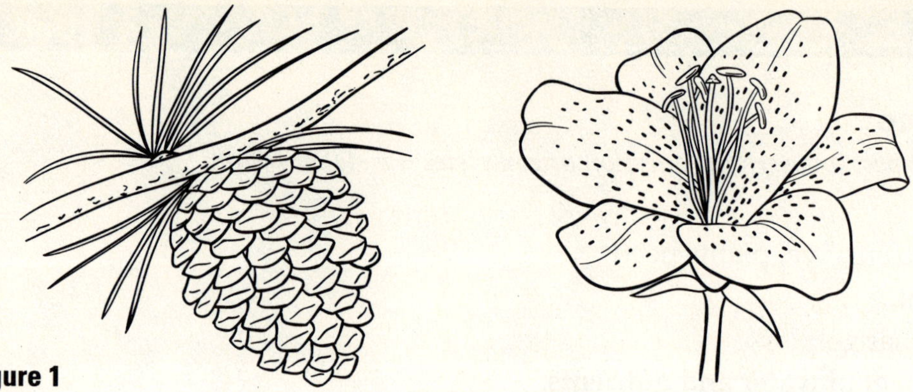
Figure 1

A B

____**10.** Which reproductive structure in Figure 1 is pollinated by animals?
- a. A only
- b. B only
- c. A and B
- d. neither A nor B

____**11.** Environmental factors that cause a seed to germinate are
- a. gravity and light.
- b. lengths of daylight and darkness.
- c. temperature and moisture.
- d. wind and soil.

____**12.** Plants produced by vegetative reproduction are
- a. genetically identical to the parent plant.
- b. genetically different from the parent plant.
- c. grown from seeds.
- d. grown from spores.

____**13.** Plant hormones that stimulate cell division and the growth of lateral buds are
- a. auxins.
- b. gibberellins.
- c. cytokinins.
- d. phytochromes.

____**14.** Because of photoperiodism, plants
- a. curl and twist around objects.
- b. bend toward a light source.
- c. grow roots downward.
- d. flower in certain seasons.

____**15.** Xerophytes are characterized by
- a. thick stems that store water.
- b. growth on the bodies of other plants.
- c. cells that pump salt out of plant tissues.
- d. air-filled spaces in their tissues.

Completion

Complete each statement on the line provided.

16. Plants are multicellular eukaryotes that have ____________________ made of cellulose.

17. In a ____________________ stem, the vascular bundles are arranged in a ring.

18. Vascular tissue is made up of ____________________, which is composed of tracheids and vessel elements, and ____________________, which is composed of sieve tube elements and companion cells.

19. As angiosperm seeds mature, the ovary walls thicken to form a ____________________ that encloses the developing seeds.

20. The response of plants to touch is called ____________________.

Short Answer

In complete sentences, write the answers to the questions on the lines provided.

21. What was the significance of vascular tissue in the evolution of plants?

____

____

22. What adaptations allow seed plants to reproduce without water?

____

____

23. How do plants regulate water loss due to transpiration?

____

____

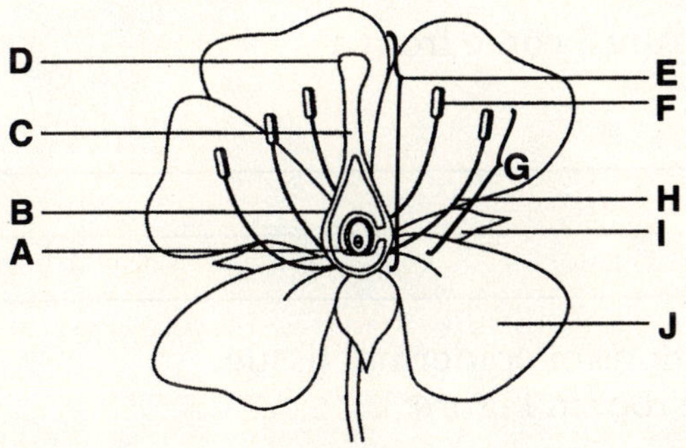

Figure 2

24. In the diagram of the flower in Figure 2, what is the structure labeled F? What is its function?

____

____

25. How do deciduous plants prepare for winter?

____

____

Name______________________ Class______________ Date __________

Using Science Skills

Use the diagram below to answer the following questions on the lines provided.

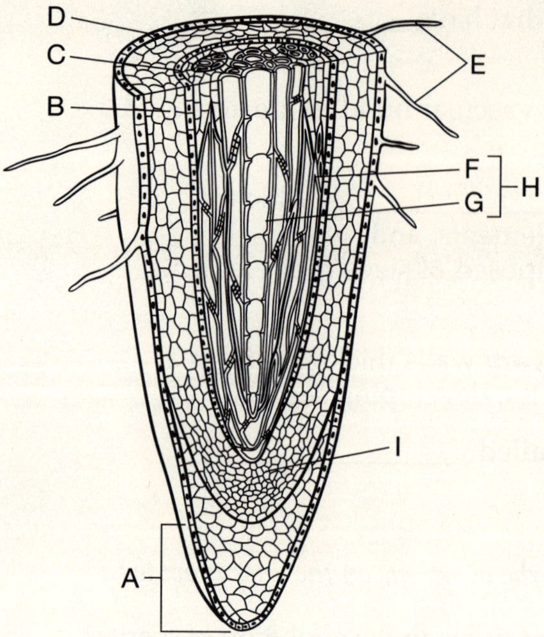

Figure 3

26. Interpreting Graphics What is the structure labeled C in Figure 3? What is its function?

____

____

____

27. Applying Concepts Where is the apical meristem in the root in Figure 3? What is its function?

____

____

28. Applying Concepts Does the root in Figure 3 come from a monocot or a dicot? Explain.

____

____

____

29. Interpreting Graphics Identify the locations of epidermal tissue, vascular tissue, and ground tissue in the root in Figure 3.

____

____

____

30. Inferring What plant hormones act on the root in Figure 3?

____

____

____

Name______________________________ Class________________ Date ____________

Multiple Choice

Write the letter that best answers the question or completes the statement on the line provided.

____ **1.** An animal is each of the following EXCEPT
a. multicellular.
b. autotrophic.
c. heterotrophic.
d. eukaryotic.

____ **2.** Animals that have no backbone, or vertebral column, are called
a. vertebrates.
b. ectoderms.
c. deuterostomes.
d. invertebrates.

____ **3.** When an animal's environment changes, sexual reproduction improves a species' ability to
a. react.
b. increase its numbers rapidly.
c. evolve.
d. produce genetically identical offspring.

____ **4.** A blastula is a stage in the development of an embryo that consists of a
a. solid ball of cells.
b. hollow ball of cells.
c. flat sheet of cells.
d. folded sheet of cells.

____ **5.** Which term is NOT associated with sponges?
a. choanocytes
b. medusa
c. spicules
d. gemmules

____ **6.** Which of the following best describes the feeding habits of sponges?
a. predators
b. filter feeders
c. parasites
d. detritivores

____ **7.** The benefit of a sponge's producing gemmules is that the gemmules
a. are a form of sexual reproduction.
b. defend the sponge against predators.
c. help to filter food for the sponge.
d. allow the sponge to survive difficult environmental conditions.

____ **8.** Which of the following animals has an asymmetrical body plan?
a. jellyfish
b. coral
c. sponge
d. none of the above

____ 9. Sponges benefit some marine animals by
a. providing a habitat.
b. eating diseased coral.
c. poisoning their predators.
d. supplying them with oxygen.

____10. Cnidocytes help a cnidarian survive by
a. storing food.
b. forming colonies.
c. paralyzing prey.
d. providing movement.

____11. The body symmetry of a cnidarian is
a. radial in the medusa stage and bilateral in the polyp stage.
b. radial in both the medusa and polyp stages.
c. bilateral in both the medusa and polyp stages.
d. bilateral in the medusa stage and radial in the polyp stage.

____12. In jellyfishes, the medusa stage reproduces
a. asexually.
b. sexually.
c. by budding.
d. by gemmules.

____13. Cnidarians in the class Anthozoa are distinguished by
a. having only colonial species.
b. having only a polyp stage.
c. reproducing only sexually.
d. obtaining food only from symbionts.

____14. The Portuguese man-of-war is a member of what class of cnidarian?
a. Hydrozoa
b. Scyphozoa
c. Anthozoa
d. Porifera

____15. Coral bleaching is a serious ecological threat because it
a. destroys the beauty of a coral reef.
b. causes corals to die.
c. pollutes the water.
d. causes coral to overpopulate the area.

Completion

Complete each statement on the line provided.

16. Animals that have a backbone, or vertebral column, are called ____________________.

17. The body mechanism in which the product or result of a process stops or limits the process is called ____________________.

18. The opening through which wastes leave a digestive tract is the ____________________.

19. The internal space of a cnidarian is called a(an) ____________________.

20. In a jellyfish, the ____________________ stage reproduces asexually.

Name______________________ Class______________ Date ____________

Short Answer

In complete sentences, write the answers to the questions on the lines provided.

21. What is the difference between a protostome and a deuterostome?

22. What are the important trends in animal evolution?

23. What is cephalization? What is an advantage of that characteristic?

24. What body form is typical of the scyphozoans? Describe this body form.

25. How are coral reefs formed?

Name________________________ Class________________ Date ____________

Using Science Skills

Use the diagram below to answer the following questions on the lines provided.

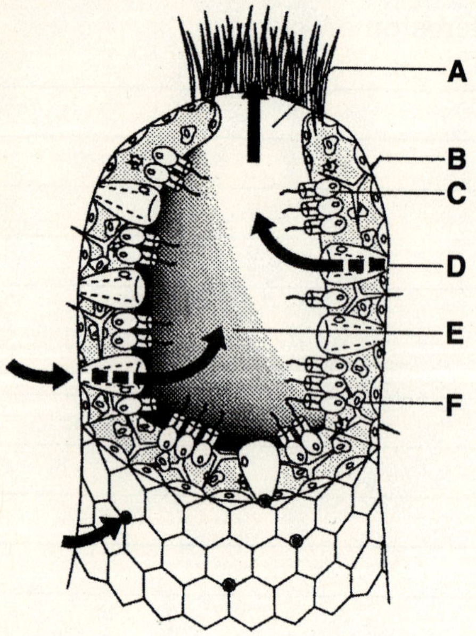

Figure 26-1

26. Inferring Why are sponges appropriately placed in the phylum Porifera? Use Figure 26-1 to explain.

____

____

____

____

____

____

27. Interpreting Graphics In Figure 26-1, which label refers to the osculum? What is this structure's function?

____

____

____

28. Interpreting Graphics Which label in Figure 26-1 represents an archaeocyte? What is the function of archaeocytes?

____

____

____

29. Applying Concepts In Figure 26-1, which label represents the structure that moves a steady current of water through a sponge? What is this structure called?

____

Name____________________ Class____________ Date __________

30. Applying Concepts What is the structure labeled B in Figure 26–1? How does this structure enable organisms within a sponge to carry out photosynthesis?

____

____

____

____

____

Essay

Write the answer to each question in the space provided.

31. How do the three germ layers form, and what does each germ layer give rise to in the adult organism?

32. Describe an animal with bilateral symmetry. In your description, use the terms *anterior, posterior, dorsal,* and *ventral.*

Name________________________________ Class____________________ Date ______________

33. Describe sereral advantages an animal receives from having a body cavity.

34. In sponges, how does asexual reproduction by budding differ from gemmule production?

35. Describe the feeding behavior of cnidarians.

Name______________________________ Class__________________ Date ______________

Multiple Choice

Write the letter that best answers the question or completes the statement on the line provided.

____ **1.** What are animals that have a backbone called?
a. invertebrates
b. heterotrophs
c. vertebrates
d. eukaryotes

____ **2.** Which of the following are trends in animal evolution as animals become more complex?
a. having a body cavity
b. bilateral body symmetry
c. high levels of cell specialization
d. all of the above

____ **3.** When an animal zygote undergoes a series of divisions to form a hollow ball of cells, it is called a(an)
a. blastula.
b. spicule.
c. ectoderm.
d. anus.

____ **4.** What is an animal called whose mouth is formed from the blastopore?
a. protostome
b. deuterstome
c. mesoderm
d. archaeocyte

____ **5.** An animal that has distinct left and right sides shows
a. radial symmetry.
b. segmentation.
c. several planes of symmetry.
d. bilateral symmetry.

____ **6.** Unlike plants, sponges are
a. heterotrophic.
b. autotrophic.
c. unicellular.
d. photosynthetic.

____ **7.** What are the specialized cells that use flagella to move a steady current of water through a sponge?
a. spicules
b. choanocytes
c. archaeocytes
d. cnidocytes

____ **8.** Sponges reproduce sexually through a process called
a. cephalization.
b. internal fertilization.
c. radial symmetry.
d. budding.

____ **9.** Some sponges are green because they
a. are more like plants than animals.
b. filter algae out of the water for food.
c. have photosynthetic organisms in their tissues.
d. are primary producers.

Name________________________________ Class____________________ Date ______________

_____**10.** Which of the following is a characteristic of cnidarians?

a. bilateral symmetry c. radial symmetry
b. cephalization d. internal fertilization

_____**11.** Cnidarians have two basic body types, a medusa and a(an)

a. larva. c. polyp.
b. tentacle. d. osculum.

_____**12.** The nerve cells of cnidarians make up a(an)

a. brain. c. hydrostatic skeleton.
b. ocellus. d. nerve net.

_____**13.** What is a digestive chamber with one opening called?

a. gastrovascular cavity c. blastula
b. central cavity d. hydrostatic skeleton

_____**14.** The class Scyphozoa contains

a. jellyfishes. c. corals.
b. hydras. d. sea anemones.

_____**15.** Coral reefs occur in areas where there are

a. high levels of sediment.
b. very deep waters.
c. high levels of light.
d. very cool waters.

Completion

Complete each statement on the line provided.

16. Animals that do not have backbones are called ____________________.

17. The body mechanism in which the product or result of a process stops or limits the process is called ____________________ inhibition.

18. All vertebrates are ____________________, which are animals whose anus is formed from the blastopore.

19. An immature stage of an organism that looks different from the adult form is a(an) ____________________.

20. The stinging cells of cnidarians are called ____________________.

Short Answer

In complete sentences, write the answers to the questions on the lines provided.

21. What are seven essential functions that animals carry out?

____

____

____

____

Name________________________ Class________________ Date ____________

22. What are the three germ layers that develop in most animal embryos?

____

____

____

23. What is cephalization?

____

____

____

24. What does the movement of water through the body provide for a sponge?

____

____

____

____

25. Describe the polyp form of cnidarians.

____

____

____

____

Using Science Skills

Use the diagram below to answer the following questions on the lines provided.

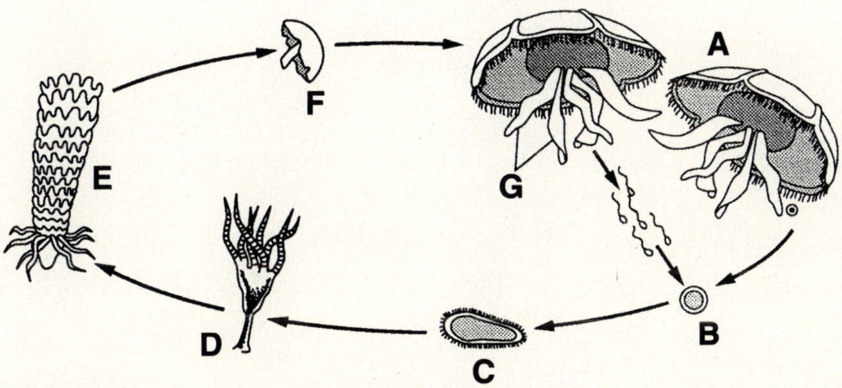

Figure 26-1

26. **Interpreting Graphics** In Figure 26-1, what letter shows a zygote, or fertilized egg?

____

Name______________________ Class______________ Date ____________

27. **Interpreting Graphics** In Figure 26-1, what is the structure labeled C?

____

____

28. **Applying Concepts** In Figure 26-1, is the structure labeled F produced sexually or asexually?

____

____

29. **Interpreting Graphics** Using Figure 26-1, determine which body form—polyp or medusa—reproduces sexually.

____

____

30. **Inferring** What evidence do you see in Figure 26-1 that this organism, *Aurelia*, is not a hydra?

____

____

____

____

____

____

Name______________________________ Class__________________ Date ______________

Multiple Choice

Write the letter that best answers the question or completes the statement on the line provided.

____ 1. Which of the following is NOT true about flatworms?
a. They have a fluid-filled body cavity called a coelom.
b. They are the simplest animals to have three embryonic germ layers.
c. They are acoelomates.
d. They are bilaterally symmetrical.

____ 2. In free-living flatworms, what organ pumps food into the digestive cavity?
a. coelom
b. ganglia
c. pharynx
d. flame cells

____ 3. Many flatworms can detect changes in the amount of light in their environment using groups of cells called
a. flame cells. c. ganglia.
b. nerve cords. d. eyespots.

____ 4. In the tapeworm, both male and female reproductive organs are contained in each mature
a. scolex. c. cyst.
b. proglottid. d. egg.

____ 5. In a pseudocoelom, mesoderm partially lines the
a. germ layer. c. blood vessels.
b. body cavity. d. pharynx.

____ 6. Which of the following is NOT true about roundworms?
a. They have a digestive system with two openings.
b. They have specialized tissues and organ systems.
c. They are segmented worms.
d. They have pseudocoeloms.

____ 7. What causes the disease called elephantiasis?
a. flukes c. hookworms
b. filarial worms d. ascarid worms

____ 8. Which of these animals has a true coelom?
a. filarial worm c. planarian
b. tapeworm d. leech

____ 9. In annelids, nitrogen-containing wastes are eliminated by
a. clitella. c. nephridia.
b. parapodia. d. gills.

Name________________________________ Class____________________ Date _______________

_____10. The muscular extension of a leech that penetrates the tissue of its host is the

a. septum. c. proboscis.
b. radula. d. ganglion.

_____11. The larvae of many marine annelids are ecologically important because they

a. poison coral reefs.
b. are eaten by fishes and other marine animals.
c. feed on earthworms.
d. none of the above

_____12. Which of the following is NOT true about mollusks?

a. They usually have an external or internal shell.
b. They are all filter feeders.
c. They are soft-bodied animals.
d. They have tube-shaped nephridia to remove ammonia from blood.

_____13. The tubelike structure through which water enters and leaves a mollusk's body is the

a. sinus. c. coelom.
b. siphon. d. mantle cavity.

_____14. The most active mollusks are the

a. gastropods. c. bivalves.
b. cephalopods. d. nudibranchs.

_____15. Which cephalopods have external shells?

a. nautiluses c. octopi
b. squid d. cuttlefish

Completion

Complete each statement on the line provided.

16. A coelom is a fluid-filled body cavity that is lined with tissue derived from the ____________________ .

17. A(an) ____________________ is an individual organism that has both male and female reproductive organs.

18. In an earthworm's digestive system, the ____________________ grinds food into small pieces.

19. Sandworms and bloodworms are members of the class of marine annelids called ____________________ .

20. A cephalopod's ____________________ is divided into tentacles, or arms.

Name________________________ Class________________ Date ____________

Short Answer

In complete sentences, write the answers to the questions on the lines provided.

21. How can washing vegetables before you eat them help prevent the severe malnutrition associated with ascarid worms?

22. Compare the body cavities of roundworms and annelids.

23. In what ways do earthworms improve soil quality?

24. How do most aquatic mollusks breathe?

25. What are the three major classes of mollusks? Provide three examples of organisms in each class.

Name______________________ Class______________ Date ____________

Using Science Skills

Use the diagram below to answer the following questions on the lines provided.

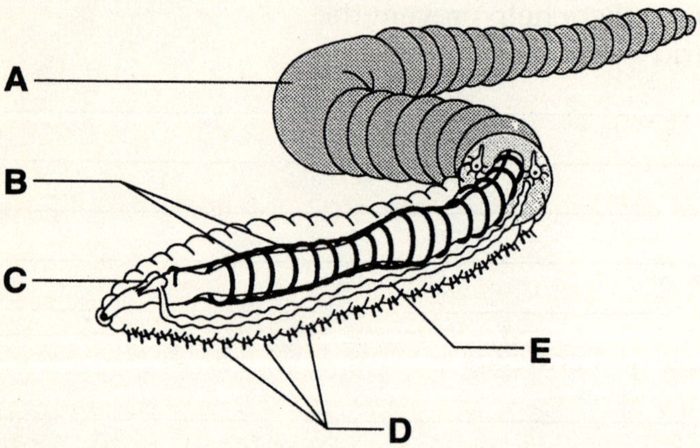

Figure 27-1

26. **Applying Concepts** What animal is shown in Figure 27-1, and in what class and phylum does it belong?

27. **Interpreting Graphics** In Figure 27-1, what is the structure labeled A? What is the function of this structure?

28. **Interpreting Graphics** In Figure 27-1, what is the structure labeled B? What is its function?

29. **Applying Concepts** Identify the structures labeled D in Figure 27-1. How are these structures useful to a land-dwelling animal?

Name______________________________ Class__________________ Date ______________

30. **Inferring** Identify structure C in Figure 27-1, and infer its relationship to structure E.

____

____

____

____

Essay

Write the answer to each question in the space provided.

31. Describe the life cycle of the blood fluke.

32. Compare excretion in flatworms and annelids.

Name________________________ Class________________ Date______________

33. What is a trochophore? Why is it important in inferring evolutionary relationships?

34. How do filter-feeding bivalves obtain their food?

35. What is the difference between an open circulatory system and a closed circulatory system? Provide an example of a group that has each type of system.

Name________________________________ Class____________________ Date ______________

Multiple Choice

Write the letter that best answers the question or completes the statement on the line provided.

____ **1.** Asexual reproduction in free-living flatworms takes place by the process known as

a. budding. c. fertilization.
b. fission. d. hermaphroditism.

____ **2.** Some flatworms have clusters of nerve cells that control the nervous system. Each cluster is called a(an)

a. ganglion. c. eyespot.
b. brain. d. flame cell.

____ **3.** Free-living flatworms, most of which live in marine environments or fresh water, are

a. flukes. c. tapeworms.
b. turbellarians. d. roundworms.

____ **4.** An intermediate host is an organism in which a parasite

a. reproduces asexually.
b. reproduces sexually.
c. causes tissue decay.
d. clogs blood vessels.

____ **5.** Which of the following is a parasitic roundworm?

a. *Trichinella*
b. *Schistosoma*
c. a tapeworm
d. a turbellarian

____ **6.** Roundworms have a digestive system

a. with two openings.
b. with one opening.
c. within a true coelom.
d. none of the above

____ **7.** To move, roundworms use their

a. proglottids.
b. hydrostatic skeleton.
c. tentacles.
d. pharynx.

____ **8.** In earthworms, food is ground into small pieces in the

a. crop. c. pharynx.
b. gizzard. d. esophagus.

____ 9. The body of an annelid has
a. a backbone. c. segments.
b. an external shell. d. stinging tentacles.

____10. A type of annelid that is an external parasite is the
a. tapeworm. c. leech.
b. polychaete. d. earthworm.

____11. The bristlelike structures on some annelids' bodies are called
a. setae. c. nephridia.
b. suckers. d. ganglia.

____12. Earthworm tunnels provide passageways for
a. leeches. c. plant roots and water.
b. polychaetes. d. planarians.

____13. The thin layer of tissue that covers a mollusk's body is called the
a. mantle. c. visceral mass.
b. foot. d. shell.

____14. A pond snail is an example of a(an)
a. gastropod. c. roundworm.
b. flatworm. d. annelid.

____15. Filter-feeding bivalves can be used to monitor the environmental health of a habitat because
a. the bivalves reproduce rapidly in polluted water.
b. the bivalves concentrate pollutants and microorganisms in their tissues.
c. the bivalves live near deep-sea vents.
d. some bivalves never get cancer.

Completion

Complete each statement on the line provided.

16. The simplest animals to have three embryonic germ layers, bilateral symmetry, and cephalization are the soft worms called ____________________.

17. Many free-living roundworms are ____________________, which are animals that eat other animals.

18. Annelids have a(an) ____________________ circulatory system in which blood is contained in a network of blood vessels.

19. The type of embryonic tissue that lines a true coelom is called ____________________.

20. The foot of most cephalopods is divided into eight or more arms called ____________________.

Name________________________________ Class____________________ Date ______________

Short Answer

In complete sentences, write the answers to the questions on the lines provided.

21. What are the characteristics of flatworms?

22. What role do freshwater snails play in the life cycle of the blood fluke *Schistosoma?*

23. Describe the tube-within-a-tube body plan of roundworms.

24. How do hookworms infect humans?

25. What are the four parts of the body plan of most mollusks?

Name______________________________ Class__________________ Date______________

Using Science Skills

Use the diagram below to answer the following questions on the lines provided.

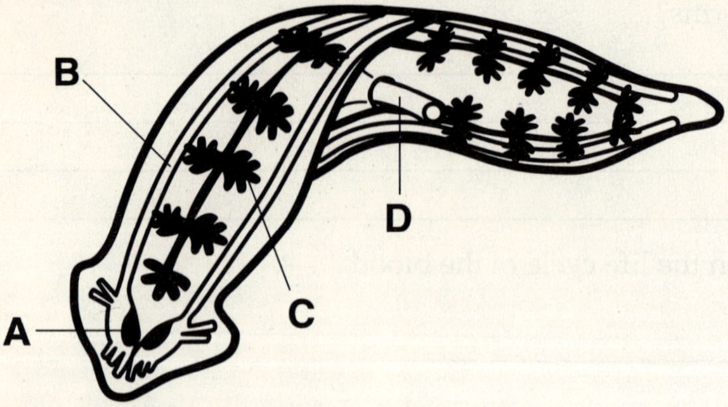

Figure 27-1

26. Applying Concepts To which phylum does the animal shown in Figure 27-1 belong?

____

____

27. Interpreting Graphics Which structure shown in Figure 27-1 (A, B, C, or D) takes food into the animal's body cavity? What is this structure called?

____

28. Interpreting Graphics Which label in Figure 27-1 points to a nerve cord? At the anterior end of the worm, what structures connect to nerve cords?

____

____

29. Interpreting Graphics What structure shown in Figure 27-1 allows the animal to sense light? What is this structure called?

____

30. Inferring The animal shown in Figure 27-1 uses diffusion to transport oxygen, nutrients, and wastes through its body. How does the body shape of the animal enable it to depend upon diffusion for transport?

____

____

____

Name________________________ Class________________ Date ____________

Multiple Choice

Write the letter that best answers the question or completes the statement on the line provided.

____ **1.** The appendages of arthropods are
a. found only on the head.
b. hard and immovable.
c. jointed and extend from the body wall.
d. divided into six branches.

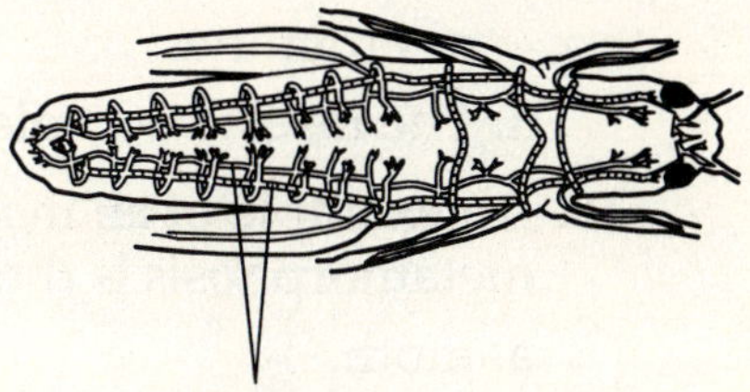

Figure 28-1

____ **2.** The structures labeled X in Figure 28-1 are filled with
a. water. c. air.
b. blood. d. nitrogenous wastes.

____ **3.** All of the following are true about arthropods EXCEPT that
a. they have a closed circulatory system.
b. they have an exoskeleton made of chitin.
c. they include herbivores, carnivores, and omnivores.
d. they move using muscles controlled by the nervous system.

____ **4.** An arthropod is vulnerable to predators during the molting period because
a. it must come out of hiding to molt.
b. its new exoskeleton is soft.
c. molting cannot occur without the assistance of predators.
d. predators are more numerous during this period.

____ **5.** The function of mandibles is to
a. bite and grind food.
b. sense the environment.
c. propel an arthropod when it swims.
d. support an arthropod when it walks.

____ **6.** An example of a chelicerate is a
a. lobster. c. crayfish.
b. centipede. d. spider.

____ **7.** Spiders feed by
a. swallowing their prey whole.
b. biting off and swallowing pieces of their prey.
c. sucking up prey tissues that have been liquefied by enzymes.
d. sipping nectar through a tubelike mouthpart.

____ **8.** An insect can detect minute movements in its environment by using its compound eyes and its
a. Malpighian tubules. c. chelicerae.
b. tracheal tubes. d. sensory hairs.

____ 9. Which of the following explains why flying has been beneficial to insects?
a. Nymphs can escape most predators.
b. Insects can disperse long distances and colonize a variety of habitats.
c. Insect societies can communicate without using pheromones.
d. The insect body plan can function with only two sections.

____10. An example of an insect that undergoes incomplete metamorphosis is the
a. moth. c. beetle.
b. bee. d. chinch bug.

____11. Which of the following is NOT true about insect societies?
a. All species use the same "language" of visual and chemical signals.
b. Individuals may be specialized to perform particular roles.
c. Every individual in a society works for the benefit of the whole group.
d. Each caste has a body form specialized for its role.

____12. In most species of echinoderms, respiration occurs mainly in the
a. tube feet. c. ring canal.
b. madreporite. d. radial canals.

____13. The skeleton of an echinoderm is an
a. exoskeleton made of calcium carbonate.
b. exoskeleton made of chitin.
c. endoskeleton made of calcium carbonate.
d. endoskeleton made of chitin.

____14. The water vascular system of echinoderms is involved with each of the following body functions EXCEPT
a. respiration. c. movement.
b. circulation. d. reproduction.

____15. In the Great Barrier Reef of Australia, extensive areas of coral have been destroyed by the crown of thorns, which is a type of
a. sea cucumber. c. sea star.
b. sea urchin. d. sea lily.

Completion

Complete each statement on the line provided.

16. In arthropods, ganglia are located along a ventral ____________________.

17. If an arthropod has two pairs of antennae and two or three body sections, it is classified in the group called ____________________.

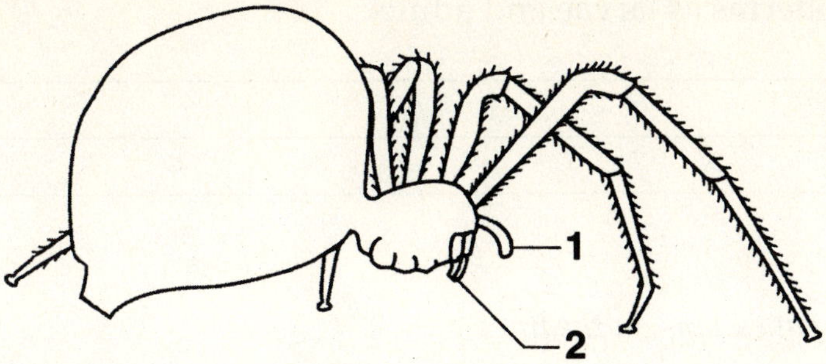

Figure 28-2

18. In the arthropod in Figure 28-2, the appendages labeled 1 are called ____________________.

19. In the arthropod in Figure 28-2, the appendages labeled 2, which contain mouthparts, are called ____________________.

20. In an echinoderm, oxygen is carried throughout the body by the ____________________ system.

Short Answer

In complete sentences, write the answers to the questions on the lines provided.

21. Name the three sections of an insect's body and the two sections of a chelicerate's body.

____

____

22. Identify the respiratory organs used by spiders and by horseshoe crabs.

____

____

23. Name two ways in which insects are beneficial to humans.

____

____

____

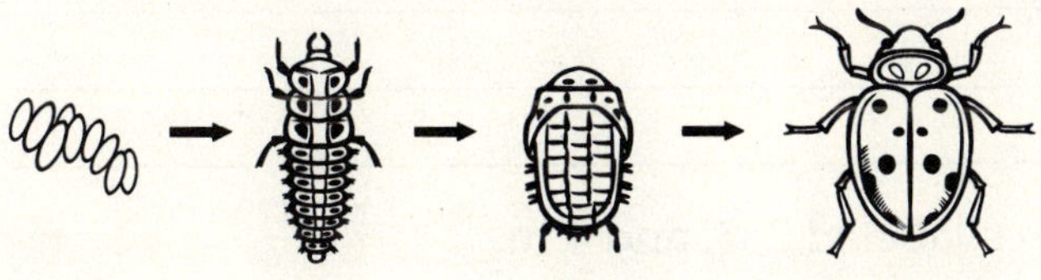

1 2 3 4

Figure 28-3

24. Figure 28-3 shows the development of a ladybug. Identify the stages labeled 1 through 4.

____

____

Name______________________ Class__________________ Date ____________

25. Contrast the body symmetry of echinoderms as larvae and adults.

____

____

____

Using Science Skills

Use the diagram below to answer the following questions on the lines provided.

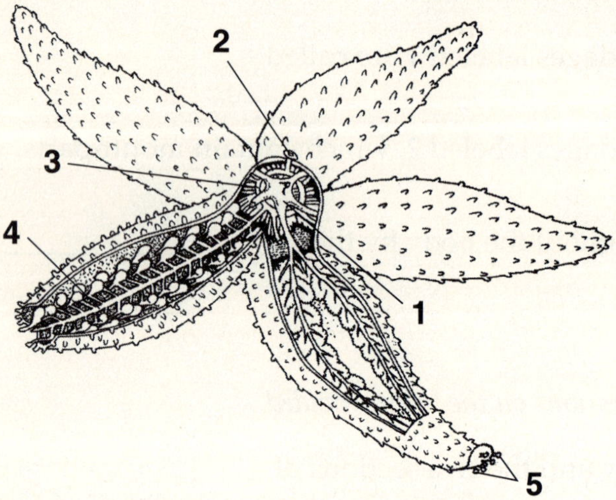

Figure 28-4

26. **Interpreting Graphics** Name the type of body symmetry shown by the animal in Figure 28-4. What features visible in the figure demonstrate this type of symmetry?

____

____

____

____

27. **Applying Concepts** What is the structure labeled 1 in Figure 28-4? Explain how the animal uses this structure when it feeds.

____

____

____

28. **Interpreting Graphics** Identify the structures labeled 2, 3, and 4 in Figure 28-4. What do these three structures have in common?

____

____

____

____

Name______________________________ Class__________________ Date ______________

29. **Applying Concepts** What are the structures labeled 5 in Figure 28-4? Describe the role of these structures in movement.

30. **Applying Concepts** The structures labeled 5 in Figure 28-4 help the animal obtain food. How do they do this?

Essay

Write the answer to each question in the space provided.

31. Describe the events that take place when an arthropod molts.

32. Describe the external structural differences between spiders and insects.

33. Compare and contrast centipedes and millipedes.

34. Explain how a worker honeybee communicates the location of food that is more than 50 meters from the hive.

35. Describe reproduction and development in sea stars.

Name______________________ Class________________ Date ____________

Multiple Choice

Write the letter that best answers the question or completes the statement on the line provided.

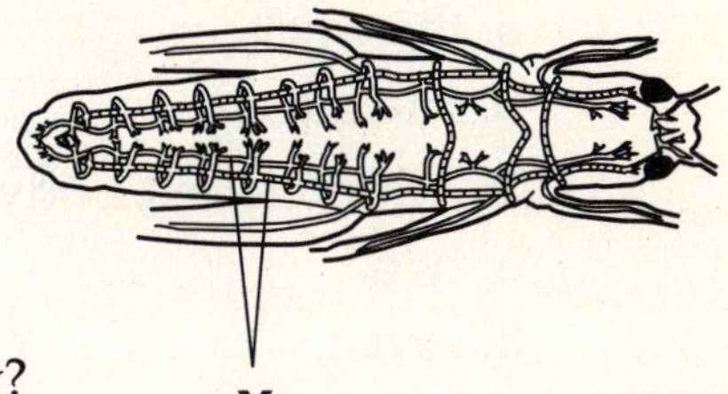

Figure 28-1

____ **1.** Figure 28-1 shows the respiratory system of a grasshopper. The structures labeled X are called
a. tracheal tubes. c. book lungs.
b. Malpighian tubules. d. book gills.

____ **2.** Which of the following habitats do arthropods occupy?
a. the sea
b. the land
c. the air
d. all of the above

____ **3.** What does molting enable arthropods to do?
a. to breathe c. to grow
b. to reproduce d. to eat

____ **4.** Arthropods are classified based on the number and structure of their
a. eyes and wings.
b. body segments and appendages.
c. muscles and bones.
d. gills and hearts.

____ **5.** Insects are classified as
a. crustaceans. c. uniramians.
b. arachnids. d. chelicerates.

____ **6.** How many pairs of legs are there on most body segments of a centipede?
a. one c. five
b. two d. six

____ **7.** The compound eyes of insects
a. are made of many lenses.
b. are located on the thorax.
c. produce an image that is more detailed than what humans see.
d. are not very good at detecting movement.

____ **8.** The body of an insect is divided into a
a. head and a thorax.
b. head and a cephalothorax.
c. head, a thorax, and an abdomen.
d. cephalothorax and an abdomen.

Name_______________________________ Class___________________ Date ______________

____ 9. Which of the following is NOT a stage of complete metamorphosis?

a. nymph
b. egg
c. larva
d. pupa

____10. In most ant societies, egg laying is performed by

a. the workers.
b. one queen.
c. hundreds of queens.
d. reproductive males.

____11. In echinoderms, body parts usually occur in multiples of

a. two.
b. seven.
c. three.
d. five.

____12. In an echinoderm, the structure that operates like a living suction cup is the

a. madreporite.
b. tube foot.
c. stomach.
d. nerve ring.

____13. Which structure is part of an echinoderm's water vascular system?

a. skin gill
b. anus
c. madreporite
d. stomach

____14. The echinoderms that look like warty, moving pickles are

a. sea urchins.
b. sea cucumbers.
c. sea stars.
d. feather stars.

____15. Two echinoderms that are in the same class are

a. brittle stars and sea stars.
b. sea stars and sea cucumbers.
c. sea cucumbers and sea urchins.
d. sea urchins and sand dollars.

Completion

Complete each statement on the line provided.

16. A typical primitive arthropod had many identical ____________________, each with a pair of appendages.

17. An arthropod's exoskeleton is made of____________________.

18. Chelicerae contain ____________________ that stab and paralyze prey.

19. Insects have ____________________ pairs of legs.

20. Within a society, particular tasks or roles are performed by groups of individuals called ____________________.

Name______________________ Class______________ Date__________

Short Answer

In complete sentences, write the answers to the questions on the lines provided.

21. List three features that are present in the body plan of arthropods.

22. Suppose you observed an arthropod whose body was divided into a cephalothorax and an abdomen. Which of the three major groups of arthropods could this specimen belong to?

23. How are mites and ticks harmful to humans?

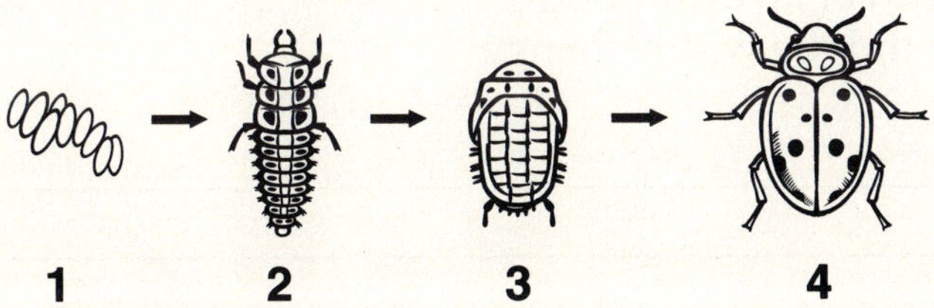

Figure 28-2

24. What type of metamorphosis is shown in Figure 28-2?

25. What will happen if a sea star is pulled into pieces?

Name________________________ Class________________ Date ____________

Using Science Skills

Use the chart below to answer the following questions on the lines provided.

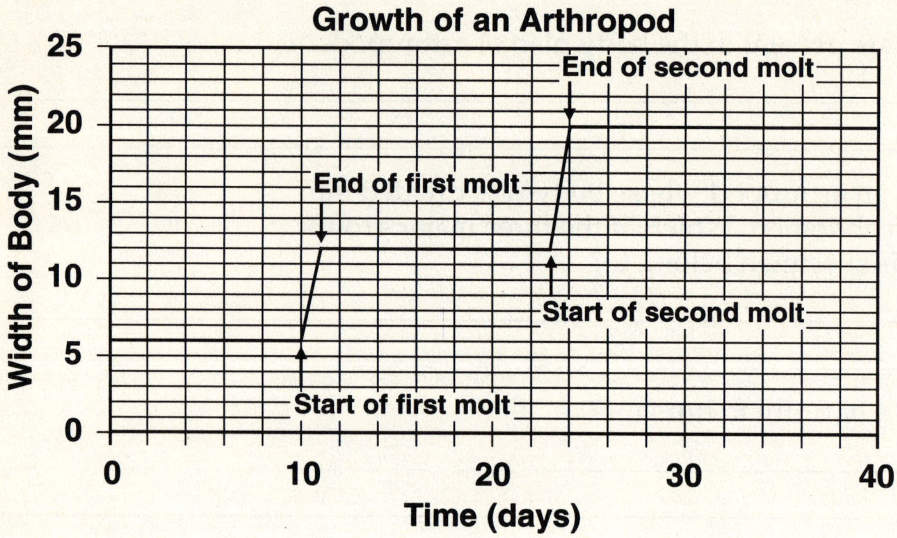

Figure 28-3

26. Calculating Figure 28-3 shows how the width of an arthropod's body changed over time. By what amounts did the width change during the first and second molts?

____

____

27. Applying Concepts Explain why the width of the arthropod's body changed during each molt in Figure 28-3.

____

____

28. Using Tables and Graphs In Figure 28-3, what was the width of the arthropod's body between the end of the first molt and the start of the second molt?

____

29. Applying Concepts Explain why the width of the arthropod's body did not change between the end of the first molt and the start of the second molt in Figure 28-3.

____

____

____

30. Using Tables and Graphs In Figure 28-3, how long did it take for the width of the arthropod's body to increase from 6 mm to 20 mm?

____

____

Name______________________________ Class__________________ Date ______________

Multiple Choice

Write the letter that best answers the question or completes the statement on the line provided.

____ **1.** Which of the following did NOT occur during the Cambrian Explosion?

a. Animals acquired specialized cells, tissues, and organ systems.

b. There was an extraordinary growth in animal diversity.

c. Animals evolved simpler body plans.

d. Animal appendages became specialized for a variety of functions.

____ **2.** One characteristic that made early animals different from all animals of today was their

a. habitat.
c. body plan.
b. body segmentation.
d. bilateral symmetry.

____ **3.** Animals of the Cambrian Period typically had all of the following EXCEPT

a. body symmetry.
c. some type of skeleton.
b. segmentation.
d. a backbone.

____ **4.** The classification of an animal as a deuterostome or a protostome is based on

a. its body symmetry.

b. whether or not it has a coelom.

c. what happens to the blastopore.

d. the number of germ layers it has.

____ **5.** An acoelomate is an animal that has

a. a body cavity lined with endoderm and ectoderm.

b. a body cavity partially lined with mesoderm.

c. a body cavity completely lined with mesoderm.

d. no body cavity between the germ layers.

____ **6.** Which invertebrates exhibit radial symmetry?

a. cnidarians and echinoderms

b. sponges and flatworms

c. roundworms and annelids

d. mollusks and arthropods

____ **7.** Cephalization refers to the

a. division of the body into upper and lower sides.

b. concentration of sense organs and nerve cells in the front of the body.

c. joining together of specialized cells to form tissues.

d. formation of a body cavity between the germ layers.

Name________________________ Class__________________ Date ____________

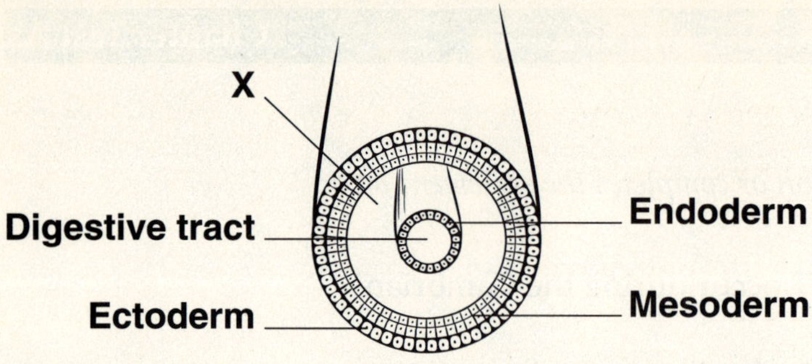

Figure 29-1

____ **8.** One animal that has a body construction like that shown in Figure 29-1 is a

a. flatworm.
c. sponge.
b. roundworm.
d. jellyfish.

____ **9.** In Figure 29-1, the space labeled X is called a

a. protostome.
b. deuterostome.
c. coelom.
d. pseudocoelom.

____**10.** The distinguishing feature of a closed circulatory system is that

a. it does not include a heart.
b. blood is contained within vessels that extend throughout the body.
c. blood is kept at low pressure.
d. blood is circulated less efficiently than in an open circulatory system.

____**11.** Which structures are NOT part of an excretory system?

a. flame cells
c. Malpighian tubules
b. spiracles
d. nephridia

____**12.** Which of the following best describes uric acid?

a. more toxic than ammonia, leaves the body through excretory pores
b. more toxic than ammonia, leaves the body through the rectum
c. less toxic than ammonia, leaves the body through excretory pores
d. less toxic than ammonia, leaves the body through the rectum

____**13.** An example of an invertebrate with a hydrostatic skeleton is a(an)

a. spider.
c. sea star.
b. sponge.
d. earthworm.

Name______________________ Class________________ Date ____________

____**14.** An endoskeleton is a

a. shell of a mollusk.
b. fluid-filled body cavity that supports the muscles.
c. structural support located inside the body.
d. hard body covering made of chitin.

____**15.** Which statement refers to sexual reproduction?

a. All offspring are genetically identical to the parent.
b. Offspring are produced from the fusion of male and female gametes.
c. An organism breaks into pieces that grow into new individuals.
d. New individuals are produced from outgrowths of the parent's body wall.

Completion

Complete each statement on the line provided.

16. Over the course of evolution, ________________ allowed animals to increase in body size with a minimum of new genetic material.

17. Echinoderms differ from cnidarians in that echinoderms develop from three embryonic ________________ layers.

18. If an invertebrate has gills, it most likely lives in a(an) ________________ environment.

19. The function of the structure labeled X in Figure 29-2 is to rid the body of excess ________________ .

20. If an animal moves by contracting muscles that surround a body cavity filled with fluid, the animal has a(an) ________________ skeleton.

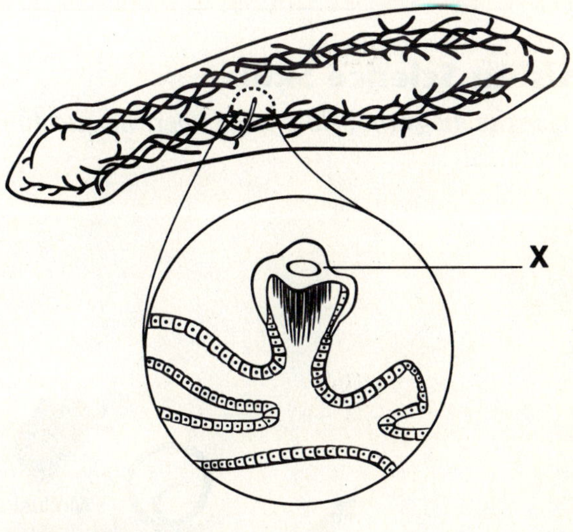

Figure 29-2

Short Answer

In complete sentences, write the answers to the questions on the lines provided.

21. Why are animal fossils more abundant from the Cambrian Period than from earlier periods?

____

____

____

22. Name three invertebrate phyla with bilateral symmetry.

____

____

23. Explain the advantage of cephalization.

____

____

____

____

24. Why must the surfaces of respiratory systems be moist?

____

____

____

____

25. Identify each circulatory system in Figure 29-3 as open or closed.

____

____

____

____

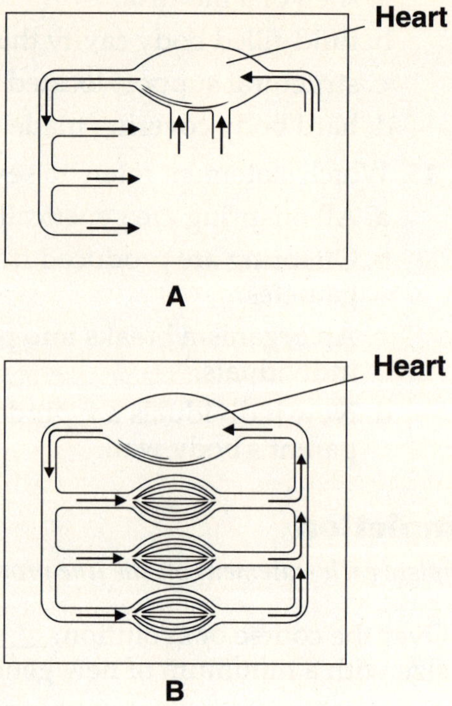

Figure 29-3

Using Science Skills

Use the diagram below to answer the questions on page 361 on the lines provided.

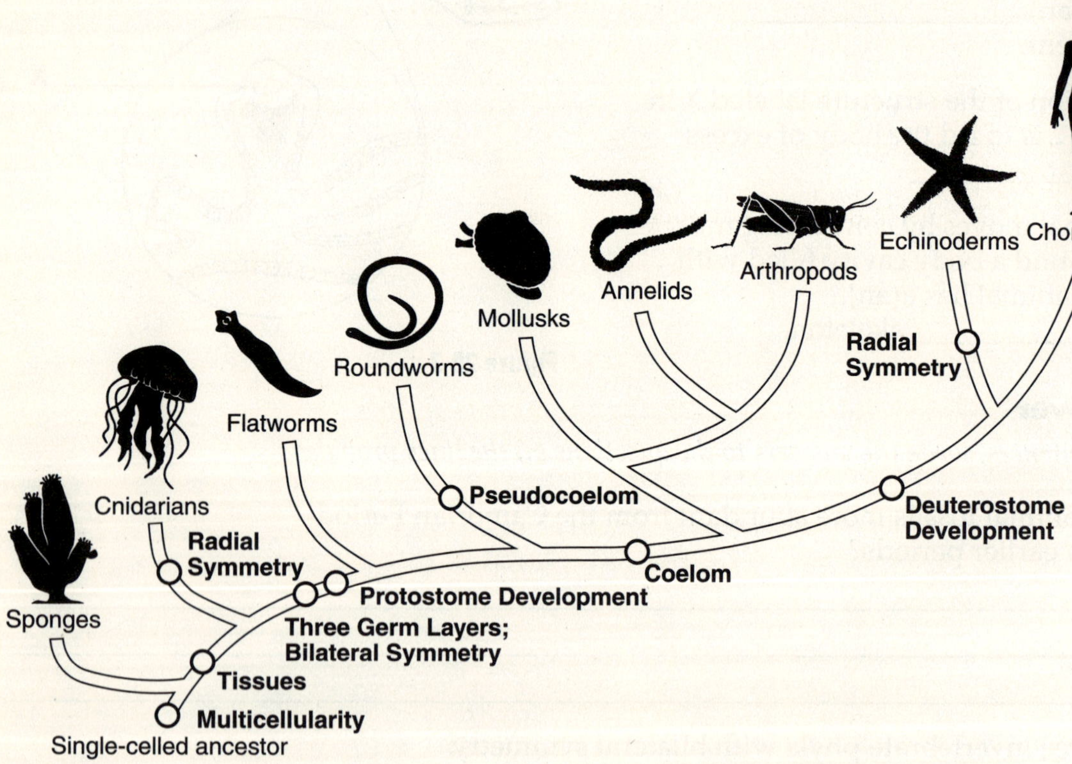

Figure 29-4

26. **Interpreting Graphics** Figure 29-4 shows one theory about the evolutionary relationships among groups of living invertebrates and chordates. Which feature evolved twice during the course of invertebrate evolution?

____

27. **Interpreting Graphics** According to Figure 29-4, which group(s) of invertebrates develop from three germ layers?

____

____

28. **Interpreting Graphics** Look at Figure 29-4. Which group(s) of invertebrates do NOT have tissues?

____

29. **Interpreting Graphics** Use Figure 29-4 to determine to which two groups of invertebrates annelids are most closely related.

____

30. **Interpreting Graphics** Which of the two types of development shown in Figure 29-4 do mollusks undergo?

____

Essay

Write the answer to each question in the space provided.

31. Describe the changes in internal specialization that occurred during the evolution of animals.

32. Compare and contrast cnidarians, mollusks, and echinoderms in terms of body symmetry, presence of a body cavity, and cephalization.

33. To remove nitrogenous wastes, some terrestrial invertebrates convert ammonia to urea. Why is that advantageous for these invertebrates?

34. Describe the organization of nervous systems in cnidarians, flatworms, and cephalopod mollusks.

35. Contrast hydrostatic skeletons, exoskeletons, and endoskeletons in terms of their structures.

Name________________________ Class ________________ Date ____________

Multiple Choice

Write the letter that best answers the question or completes the statement on the line provided.

____ **1.** The diversity of invertebrate phyla underwent its greatest increase
- a. before the Cambrian Period.
- b. during the Cambrian Period.
- c. after the Cambrian Period.
- d. both before and after the Cambrian Period.

____ **2.** Which of the following invertebrates are deuterostomes?
- a. worms
- b. arthropods
- c. mollusks
- d. echinoderms

____ **3.** In a protostome, the blastopore becomes a(an)
- a. mouth.
- b. anus.
- c. zygote.
- d. blastula.

____ **4.** A body cavity that forms between the germ layers is called a(an)
- a. coelom.
- b. blastopore.
- c. mesoderm.
- d. ectoderm.

____ **5.** Some type of body symmetry is found in all invertebrates EXCEPT
- a. cnidarians.
- b. mollusks.
- c. sponges.
- d. flatworms.

____ **6.** During development, a zygote divides repeatedly to form a(an)
- a. endoderm.
- b. endoskeleton.
- c. blastula.
- d. germ layer.

____ **7.** Invertebrates that break down their food through intracellular digestion include
- a. annelids.
- b. mollusks.
- c. arthropods.
- d. sponges.

____ **8.** Which invertebrate has a gastrovascular cavity?
- a. arthropod
- b. mollusk
- c. cnidarian
- d. roundworm

Name______________________________ Class____________________ Date ______________

____ **9.** A true digestive tract is found in
a. annelids. c. cnidarians.
b. sponges. d. flatworms.

____**10.** Gases diffuse most efficiently across a respiratory membrane if the membrane is
a. thick and dry. c. thick and moist.
b. thin and dry. d. thin and moist.

____**11.** In insects, gas exchange takes place through a network of
a. tracheal tubes. c. book lungs.
b. mantle cavities. d. blood vessels.

____**12.** In an open circulatory system, blood
a. never leaves the heart.
b. does not come in direct contact with the tissues.
c. is always contained within a system of blood vessels.
d. is pumped through a system of sinuses.

____**13.** An example of an animal with an open circulatory system is a(an)
a. sponge. c. arthropod.
b. cnidarian. d. annelid.

____**14.** Which of the following is a function of an excretory system?
a. eliminating nitrogenous wastes from the body
b. exchanging oxygen and carbon dioxide with the environment
c. gathering and processing information from the environment
d. obtaining and digesting food

____**15.** The eyespots of flatworms can
a. detect the presence of light.
b. detect motion.
c. detect color.
d. form images.

Completion

Complete each statement on the line provided.

16. ____________________ fossils are tracks and burrows made by soft-bodied animals whose bodies were not fossilized.

17. A body cavity lined partially with mesoderm is called a(an) ____________________.

18. The concentration of sense organs and nerve cells in the front of the body is called ____________________.

Name________________________ Class________________ Date ____________

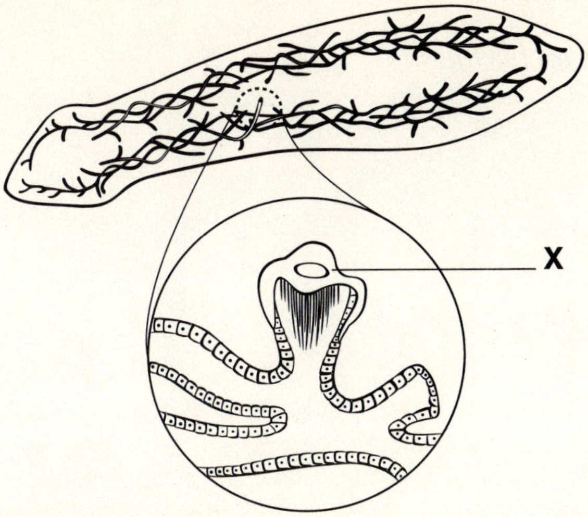

Figure 29-1

19. In Figure 29-1, the structure labeled X is called a(an) ________________ .

20. The simplest nervous systems are called ________________ .

Short Answer

In complete sentences, write the answers to the questions on the lines provided.

21. What is a blastopore?

____

____

22. What is the definition of a coelomate?

____

____

____

23. What is intracellular digestion?

____

____

____

24. What are Malpighian tubules?

____

____

____

25. Why must all animals eliminate ammonia from their bodies?

____

____

____

Name______________________________ Class____________________ Date ______________

Using Science Skills

Use the diagram below to answer the following questions on the lines provided.

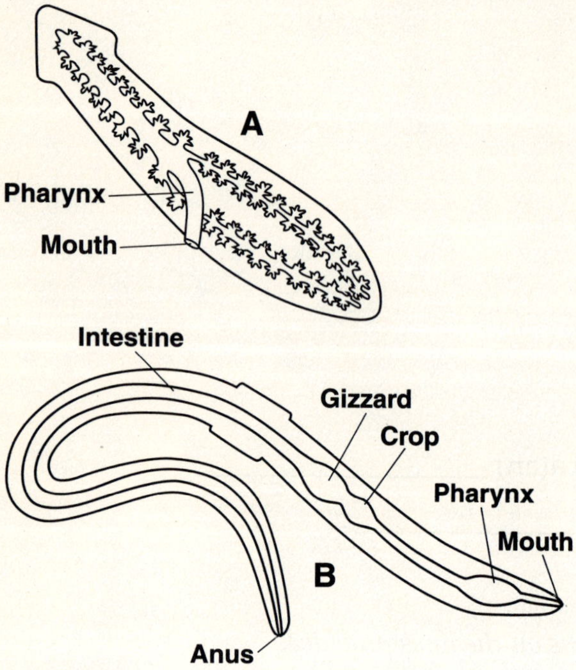

Figure 29-2

26. **Interpreting Graphics** Figure 29-2 shows two main types of digestive systems found in invertebrates. Identify the type of digestive system shown in A.

____

27. **Classifying** Name two phyla of invertebrates that have the type of digestive system shown in A in Figure 29-2.

____

28. **Interpreting Graphics** Identify the type of digestive system shown in B in Figure 29-2.

____

29. **Classifying** Name two phyla of invertebrates that have the type of digestive system shown in B in Figure 29-2.

____

30. **Applying Concepts** Describe the passage of food through the digestive system shown in B in Figure 29-2.

____

____

____

____

Multiple Choice

Write the letter that best answers the question or completes the statement on the line provided.

____ **1.** Each of the following is one of the essential functions animals carry out EXCEPT

a. respiration. c. cephalization.

b. excretion. d. circulation.

____ **2.** Which of the following is NOT a reason why sponges are classified as animals?

a. Sponges are multicellular.

b. Sponges have no cell walls.

c. Sponges are heterotrophic.

d. Sponges are asymmetrical.

____ **3.** What are the simplest animals to have body symmetry and specialized cells?

a. flatworms c. trilobites

b. sponges d. cnidarians

____ **4.** Which of the following is true about corals?

a. Corals have only the medusa stage in their life cycles.

b. Coral polyps secrete an underlying skeleton of calcium carbonate.

c. The polyp form of corals is restricted to a small larval stage.

d. One coral polyp forms a balloonlike float that keeps the entire colony afloat.

____ **5.** Which of the following is true about flatworms?

a. Flatworms are worms with segmented bodies.

b. Most free-living flatworms reproduce asexually.

c. All flatworms rely on diffusion for some essential body functions.

d. Flatworms are unsegmented worms with pseudocoeloms.

____ **6.** What kind of organism causes trichinosis?

a. flatworm c. roundworm

b. annelid d. trochophore

____ **7.** Which of the following is NOT true about earthworms?

a. Their tunnels allow for the growth of oxygen-requiring soil bacteria.

b. Earthworms have a pseudocoelom.

c. Earthworms have a digestive tract that includes a mouth and an anus.

d. Earthworms are hermaphrodites that reproduce sexually.

_____ 8. Which of the following describes a mollusk mantle?
   a. a thin layer of tissue that covers most of the body
   b. a system of internal tubes that opens to the outside through a sievelike structure
   c. the segment that contains the brain, eyes, mouth, and legs
   d. the section that consists of the internal organs

_____ 9. All arthropods have
   a. pedipalps.
   b. a madreporite.
   c. jointed appendages.
   d. a carapace.

_____10. Which of the following is NOT true about insects?
   a. Insects have a body divided into three parts.
   b. Insects have four pairs of walking legs.
   c. Insects have chemical receptors for taste and smell on their mouthparts.
   d. Insect legs are attached to the thorax.

_____11. What are specific chemical receptors that insects use that affect the behavior or development of other individuals of the same species?
   a. pheromones
   b. chitin
   c. chelicerae
   d. spiracles

_____12. Which of the following is NOT true about echinoderms?
   a. Most adults exhibit five-part radial symmetry.
   b. The water vascular system carries out many essential body functions.
   c. Echinoderms have an internal skeleton.
   d. The central ring is called the cephalothorax.

_____13. The Burgess Shale animals lived during the
   a. Ediacaran Period.
   b. Trilobite Period.
   c. Canadian Period.
   d. Cambrian Period.

_____14. Which of the following do NOT exhibit bilateral symmetry?
   a. roundworms
   b. mollusks
   c. cnidarians
   d. arthropods

_____15. Which of the following is NOT true about complex animals?
   a. Most move blood through their bodies using one or more hearts.
   b. Most have decentralized nervous systems.
   c. Most have a true coelom.
   d. Most use extracellular digestion.

Name______________________________ Class________________ Date ____________

Completion

Complete each statement on the line provided.

16. An animal without a backbone is called a(an)______________.

17. Asexual reproduction in flatworms is by a process called__________________.

18. In a bivalve, water enters and leaves the body through a(an) __________________, a tubelike structure.

19. The process in which an arthropod sheds its exoskeleton and manufactures a larger one to take its place is ________________.

20. The concentration of sense organs and nerve cells in the front of the body is called ______________________________________.

Short Answer

In complete sentences, write the answers to the questions on the lines provided.

21. What are the important evolutionary trends that have occurred in animals as they've become more complex over time?

____

____

____

22. Describe the two stages in the life cycle of a cnidarian.

____

____

____

23. What is the difference between internal fertilization and external fertilization?

____

____

24. What are three subphylums of Arthropoda? For each, name two example organisms or groups.

____

____

25. Do echinoderms have a closer evolutionary relationship to mollusks or to chordates? Explain.

____

____

____

Name________________________ Class________________ Date____________

Using Science Skills

Use the diagram below to answer the following questions on the lines provided.

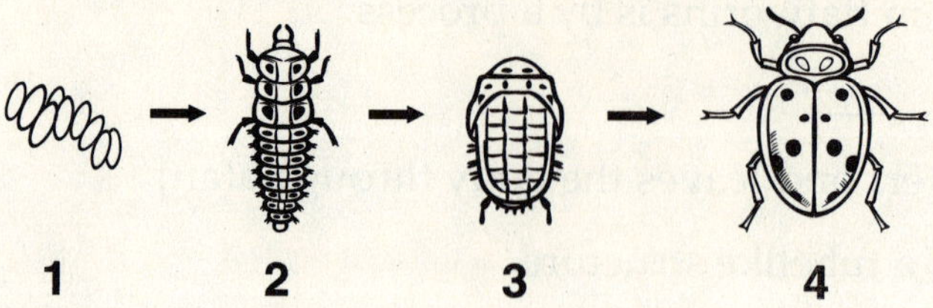

Figure 1

26. Interpreting Graphics What is the process called that is illustrated in Figure 1? Name three examples of organisms in which the process occurs.

27. Applying Concepts What does drawing 1 of Figure 1 represent? Describe the process in which these objects were formed.

28. Applying Concepts What does drawing 2 of Figure 1 represent? How is it different from an adult?

29. Applying Concepts What does drawing 3 of Figure 1 represent? Describe what occurs at this stage.

30. Comparing and Contrasting How is the process shown in Figure 1 like the other form of metamorphosis? How is it different?

Name_______________ Class_______________ Date_______________

Essay

Write the answer to each question in the space provided.

31. Compare the formation of a protostome and a deuterostome, and name the main groups of invertebrates that each type of embryological development includes.

32. What is the difference between a primary host and an intermediate host? Give an example of each.

33. How are modern arthropods different from primitive arthropods of hundreds of millions of years ago? Describe the trends in the evolution of arthropods.

34. Describe three types of skeletons found in invertebrates.

35. Describe how terrestrial invertebrates solve the problem caused by having to remove nitrogenous wastes from the body.

Name______________________________ Class__________________ Date ______________

Multiple Choice

Write the letter that best answers the question or completes the statement on the line provided.

____ **1.** Animals that have no backbone, or vertebral column, are called

a. vertebrates. c. heterotrophs.
b. invertebrates. d. prokaryotes.

____ **2.** What is a protostome?

a. an animal whose anus is formed from the blastopore
b. an embryo just after fertilization
c. the outermost germ layer
d. an animal whose mouth is formed from the blastopore

____ **3.** What is one reason why sponges are classified as animals?

a. Their cells have cell walls.
b. Sponges are heterotropic.
c. Sponges form partnerships with photosynthetic organisms.
d. Sponges are autotrophic.

____ **4.** Cnidarians have sting cells called

a. choanocytes. c. cnidocytes.
b. archaeocytes. c. flame cells.

____ **5.** In flatworms, what structure pumps food into the digestive cavity?

a. ganglia c. pharynx
b. gizzard d. scolex

____ **6.** Which of the following describes a pseudocoelom?

a. body cavity that is lined only partially with mesoderm tissue
b. body cavity between the mesoderm and the endoderm
c. body cavity that is fully lined with mesoderm tissue
d. body cavity between the mantle and the thorax

____ **7.** Which of the following is true about annelids?

a. Annelids are acoelomates.
b. Each body segment contains several pairs of antennae.
c. Annelids rely on diffusion to transport oxygen and nutrients to their tissues.
d. Annelids have segmented bodies and a true coelom.

____ **8.** Which of the following animals is classified as a cephalopod?

a. octopus c. mussel
b. nudibranch d. scallop

Name______________________________ Class__________________ Date ______________

____ **9.** Arthropods have

a. a soft body and an internal shell.

b. a segmented body and an exoskeleton.

c. a flattened body with an external shell.

d. spiny skin and an internal skeleton.

____ **10.** Which of the following are MOST closely related to spiders?

a. insects

b. horseshoe crabs

c. centipedes

d. echinoderms

____ **11.** What is a process of changing shape and form that occurs in the development of insects?

a. cephalization

b. feedback inhibition

c. internal fertilization

d. metamorphosis

____ **12.** Which of the following is true about echinoderms?

a. Most adult echinoderms exhibit radial symmetry.

b. An echinoderm body is divided into a head, a thorax, and an abdomen.

c. Echinoderms have one pair of antennae and unbranched appendages.

d. Adult echinoderms have an exoskeleton.

____ **13.** Which of the following animals are deuterostomes?

a. cnidarians and mollusks

b. annelids and arthropods

c. echinoderms and chordates

d. mollusks and arthropods

____ **14.** Complex animals break down food through the process of

a. complete metamorphosis.

b. extracellular digestion.

c. cephalization.

d. intracellular digestion.

____ **15.** In terrestrial animals, respiratory surfaces are covered with

a. layers of chitin.

b. uric acid.

c. water or mucus.

d. pedipalps.

Name________________________ Class________________ Date ____________

Completion

Complete each statement on the line provided.

16. A(an) ________________ is an immature stage of an organism that looks different from the adult form.

17. The simplest animal to have bilateral symmetry is a(an) ______________________.

18. Mollusks called ______________________ have two shells that are held together by one or two powerful muscles.

19. A spider's two body sections are the abdomen and the ______________________.

20. Most adult echinoderms exhibit five-part ______________________ symmetry.

Short Answer

In complete sentences, write the answers to the questions on the lines provided.

21. What are the two stages of a cnidarian called?

____

____

22. What are the four parts of a mollusk body plan?

____

____

23. What is an open circulatory system?

____

____

____

24. What are the three body sections of an insect? To which section(s) are the legs attached?

____

____

____

25. What is a coelom? Which invertebrate groups have a true coelom?

____

____

____

Name______________________ Class__________________ Date ______________

Using Science Skills

Use the diagram below to answer the following questions on the lines provided.

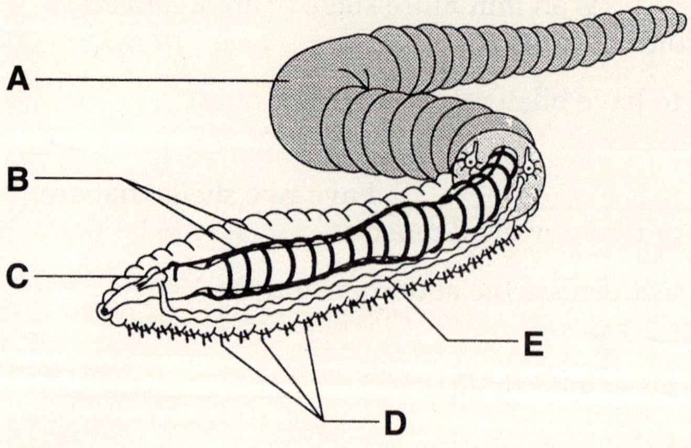

Figure 1

26. Classifying What animal is shown in Figure 1? What phylum contains this animal?

____

____

27. Interpreting Graphics What is the name of structure A in Figure 1? What function is it involved in?

____

____

28. Interpreting Graphics Structure E in Figure 1 is a nerve cord that runs the length of the body. What structure is it attached to, and what is that structure called?

____

____

29. Inferring Structure B in Figure 1 is the dorsal blood vessel, which contains blood that moves from the head to the tail. What type of circulation system does this indicate that the organism has?

____

____

30. Inferring The structures indicated by D are brushlike structures on the outer surface of the animal. What is the name of these structures, and how could they help as this animal moves?

____

____

____

Name______________________________ Class__________________ Date ______________

Multiple Choice

Write the letter that best answers the question or completes the statement on the line provided.

____ **1.** Which of these chordate characteristics exists as paired structures?

a. tail
b. notochord
c. pharyngeal pouch
d. nerve cord

____ **2.** Which of the following statements about a vertebrate's skeleton is INCORRECT?

a. It supports and protects the body.
b. It is an endoskeleton.
c. It grows as the animal grows.
d. It is made entirely of nonliving material.

____ **3.** Lancelets belong to the phylum

a. Urochordata.
b. Cephalochordata.
c. Vertebrata.
d. Chordata.

____ **4.** In most fishes, the structures that are most important for obtaining oxygen from water are the

a. scales.
b. gills.
c. lungs.
d. vertebrae.

____ **5.** Fishes that lived during the late Cambrian Period

a. lacked paired fins.
b. had powerful jaws.
c. had limbs.
d. had soft bodies with little or no armor.

____ **6.** After passing through the gills of a fish, blood circulates through the rest of the body and then collects in the

a. atrium.
b. ventricle.
c. bulbus arteriosus.
d. sinus venosus.

____ **7.** Most fishes get rid of nitrogenous wastes by

a. taking in ammonia through the gills and eliminating it from the kidneys.
b. taking in water through the kidneys and eliminating ammonia from the gills.
c. eliminating ammonia from the gills and from the kidneys.
d. eliminating urine from the gills and ammonia from the kidneys.

____ 8. All fishes in the class Chondrichthyes are alike in the
- a. foods that they eat.
- b. size and form of their teeth.
- c. shape of their bodies.
- d. composition of their skeletons.

____ 9. Which feature distinguishes most fishes from most amphibians?
- a. a vertebral column
- b. scales
- c. breathing with gills during at least part of the life cycle
- d. living in water during at least part of the life cycle

____10. The dominance of amphibians during the Carboniferous Period ended because
- a. many of their habitats disappeared due to climate changes.
- b. swampy fern forests became more widespread.
- c. amphibians were never very numerous during that period.
- d. amphibians did not evolve from the first forms that climbed onto land.

____11. Which of the following is NOT an amphibian adaptation that provides support against the pull of gravity?
- a. strong bones in the limbs
- b. strong bones in the pelvic girdle
- c. a bony cage around the internal organs
- d. lungs for breathing air

____12. Which of the following structures are missing from many salamanders that live on land?
- a. lungs
- b. kidneys
- c. legs
- d. eyes

____13. Which of the following is NOT a characteristic of an amphibian's circulatory system?
- a. two loops
- b. heart with three chambers
- c. right atrium and left atrium
- d. complete separation of oxygen-rich and oxygen-poor blood

____14. A frog's tympanic membranes would be most useful for
- a. enabling the frog to jump long distances.
- b. filtering wastes from the frog's blood.
- c. listening to the mating calls of other frogs.
- d. keeping the frog's eyes from drying out on land.

Name____________________ Class________________ Date ____________

Figure 30-1

____**15.** Amphibians like the one in Figure 30-1 are

a. herbivores as larvae and carnivores as adults.

b. carnivores as larvae and herbivores as adults.

c. herbivores as larvae and adults.

d. carnivores as larvae and adults.

Completion

Complete each statement on the line provided.

16. In fishes and amphibians, gills develop from slits that form in the __________________ .

17. One basic characteristic of fishes is the presence of __________________ , which they use to obtain oxygen from water.

18. If a vertebrate is aquatic as a larva and terrestrial as an adult, it is a(an) __________________ .

19. One adaptation of amphibians for life on land is the presence of mucous glands in the skin, which can help protect amphibians from __________________ .

20. In a frog's heart, blood from the body enters the __________________ .

Short Answer

In complete sentences, write the answers to the questions on the lines provided.

21. In vertebrates, how does the front end of the spinal cord differ from the rear end?

____

____

____

22. Identify one feature of early fishes from the Cambrian Period that was probably useful as a defense against predators.

____

____

____

Name________________________ Class__________________ Date______________

23. Describe the structure and list two functions of the pyloric ceca of fishes.

___

___

___

___

24. Distinguish between anadromous and catadromous fishes.

___

___

___

___

___

25. Describe the composition of the skeleton of each fish in Figure 30-2.

___

___

___

___

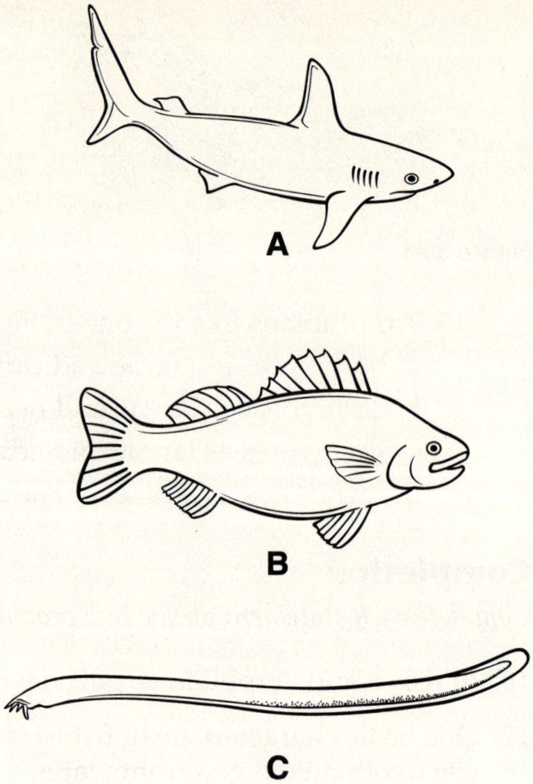

Figure 30-2

Using Science Skills

Use the diagram below to answer the following questions on the lines provided.

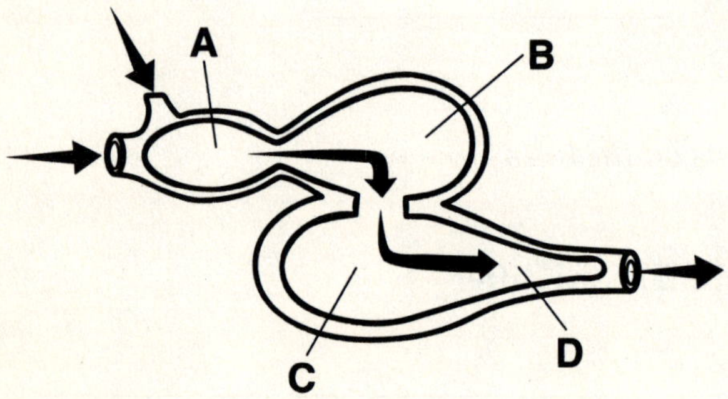

Figure 30-3

26. **Interpreting Graphics** Identify the heart chambers labeled A through D in Figure 30-3.

___

___

___

27. **Applying Concepts** In Figure 30-3, is the blood that enters the chamber labeled A rich in oxygen or poor in oxygen?

___

28. **Interpreting Graphics** In Figure 30-3, which chamber is the actual pumping portion of the heart?

____

____

29. **Applying Concepts** To what vessel does the chamber labeled D connect? (This vessel is not shown in Figure 30-3.)

____

____

30. **Applying Concepts** In Figure 30-3, what are the first organs that blood passes through after passing through the chamber labeled D?

____

____

Essay

Write the answer to each question in the space provided.

31. Contrast the structure of an adult lancelet and an adult tunicate.

32. Explain how having jaws was a useful adaptation for early fishes.

Name_______________________ Class_______________ Date ___________

33. Name the five main parts of a fish's brain, and describe the function of each part.

34. How are worldwide amphibian populations changing today? What possible explanations have been proposed to explain the change?

35. Describe the circulation of blood through the body of an adult amphibian. Include the chambers of the heart in your description.

Name______________________________ Class ________________ Date ____________

Multiple Choice

Write the letter that best answers the question or completes the statement on the line provided.

____ **1.** In chordates, the long supporting rod that runs through the body is called the
a. nerve cord. c. pharyngeal pouch.
b. notochord. d. tail.

____ **2.** A vertebrate is any chordate that has a
a. backbone. c. hollow nerve cord.
b. notochord. d. tail that extends beyond the anus.

____ **3.** The two groups of nonvertebrate chordates are
a. tunicates and lancelets.
b. skates and rays.
c. frogs and toads.
d. lungfishes and coelacanths.

____ **4.** Most fishes are characterized by each of the following EXCEPT
a. a cartilaginous skeleton.
b. scales.
c. paired fins.
d. gills.

____ **5.** The first vertebrates to evolve were
a. amphibians. c. tunicates.
b. lancelets. d. fishes.

____ **6.** In fishes with gills, oxygen-rich water enters through the
a. mouth and leaves through openings in the pharynx.
b. mouth and leaves through the bladder.
c. openings in the pharynx and leaves through the mouth.
d. openings in the pharynx and leaves through the anus.

____ **7.** Blood flows through the body of a fish in a
a. single-loop, open circulatory system.
b. single-loop, closed circulatory system.
c. double-loop, open circulatory system.
d. double-loop, closed circulatory system.

____ **8.** The organ that adjusts the buoyancy of many bony fishes is the
a. swim bladder. c. ventricle.
b. cerebellum. d. kidney.

Name______________________________ Class__________________ Date ______________

____ **9.** Modern jawless fishes include
a. skates.
b. sharks.
c. lampreys.
d. lungfishes.

____ **10.** Which of the following is NOT a characteristic of most amphibians?
a. They live on land as adults.
b. They breathe with lungs as adults.
c. They have moist skin that contains mucous glands.
d. They have scales and claws.

____ **11.** The first amphibians probably resembled
a. jawless fishes, like the lamprey.
b. cartilaginous fishes, like the skate.
c. lobe-finned fishes, like the coelacanth.
d. ray-finned fishes, like the salmon.

____ **12.** In the digestive system of a frog, where does food go after it leaves the mouth?
a. to the gallbladder
b. to the esophagus
c. to the cloaca
d. to the pancreas

____ **13.** In a frog, the cavity through which digestive wastes, urine, and eggs or sperm leave the body is the
a. cloaca.
b. colon.
c. gallbladder.
d. pancreas.

____ **14.** The eggs of amphibians can dry out easily because they
a. are never encased in jelly.
b. do not have shells.
c. are usually laid on land.
d. are always fertilized externally.

____ **15.** The amphibian in Figure 30-1 is a
a. caecilian.
b. frog.
c. salamander.
d. toad.

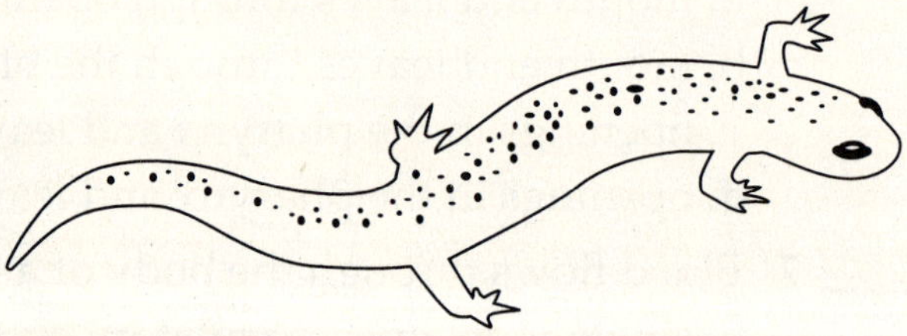

Figure 30-1

Completion

Complete each statement on the line provided.

16. Individual segments of the backbone are called ___________________.

17. The skeletons of some ancient fishes were made of ___________________, which is softer and more flexible than bone.

18. Adult salmon can distinguish their home stream from other streams by using their sense of ___________________.

Name________________________ Class__________________ Date ______________

19. The Carboniferous Period is sometimes referred to as the Age of __________________.

20. An adult amphibian's heart has __________________ chambers.

Short Answer

In complete sentences, write the answers to the questions on the lines provided.

21. List the four characteristics of chordates.

22. How do lancelets move?

23. To which of the three main groups of fishes do each of the animals in Figure 30-2 belong?

24. List two adaptations of early amphibians that helped them live on land.

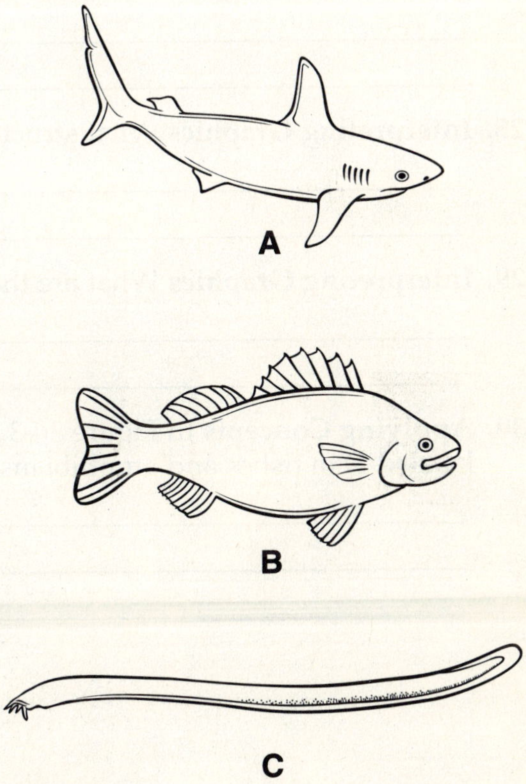

Figure 30-2

25. What are the three groups of modern amphibians?

____

____

____

____

Using Science Skills

Use the diagram below to answer the following questions on the lines provided.

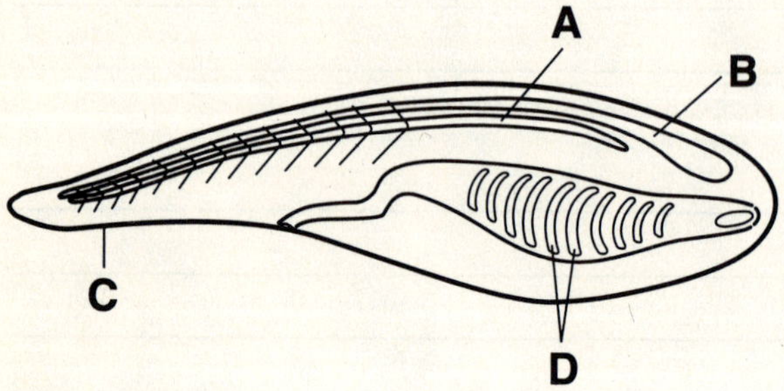

Figure 30-3

26. **Interpreting Graphics** In Figure 30-3, what is the structure labeled A?

____

____

27. **Interpreting Graphics** In Figure 30-3, what is the structure labeled B?

____

____

28. **Interpreting Graphics** What structure is labeled C in Figure 30-3?

____

____

29. **Interpreting Graphics** What are the structures labeled D in Figure 30-3?

____

____

30. **Applying Concepts** In Figure 30-3, what happens to the structures labeled D in fishes and amphibians?

____

____

____

Name______________________________ Class__________________ Date ______________

Multiple Choice

Write the letter that best answers the question or completes the statement on the line provided.

____ **1.** Reptile hearts contain
a. one atrium and one ventricle.
b. one atrium and either one or two ventricles.
c. two atria and two ventricles.
d. two atria and either one or two ventricles.

____ **2.** One reason for the first great adaptive radiation of reptiles was that
a. reptiles could reproduce without water.
b. reptiles were strictly aquatic.
c. Earth's climate became warmer and more humid.
d. lakes and swamps became more numerous.

____ **3.** One characteristic that all dinosaurs shared was
a. a similar body size.
b. the number of legs they used for moving around on land.
c. what they ate.
d. how the embryos developed.

____ **4.** Which of the following is NOT an adaptation for a completely terrestrial life?
a. limb girdles
b. lungs
c. scaly skin
d. amniotic egg

____ **5.** Compared with the limbs of most amphibians, the limbs of reptiles tend to be
a. smaller and weaker.
b. capable of carrying less weight.
c. located farther under the body.
d. less specialized in form and function.

____ **6.** All reptiles reproduce by
a. internal fertilization, and most are oviparous.
b. internal fertilization, and most are viviparous.
c. external fertilization, and most are viviparous.
d. external fertilization, and all are ovoviviparous.

____ **7.** A four-chambered heart is found in
a. tuataras. c. turtles.
b. snakes. d. alligators.

_____ 8. Caimans and gavials belong to the order

a. Squamata. c. Testudines.

b. Crocodilia. d. Rhynchocephalia.

_____ 9. What is the order in which food passes through the digestive system of a seed-eating bird?

a. large intestine → crop → stomach → small intestine

b. stomach → large intestine → crop → small intestine

c. crop → stomach → small intestine → large intestine

d. crop → large intestine → small intestine → stomach

_____10. Which of the following statements about temperature regulation in birds is NOT correct?

a. Birds are endotherms.

b. A bird's body temperature depends on the temperature of the outside environment.

c. Bird's can generate their own body heat.

d. Feathers help birds conserve body heat.

_____11. The main function of contour feathers is to

a. trap air close to a bird's body.

b. provide the lifting force and balance needed for flight.

c. make a bird's respiratory system more efficient.

d. release a powder that repels water.

_____12. A bird's respiratory system is more efficient than that of other land vertebrates because

a. inhaled air passes directly into the lungs.

b. oxygen-poor air is inhaled along with oxygen-rich air.

c. air flows through the lungs in only one direction.

d. none of the space in the lungs is occupied by tubes.

_____13. The structure of a bird's heart ensures that

a. oxygen-rich blood never mixes with oxygen-poor blood.

b. oxygen-rich blood is always pumped to the lungs.

c. oxygen-poor blood is always pumped to the body.

d. oxygen-poor blood never enters the heart.

_____14. If a bird has a hooked bill and sharp talons, it might be a(an)

a. water bird. c. bird of prey.

b. ostrich. d. perching bird.

_____15. Which of the following is NOT a likely mechanism by which migrating birds find their way?

a. using stars as guides

b. following landmarks

c. following the scent of certain flowers

d. sensing Earth's magnetic field

Name________________________________ Class____________________ Date ______________

Completion

Complete each statement on the line provided.

16. A mass ____________________ at the end of the Cretaceous Period was caused by a series of natural disasters and marked the end of the dinosaurs.

17. An egg in which the embryo is protected by several membranes is called a(an) ____________________ egg.

18. Birds require an enormous amount of ____________________ for flight and for maintaining a constant body temperature.

19. Birds have a(an) ____________________ - chambered heart.

20. On a dark, moonless night with overcast skies, some migrating birds may still find their way by following Earth's ____________________ field.

Short Answer

In complete sentences, write the answers to the questions on the lines provided.

21. What disadvantage for reptiles does their skin create? How do reptiles overcome this disadvantage?

22. Compare the lungs of amphibians and reptiles.

23. Describe a defensive mechanism, other than escape, that reptiles in the order Testudines can use to protect themselves.

24. How does structure A in Figure 31-1 differ from the corresponding structure in other vertebrates?

25. Explain how the shape of the structure labeled B in Figure 31-1 is an adaptation for flight.

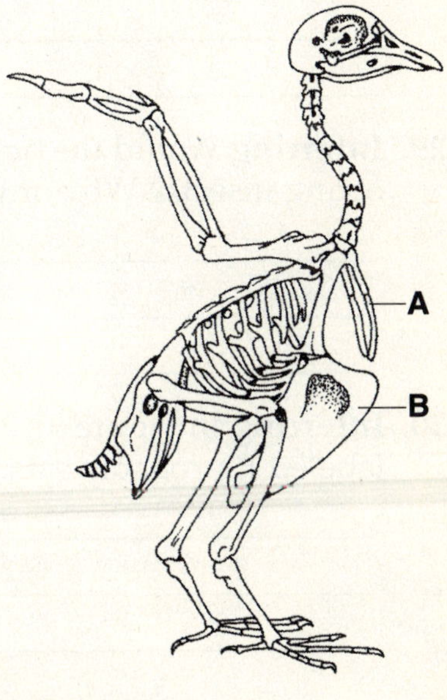

Figure 31-1

Name______________________ Class______________ Date ____________

Using Science Skills

Use the diagram below to answer the following questions on the lines provided.

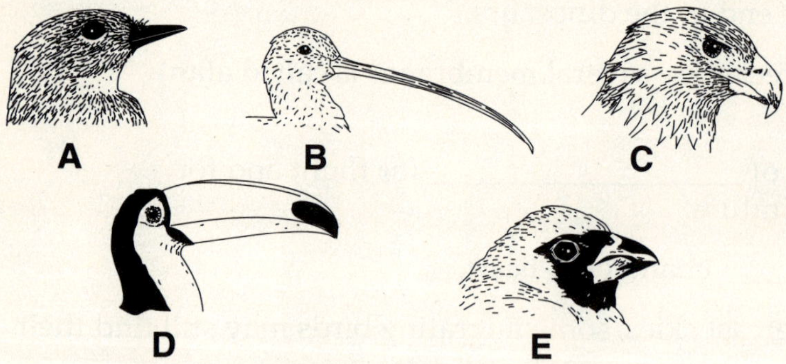

Figure 31-2

26. Applying Concepts Identify the type of food that the bird labeled A in Figure 31-2 probably eats. Explain your answer.

____

____

____

27. Inferring How is the bird labeled B in Figure 31-2 adapted to feeding in a shallow, wet environment?

____

____

____

28. Applying Concepts In Figure 31-2, identify the type of food that the bird labeled C probably eats. Explain your answer.

____

____

____

29. Inferring Would the bird labeled D in Figure 31-2 be successful at eating insects? Why or why not?

____

____

____

30. Inferring In Figure 31-2, identify the type of food that the bird labeled E probably eats. Explain your answer.

____

____

____

____

Name_______________ Class_______________ Date_______________

Essay

Write the answer to each question in the space provided.

31. Describe the range of characteristics exhibited by dinosaurs.

32. Compare and contrast the circulatory systems of a lizard and an alligator.

33. Describe the evidence relating to whether *Archaeopteryx* should be classified as a dinosaur or a bird.

34. Describe the flow of air through the respiratory system of a bird. Explain why this pattern of airflow is an advantage for birds.

35. Describe three ways in which birds interact with natural ecosystems and human society.

Name______________________________ Class__________________ Date ______________

Multiple Choice

Write the letter that best answers the question or completes the statement on the line provided.

____ **1.** The only places on Earth where most reptiles cannot live are very
a. hot areas. c. dry areas.
b. cold areas. d. wet areas.

____ **2.** One way an ectotherm can warm its body is to
a. move into the shade.
b. bask in the sun.
c. speed up its metabolism.
d. crawl into an underground burrow.

____ **3.** The earliest known reptiles lived during the
a. Permian Period.
b. Triassic Period.
c. Carboniferous Period.
d. Cretaceous Period.

____ **4.** Which of the following is NOT thought to have contributed to the extinction of the dinosaurs?
a. massive volcanic eruptions
b. the collision of an asteroid or a comet
c. the rising of sea level
d. major forest fires

____ **5.** Most reptiles exchange gases through their
a. gills. c. lungs.
b. skin. d. limbs.

____ **6.** The difference between reptilian eggs and amphibian eggs is that reptilian eggs
a. must develop in water.
b. always hatch inside the mother's body.
c. are surrounded by a protective shell and membranes.
d. do not contain an embryo.

____ **7.** The tuatara belongs to the order
a. Squamata. c. Testudines.
b. Crocodilia. d. Rhynchocephalia.

____ **8.** Most lizards have all of the following EXCEPT
a. clawed toes. c. movable eyelids.
b. a third eye. d. external ears.

Name____________________ Class______________ Date ____________

____ 9. The dorsal part of a turtle's shell is called the
a. carapace. c. amnion.
b. plastron. d. allantois.

____10. Which of the following is NOT a characteristic of all birds?
a. They have an outer covering of feathers.
b. They maintain a constant internal body temperature.
c. They have two legs.
d. They can fly.

____11. An endotherm is an animal that
a. has a low rate of metabolism.
b. can generate its own body heat.
c. stores large amounts of food in its stomach.
d. can see color very well.

____12. In which part of a bird's digestive system is food stored and moistened before it enters the stomach?
a. cloaca c. gizzard
b. small intestine d. crop

____13. One characteristic that *Archaeopteryx* had but most dinosaurs lacked was
a. feathers. c. a tail.
b. teeth. d. claws.

____14. The largest order of birds consists of
a. birds of prey. c. parrots.
b. flightless birds. d. perching birds.

____15. Insect-eating birds are beneficial to human society because they
a. pollinate flowers.
b. help keep mosquito populations under control.
c. disperse seeds.
d. provide food for farm animals.

Completion

Complete each statement on the line provided.

16. In reptiles that live entirely on land, excess water is absorbed from urine in the ________________.

17. The ________________ is the ventral part of a turtle's shell.

18. The structure labeled A in Figure 31-1 is the ________________.

19. The structure labeled B in Figure 31-1 is the ________________.

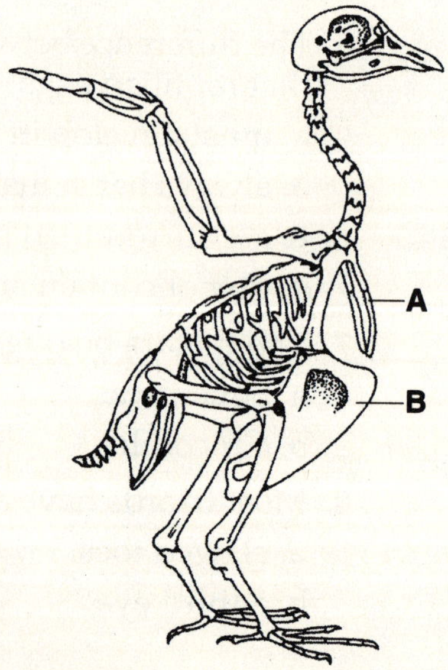

Figure 31-1

Name______________________ Class______________ Date ____________

20. The droppings of fruit-eating birds disperse ______________ over great distances.

Short Answer

In complete sentences, write the answers to the questions on the lines provided.

21. Why is it better for a reptile that lives entirely on land to excrete nitrogenous wastes as uric acid instead of ammonia?

22. Describe one threat to the survival of modern reptiles.

23. How do birds heat their bodies?

24. State the two alternative explanations for the evolutionary relationship between birds and dinosaurs.

25. What are two common characteristics of birds of prey?

Name______________________ Class______________ Date ____________

Using Science Skills

Use the diagram below to answer the following questions on the lines provided.

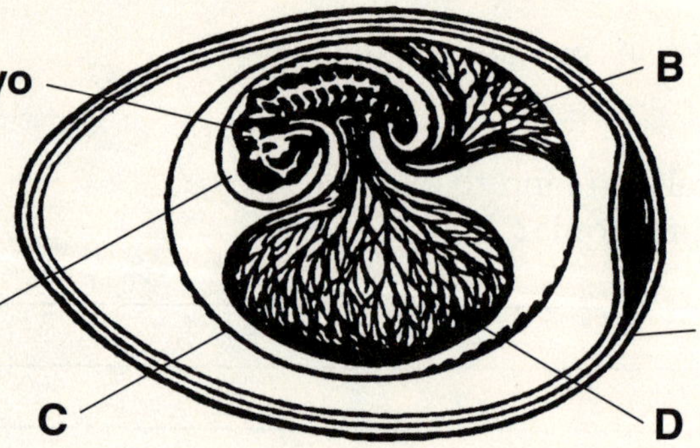

Figure 31-2

26. Interpreting Graphics In Figure 31-2, identify the structure labeled A, and describe its function.

____

____

____

27. Interpreting Graphics Identify the structure labeled B in Figure 31-2, and describe its function.

____

____

____

28. Interpreting Graphics In Figure 31-2, what are the name and the function of the structure labeled C?

____

____

____

29. Interpreting Graphics In Figure 31-2, what are the name and the function of the structure labeled D?

____

____

____

30. Applying Concepts Identify the structure labeled E in Figure 31-2, and describe its physical characteristics in most reptile eggs.

____

____

____

Name________________________ Class________________ Date ____________

Multiple Choice

Write the letter that best answers the question or completes the statement on the line provided.

____ **1.** A feature that is present in most but NOT all female mammals is

a. the ability to generate body heat.

b. viviparous development.

c. the ability to nourish young with milk.

d. the ability to breathe air.

____ **2.** Which organ is used by a cow to feed her newborn calf?

a. mammary gland c. marsupium

b. placenta d. rumen

____ **3.** The first true mammals appeared

a. about the same time dinosaurs became dominant.

b. long after dinosaurs became extinct.

c. about the same time amphibians became extinct.

d. long after birds appeared.

____ **4.** What so some mammals have to release heat from their bodies?

a. external body hair c. sweat glands

b. subcutaneous fat d. high rate of metabolism

____ **5.** The mammalian circulatory system consists of

a. one loop powered by a four-chambered heart.

b. two separate loops powered by a four-chambered heart.

c. two separate loops powered by a three-chambered heart.

d. four separate loops powered by a two-chambered heart.

____ **6.** Which part of a mammal's brain contains a well-developed cerebral cortex—the center of thinking?

a. medulla oblongata c. cerebrum

b. spinal cord d. cerebellum

____ **7.** One group of mammals is reproductively different from the other two groups because its members

a. are live-bearing.

b. reproduce by external fertilization.

c. are oviparous.

d. do not nourish their young with milk.

____ **8.** In which of the following mammals does an embryo spend the most time developing inside the mother?

a. echidna c. kangaroo

b. elephant d. koala

____ **9.** Proboscideans are mammals that

a. have sharp teeth and claws.

b. have trunks.

c. spend most of their time underwater.

d. have a single pair of long, curved incisors.

____ **10.** The similarities in body structure between an anteater and an aardvark are the result of

a. climate changes.

b. habitat destruction.

c. convergent evolution.

d. adaptive radiation.

____ **11.** Having a thumb that can move against the other fingers makes it possible for a primate to

a. hold objects firmly.

b. merge visual images.

c. display elaborate social behaviors.

d. judge the locations of tree branches.

____ **12.** A primate that uses its prehensile tail to swing from branch to branch is the

a. gibbon.

b. squirrel monkey.

c. macaque.

d. orangutan.

____ **13.** Hominids differ from other primates on the basis of all of the following EXCEPT

a. brain size.

b. method of reproduction.

c. method of locomotion.

d. shape of the hip bones.

____ **14.** Fossil evidence indicates that *Australopithecus afarensis*

a. was primarily a meat-eater.

b. had a large brain.

c. was bipedal.

d. appeared later than *Homo ergaster*.

____ **15.** The earliest hominid that belonged to the same genus as modern humans was probably

a. *Homo habilis.*

b. *Homo neanderthalensis.*

c. *Homo afarensis.*

d. *Homo ergaster.*

Name______________________________ Class__________________ Date ______________

Completion

Complete each statement on the line provided.

16. Two features that are found in all mammals but not in any other chordates are hair and __________________ .

17. Early mammals were probably nocturnal and ate __________________ .

18. In placental mammals, nutrients, oxygen, carbon dioxide, and wastes are exchanged between the embryo and the mother through the __________________ .

19. Hominids walk on two feet. This manner of walking is called __________________ locomotion.

20. The group of primates called __________________ includes New World monkeys, Old World monkeys, and hominoids.

Short Answer

In complete sentences, write the answers to the questions on the lines provided.

21. How can mammals be distinguished from other vertebrates on the basis of the fossils they leave behind?

____

____

____

22. Explain the function of the rumen. Why is this organ absent in carnivores?

____

____

____

23. What are two ways in which mammals maintain homeostasis?

____

____

____

24. Explain how young marsupials complete their development.

____

____

____

25. Why is binocular vision important to a primate that moves through the branches of trees?

____

____

____

Name________________________ Class________________ Date ____________

Using Science Skills

Use the diagram below to answer the following questions on the lines provided.

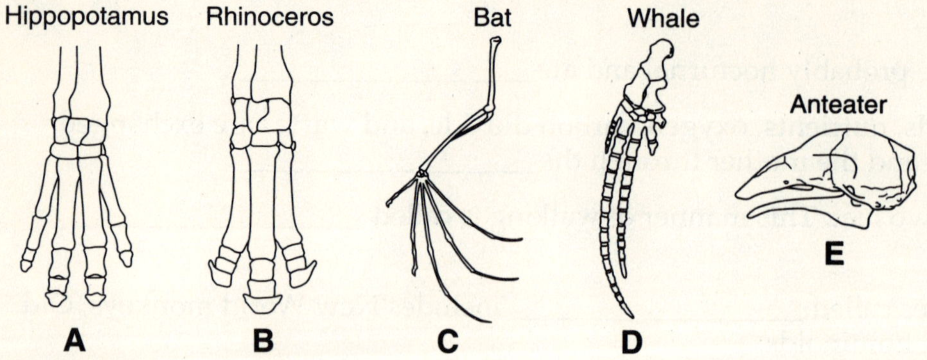

Figure 32-1

26. **Applying Concepts** In Figure 32-1, drawing A shows bones in the hoof of a placental mammal. To which order does this mammal belong?

27. **Applying Concepts** In Figure 32-1, drawing B shows bones in the hoof of a placental mammal. To which order does this mammal belong?

28. **Inferring** In Figure 32-1, drawing C shows bones in the wing of a placental mammal. How is the structure of the limb adapted to flight?

29. **Observing** In Figure 32-1, drawing D shows bones in the flipper of a carnivorous placental mammal. How does its bone structure differ from that of a bat?

30. **Inferring** In Figure 32-1, drawing E shows the skull of a placental mammal. How is the structure of this skull adapted to the anteater's diet?

Name_______________ Class_______________ Date_______________

Essay

Write the answer to each question in the space provided.

31. What adaptations do mammals have that enable them to maintain a constant body temperature?

32. Why is it important for mammals to have specialized teeth and digestive systems based on their diet?

33. Why are monotremes sometimes said to be intermediate between reptiles and most mammals in their manner of reproduction?

Name________________________ Class________________ Date ____________

34. Compare and contrast New World monkeys and Old World monkeys.

35. Explain how paleontologists have changed their views on early hominid evolution. What caused them to change their way of thinking?

Name____________________________ Class__________________ Date ______________

Multiple Choice

Write the letter that best answers the question or completes the statement on the line provided.

____ **1.** Mammals are characterized by each of the following EXCEPT
a. mammary glands.
b. hair.
c. endothermy.
d. three-chambered hearts.

____ **2.** The first true mammals appeared during the
a. Permian Period.
b. late Triassic Period.
c. Cretaceous Period.
d. Cenozoic Era.

____ **3.** Mammals that have broad, flattened molars feed on
a. insects.
b. meat.
c. plankton.
d. plants.

____ **4.** Which body system includes the rumen?
a. respiratory
b. circulatory
c. excretory
d. digestive

____ **5.** In mammals, the amount of water in the body is controlled mainly by the
a. kidneys.
b. diaphragm.
c. heart.
d. lungs.

____ **6.** The egg-laying mammals are called
a. marsupials.
b. monotremes.
c. placental mammals.
d. insectivores.

____ **7.** An example of an animal with a marsupium is a(an)
a. elephant.
b. sea lion.
c. mouse.
d. wombat.

____ **8.** The function of a placenta is to
a. exchange materials between an embryo and its mother.
b. store and process newly swallowed plant food.
c. store urine until it is eliminated from the body.
d. increase the volume of the chest cavity during breathing.

_____ **9.** Which of the following mammals is a chiropteran?
a. shrew
b. rabbit
c. bat
d. walrus

_____ **10.** The similar appearance of armadillos and aardvarks is an example of
a. continental drift.
b. convergent evolution.
c. social interaction.
d. homeostasis.

_____ **11.** Which of the following is NOT a characteristic of most primates?
a. opposable digits
b. a well-developed cerebrum
c. binocular vision
d. a cloaca

_____ **12.** An example of a prosimian is a(an)
a. spider monkey.
b. baboon.
c. orangutan.
d. lemur.

_____ **13.** Bipedal locomotion consists of
a. swinging from branch to branch.
b. using the tail to grasp branches during walking.
c. moving the big toe against the other digits.
d. walking on two feet.

_____ **14.** Which of the following was a characteristic of *Homo habilis*?
a. using tools made of stone and bone
b. producing cave paintings
c. having a skeleton similar to that of a gorilla
d. burying their dead with elaborate rituals

_____ **15.** Which statement is true of the Neanderthals?
a. They evolved after the Cro-Magnons.
b. They made stone tools.
c. They became extinct about 1 million years ago.
d. They replaced *Homo sapiens* in Europe and the Middle East.

Name______________________________ Class__________________ Date ______________

Completion

Complete each statement on the line provided.

16. Mammals have ____________________ and body fat that help conserve body heat.

17. The only monotremes that exist today are the ____________________ and the spiny anteaters.

18. Primates can display more complex behaviors than many other mammals because they have a well-developed ____________________ .

19. The anthropoids that are native to Central and South America are called ____________________ monkeys.

20. A(An) ____________________ thumb enables primates to grasp objects.

Short Answer

In complete sentences, write the answers to the questions on the lines provided.

21. Name the four specialized types of mammalian teeth.

22. What is the function of the diaphragm?

23. What are the most important characteristics that scientists use to classify mammals into subgroups?

24. What is a marsupium?

25. What is a prosimian?

Name______________________ Class________________ Date ____________

Using Science Skills

Use the diagrams below to answer the following questions on the lines provided.

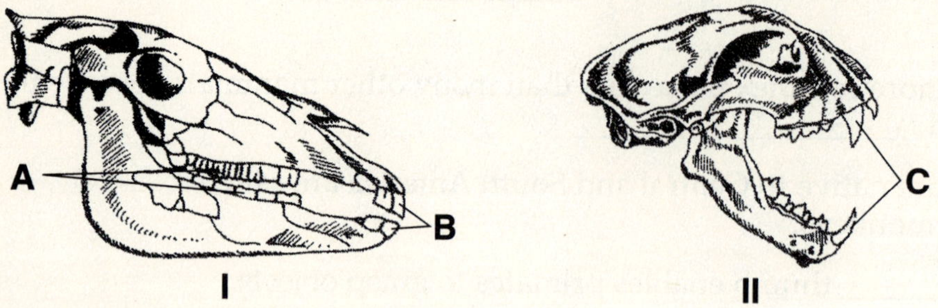

Figure 32-1

26. Inferring Which skull in Figure 32-1 is that of a carnivore and which is that of a herbivore?

27. Interpreting Graphics In Figure 32-1, identify the teeth labeled A and describe their function in this mammal.

28. Interpreting Graphics In Figure 32-1, identify the teeth labeled B and describe their function in this mammal.

29. Interpreting Graphics In Figure 32-1, identify the teeth labeled C and describe their function in this mammal.

30. Inferring Describe how the jaws move when each mammal in Figure 32-1 chews its food.

Name_______________________________ Class__________________ Date ______________

Multiple Choice

Write the letter that best answers the question or completes the statement on the line provided.

____ **1.** A phylogenetic tree shows that all chordates evolved from
   a. tunicates and lancelets.
   b. dinosaur ancestors.
   c. fishes.
   d. a common ancestor.

____ **2.** Convergent evolution is the process that produces
   a. species that are similar in appearance and behavior but not closely related.
   b. unrelated species that are different in appearance and behavior.
   c. closely related species that are similar in appearance and behavior.
   d. closely related species that are different in appearance and behavior.

____ **3.** Essential life functions are carried out most efficiently when an animal's internal body temperature is
   a. insulated.
   b. regulated by the environment.
   c. the same as its external temperature.
   d. within a particular operating range.

____ **4.** Panting is a behavior that is seen most often in
   a. endotherms that need to cool down.
   b. endotherms that need to warm up.
   c. ectotherms that need to warm up.
   d. ectotherms that are at their ideal body temperature.

____ **5.** The main difference between ectotherms and endotherms is
   a. the source of their body heat.
   b. how they obtain food to provide for their metabolism.
   c. whether they control their body temperature.
   d. whether they conserve or eliminate body heat.

____ **6.** The simplest chordates that have cephalization as adults are
   a. lancelets. c. amphibians.
   b. fishes. d. reptiles.

____ **7.** Which of the following vertebrates has a heart with four chambers?
   a. a salamander c. a goldfish
   b. a lizard d. a cow

____ **8.** Chordate respiratory structures include all of the following EXCEPT

a. simple air sacs.
b. the medulla oblongata.
c. the surface of the skin.
d. the lining of the mouth.

____ **9.** When a fish respires, water flows in which of the following patterns?

a. in through the gill slits, over the gill filaments, and out through the mouth
b. in through the gill filaments, over the gill slits, and out through the mouth
c. in through the mouth, over the gill slits, and out through the gill filaments
d. in through the mouth, over the gill filaments, and out through the gill slits

____ **10.** The main difference between an amphibian lung and a reptilian lung is that

a. an amphibian lung has a greater surface for gas exchange.
b. a reptilian lung has a greater surface area for gas exchange.
c. an amphibian lung contains thousands of alveoli, but a reptilian lung does not.
d. a reptilian lung is connected to air sacs, but an amphibian lung is not.

____ **11.** In a single-loop circulatory system, the atrium

a. receives the blood from the body.
b. pumps blood to the gills.
c. pumps blood to the lungs.
d. pumps blood throughout the entire body.

____ **12.** A single-loop circulatory system is characteristic of

a. fishes.
b. adult amphibians.
c. most reptiles.
d. crocodilians.

____ **13.** A function of ligaments in a backbone is to

a. generate forward thrust during swimming.
b. keep the backbone straight and rigid.
c. connect the vertebrae.
d. make the body bend back and forth.

____ **14.** The sets of bones that support the limbs of vertebrates are called

a. vertebral columns.
b. limb girdles.
c. ligaments.
d. vertebrae.

Name______________________________ Class__________________ Date ______________

____15. Embryos obtain nutrients from the yolk inside the egg in
a. oviparous and ovoviviparous animals.
b. viviparous and ovoviviparous animals.
c. viviparous and oviparous animals.
d. viviparous animals only.

Completion

Complete each statement on the line provided.

16. Over the course of evolution, the appearance of new adaptations has launched ____________________ in chordate groups.

17. A(An) ____________________ is an animal whose body temperature is controlled mainly by the transfer of heat between its body and its surroundings.

18. An animal that strains small pieces of food from the water is called a(an) ____________________ .

19. In mammalian lungs, ____________________ provide an enormous surface area for gas exchange.

20. The region of your brain that you use to determine the answers to test questions is called the ____________________ .

Short Answer

In complete sentences, write the answers to the questions on the lines provided.

21. If two extinct but unrelated species of chordates shared many adaptations, what can you infer about the ecological conditions those species encountered?

____

____

22. Animals A and B are terrestrial vertebrates of the same size. One is an ectotherm and the other is an endotherm. The resting metabolic rate of animal A is five times that of animal B. Using this information, state which animal is the ectotherm and briefly explain your conclusion.

____

____

23. Why is it illogical to say that the simple organ systems of a tunicate must be inferior to the organ systems of a mammal?

____

____

24. In what ways other than length do the digestive tracts of carnivores and herbivores differ?

____

____

25. Contrast movement in larval and adult tunicates.

____

____

Using Science Skills

Use the diagram below to answer the following questions on the lines provided.

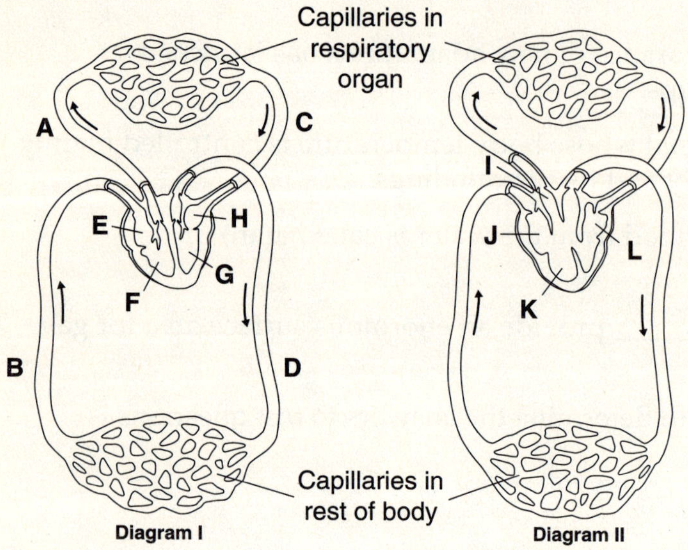

Figure 33-1

26. **Interpreting Graphics** Figure 33-1 shows two types of vertebrate circulatory systems. In Diagram I, which of the vessels labeled A, B, C, and D contain(s) oxygen-rich blood, and which contain(s) oxygen-poor blood?

____

____

____

27. **Interpreting Graphics** Name the heart chambers labeled E, F, G, and H in Diagram I of Figure 33-1.

____

____

____

28. **Interpreting Graphics** In Diagram II of Figure 33-1, where does blood go after it leaves the chamber labeled K?

____

____

____

29. Applying Concepts Which group(s) of vertebrates has(have) a circulatory system that is represented by each diagram in Figure 33-1? Choose from among the following groups: fishes, crocodilians, noncrocodilian reptiles, birds, mammals.

____

____

____

30. Comparing and Contrasting Describe the main difference between the two circulatory systems shown in Figure 33-1.

____

____

____

____

Essay

Write the answer to each question in the space provided.

31. Why is the control of body temperature important for vertebrates?

32. Describe two hypotheses concerning the evolution of endothermy in vertebrates. What does the evidence on this issue suggest?

Name______________________________ Class__________________ Date ______________

33. Contrast the respiratory systems of mammals and birds.

34. Describe the major differences among the chambers of the heart in the five main groups of vertebrates.

35. Describe the forms in which nitrogenous wastes are eliminated in different groups of chordates.

Name______________________________ Class__________________ Date ______________

Multiple Choice

Write the letter that best answers the question or completes the statement on the line provided.

____ **1.** *Pikaia* was an early

a. worm. c. chordate.
b. vertebrate. d. lancelet.

____ **2.** Which of the following is NOT an adaptation that appeared during the course of chordate evolution?

a. endothermy
b. radial symmetry
c. four limbs
d. jaws

____ **3.** The notochord is a characteristic of

a. worms. c. mammals only.
b. all chordates. d. reptiles only.

____ **4.** The rapid growth in the diversity of a group of organisms is called

a. a ladogram.
b. an ecological condition.
c. adaptive radiation.
d. convergent radiation.

____ **5.** The largest living group of chordates is the

a. amphibians. c. mammals.
b. fishes. d. birds.

____ **6.** Which of the following animals is an ectotherm?

a. bluebird c. squirrel
b. snake d. whale

____ **7.** Which of the following help mammals retain body heat?

a. hair and sweat glands
b. hair and body fat
c. bones and sweat glands
d. bones and body fat

____ **8.** Filter feeders include all of the following EXCEPT

a. lancelets. c. crocodiles.
b. flamingos. d. baleen whales.

____ **9.** Colonies of bacteria in the intestines of a cow are helpful in

a. digesting cellulose fibers.
b. producing enzymes that digest meat.
c. straining plankton from water.
d. tearing and slicing food.

____**10.** What is the general rule regarding respiratory organs in chordates?

a. Aquatic chordates use lungs, and land vertebrates use gills.

b. Aquatic chordates use lungs and gills, and land vertebrates use gills.

c. Aquatic chordates use gills, and land vertebrates use lungs.

d. Aquatic chordates use gills, and land vertebrates use lungs and gills.

____**11.** The most efficient vertebrate lungs are found in

a. amphibians. c. birds.
b. reptiles. d. mammals.

____**12.** The heart of a fish has

a. one atrium and one ventricle.

b. one atrium and two ventricles.

c. two atria and one ventricle.

d. two atria and two ventricles.

____**13.** In tunicates, some nitrogenous wastes leave the body in the form of

a. urea. c. uric acid.
b. ammonia. d. protein.

____**14.** Which of the following characteristic(s) is(are) shared by all chordates?

a. a bony skeleton c. a backbone
b. pharyngeal pouches d. fin girdles or limb girdles

____**15.** Sexual reproduction occurs in

a. fishes. c. mammals.
b. amphibians. d. all of the above

Completion

Complete each statement on the line provided.

16. The chordate family tree has its roots in ancestors that vertebrates share with tunicates and ____________________ .

17. In environments where temperatures are high and fairly constant most of the time, ____________________ is a more energy-efficient method of controlling body temperature.

18. In a mammalian lung, gas exchange occurs inside bubblelike structures called ____________________ .

19. If an animal has a four-chambered heart, it has a(an) ____________________ -loop circulatory system.

20. Animals whose embryos obtain nutrition directly from the mother's body are said to have ____________________ development.

Name______________________________ Class__________________ Date ______________

Short Answer

In complete sentences, write the answers to the questions on the lines provided.

21. Identify four features common to all chordates.

22. List three features that are included in the ways in which all vertebrates control their body temperature.

23. Describe the basic pathways of the two loops in a double-loop circulatory system.

24. List two functions of vertebrate kidneys.

25. Describe the functions of the optic lobes and olfactory bulbs in a vertebrate brain.

Name______________________ Class__________________ Date ____________

Using Science Skills

Use the diagram below to answer the following questions on the lines provided.

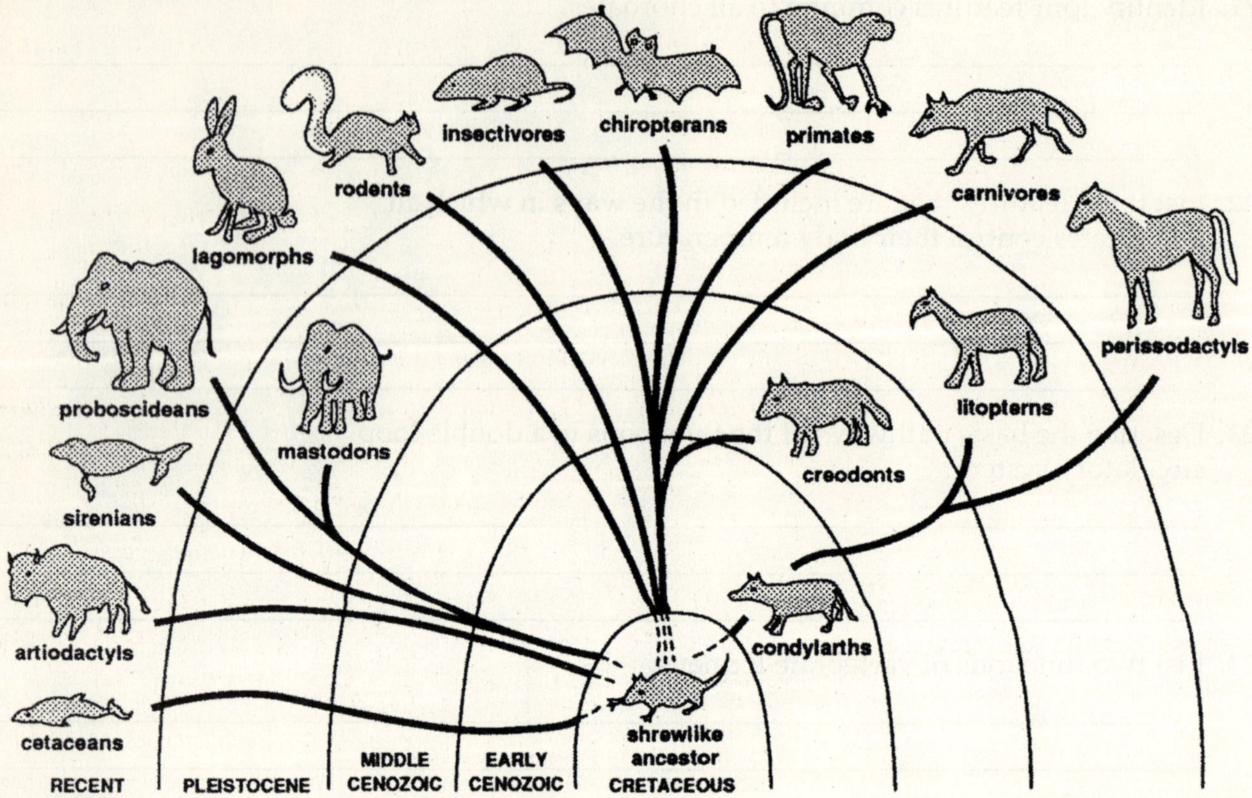

Figure 33-1

26. **Interpreting Graphics** According to Figure 33-1, which extinct group of mammals descended from the condylarths?

____

27. **Interpreting Graphics** According to Figure 33-1, during which era, period, or epoch did the ancestors of the creodonts and the carnivores split into two different lines?

____

28. **Interpreting Graphics** Use Figure 33-1 to identify two groups of mammals that are extinct.

____

29. **Classifying** According to Figure 33-1, to which living order of mammals are sirenians most closely related?

____

30. **Classifying** According to Figure 33-1, are insectivores more closely related to rodents or to chiropterans?

____

Name______________________ Class______________ Date__________

Multiple Choice

Write the letter that best answers the question or completes the statement on the line provided.

____ **1.** For a behavior to evolve under the influence of natural selection, that behavior must
a. improve survival and reproduction.
b. influence physical characteristics.
c. be acquired through learning.
d. be related to predator avoidance.

____ **2.** Animals that have very simple nervous systems cannot
a. detect external stimuli.
b. process information from their senses.
c. respond to stimuli.
d. exhibit complex behaviors.

____ **3.** Complex inborn behaviors include each of the following EXCEPT
a. web-building in spiders.
b. nest-building in birds.
c. potato-washing in macaques.
d. suckling of newborn mammals.

____ **4.** Insight learning is a form of behavior that
a. is characterized by decreased responsiveness to unimportant stimuli.
b. involves random responses that lead to either reward or punishment.
c. involves applying something already learned to a new situation.
d. is performed correctly without any previous experience.

____ **5.** In Pavlov's experiments on classical conditioning, dogs
a. learned to associate food with salivation.
b. learned to ring a bell whenever they were hungry.
c. associated a new stimulus with a reward.
d. were conditioned to produce an innate behavior in response to food.

____ **6.** When you use knowledge and experience to figure out why a lamp in your house will not light, you are using
a. classical conditioning.
b. insight learning.
c. innate behavior.
d. trial-and-error learning.

____ 7. The type of learning that begins with a random behavior that is rewarded is called
a. classical conditioning.
b. operant conditioning.
c. habituation.
d. trial-and-error learning.

____ 8. Imprinting is a form of behavior that
a. is restricted to birds.
b. is often used in the training of adult animals.
c. occurs during a specific time in young animals.
d. always involves the sense of sight.

____ 9. After a young duck imprints on a nonliving model of an adult duck, the young duck will
a. then imprint on its mother.
b. then imprint on any real adult duck.
c. follow only that model.
d. follow any other kind of model.

____10. An animal is most likely to enter into dormancy when
a. it is advantageous to be active.
b. resources are most difficult to find.
c. food is very plentiful.
d. the weather is very mild.

____11. Animals are LEAST likely to migrate out of an area where
a. there is little or no seasonal change in weather.
b. food becomes scarce during part of the year.
c. competition for resources increases during part of the year.
d. winters are very cold and snowy.

____12. It is advantageous for grazing mammals to gather in groups because groups
a. can make the available food resources last longer.
b. can migrate more easily than individuals can.
c. are more difficult for predators to locate than individuals are.
d. offer greater protection from predation.

____13. In some species of balloon flies, males spin balloons of silk and carry them while flying. If a female approaches one of the males and accepts his balloon, the two will fly off to mate. This type of behavior is an example of
a. aggression. c. courtship.
b. territorial defense. d. language.

____14. Competition for limited resources can involve each of the following EXCEPT
a. rituals. c. displays.
b. circadian rhythms. d. aggression.

____15. Nocturnal animals that have a poorly developed sense of smell are most likely to communicate by
a. sound signals. c. pheromones.
b. visual displays. d. chemical signals.

Completion

Complete each statement on the line provided.

16. A snail will withdraw into its shell if it is prodded with a sharp object. In this behavior, prodding is the __________________ and the snail's withdrawal is the response.

17. If a dog learns to ignore vehicles that drive past its house without stopping, the dog's behavior has probably been altered through the type of learning known as __________________ .

18. For an animal to become habituated to a stimulus, the stimulus must NOT be either __________________ or rewarding.

19. When an animal associates a stimulus with a reward or punishment, it learns by the process known as __________________ .

20. If an animal produces a sound and another animal obtains information by hearing the sound, the two animals are engaging in a form of __________________ .

Short Answer

In complete sentences, write the answers to the questions on the lines provided.

21. Why are innate behaviors important? Explain.

22. How is habituation useful to an animal?

23. Explain how imprinting in young geese involves both innate and learned behavior.

Name________________________ Class________________ Date ____________

24. Explain why dormancy and migration are beneficial.

____

____

____

____

25. Why is it an evolutionary advantage for an animal in a society to help other members of the society?

____

____

____

____

Using Science Skills

Use the diagram below to answer the questions on the following page on the lines provided.

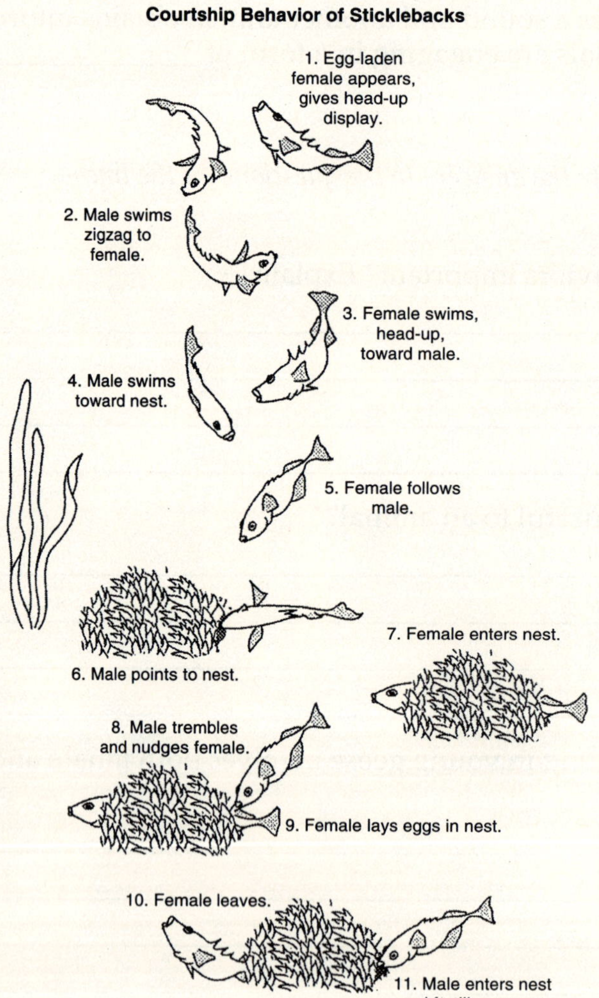

Figure 34-1

26. **Applying Concepts** Even if young sticklebacks never observe the courtship of other sticklebacks, they can perform the courtship behavior shown in Figure 34-1. Is the courtship behavior innate or learned?

____

27. **Interpreting Graphics** According to Figure 34-1, which member of the pair of stickleback fish begins the courtship?

____

28. **Inferring** In Figure 34-1, determine from the action of the sticklebacks whether the nest is built by the male or the female. Explain.

____

____

29. **Interpreting Graphics** In Figure 34-1, what stimuli does the male stickleback give to the female during courtship?

____

____

____

30. **Interpreting Graphics** In Figure 34-1, what stimuli does the female stickleback give to the male during courtship?

____

____

____

Essay

Write the answer to each question in the space provided.

31. Explain how behaviors can evolve under the influence of natural selection.

Name________________________ Class________________ Date____________

32. Contrast classical and operant conditioning.

33. Describe a Skinner box and explain what it is used for.

34. Explain the adaptive advantages of a territory to the animal that establishes it.

35. Explain how it is possible for animals to communicate without having language and what kinds of information animals communicate with nonverbal signals.

Multiple Choice

Write the letter that best answers the question or completes the statement on the line provided.

____ **1.** A stimulus is a(an)
   a. reaction to an external event.
   b. reaction to an internal event.
   c. internal or external response.
   d. internal or external signal.

____ **2.** When disturbed, certain moths lift their front wings to expose eyelike markings on their hind wings. This behavior would be most effective against predators that hunt by
   a. sound. c. sight.
   b. smell. d. touch.

____ **3.** Which of the following terms is NOT closely related to the others?
   a. innate behavior c. inborn behavior
   b. learned behavior d. instinct

____ **4.** Trial-and-error learning is also known as
   a. operant conditioning.
   b. insight learning.
   c. classical conditioning.
   d. innate behavior.

____ **5.** The process in which an animal stops responding to a repetitive stimulus is called
   a. habituation.
   b. classical conditioning.
   c. operant conditioning.
   d. instinct.

____ **6.** Any change in which a behavior is altered as a result of experience is called
   a. habituation.
   b. operant conditioning.
   c. innate behavior.
   d. learning.

____ **7.** For young geese to imprint on an object, that object must
   a. move.
   b. be some kind of animal.
   c. look like an adult goose.
   d. be the same size as an adult goose.

____ 8. The ability of salmon to recognize their home stream at spawning time is an example of
a. insight learning. c. imprinting.
b. habituation. d. communication.

____ 9. A circadian rhythm is a cycle that
a. is related to the phase of the moon.
b. is related to the temperature of the air.
c. has a daily pattern.
d. has a seasonal pattern.

____10. Migration is a behavior that is usually influenced by
a. changing seasons.
b. the phase of the moon.
c. the rise and fall of tides.
d. the time of day.

____11. Which of the following is NOT a type of social behavior?
a. courtship c. hunting in a pack
b. habituation d. communication

____12. The members of a society
a. belong to at least two species.
b. exhibit a type of social behavior.
c. act independently for each individual's benefit.
d. are usually unrelated to one another.

____13. The resources in a territory can include all of the following EXCEPT
a. food. c. potential mates.
b. nesting sites. d. predators.

____14. Dolphins communicate with one another mainly through
a. sound. c. chemical signals.
b. visual displays. d. pheromones.

____15. A pheromone is a type of
a. visual signal.
b. sound signal.
c. chemical messenger.
d. defensive display.

Completion

Complete each statement on the line provided.

16. A single, specific reaction to a stimulus is called a(an) __________________ .

17. The Russian scientist named __________________ studied conditioning in dogs.

18. The most complicated form of learning is __________________ learning.

Name________________________ Class________________ Date ____________

19. When competing for a territory, animals may show ________________ to gain control over each other.

20. Only humans are known to use __________________, a complex system of communication that uses sounds, symbols, and gestures according to a set of rules.

Short Answer

In complete sentences, write the answers to the questions on the lines provided.

21. What body systems are involved when an animal responds to a stimulus?

22. Define *innate behavior*.

23. List the four major types of learning.

24. What are the functions of courtship?

25. List the types of signals animals may use to communicate with one another.

Name________________________ Class________________ Date ____________

Using Science Skills

Use the diagram below to answer the following questions on the lines provided.

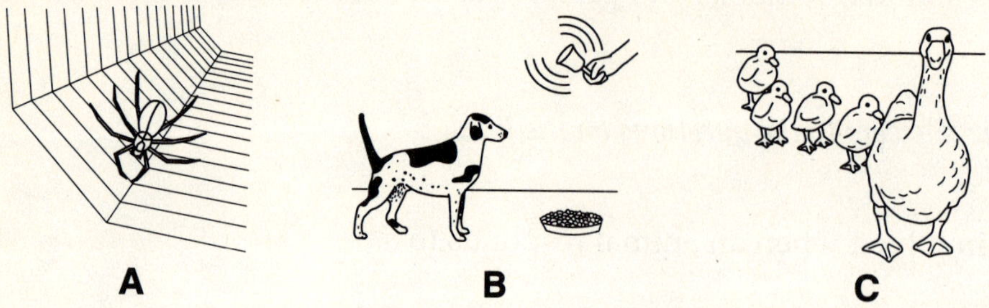

Figure 34-1

26. Interpreting Graphics In Figure 34-1, what type of behavior does drawing A show?

____

____

____

27. Interpreting Graphics In Figure 34-1, what type of learning does drawing B show?

____

____

____

28. Interpreting Graphics In Figure 34-1, identify the two stimuli that are involved in drawing B. Also, identify the response that will be produced in drawing B.

____

____

____

____

29. Applying Concepts Which of the four major types of learning is(are) NOT shown in Figure 34-1?

____

____

____

____

30. Interpreting Graphics In Figure 34-1, what behavioral process does drawing C show?

____

____

____

Name________________________ Class________________ Date ____________

Multiple Choice

Write the letter that best answers the question or completes the statement on the line provided.

____ 1. Which of the following statements about chordates is true?
a. All chordates are vertebrates.
b. All chordates have backbones.
c. All chordates have a notochord.
d. All chordates have paired appendages.

____ 2. The eggs of ovoviviparous fish are
a. released from the female immediately after being fertilized.
b. retained and nourished by the female.
c. released from the female before they are fertilized.
d. held inside the female's body as they develop.

____ 3. Which of the following structures are found in the adult frog but not in the tadpole?
a. gills
b. three-chambered heart
c. single-loop circulatory system
d. all of the above

____ 4. Reptiles are successful on land because of
a. strong limbs.
b. a single-loop circulatory system.
c. endothermy.
d. external fertilization.

____ 5. Which structure in the amniotic egg provides nutrients for the embryo as it grows?
a. amnion
b. chorion
c. allantois
d. yolk

____ 6. A characteristic of birds but not reptiles is
a. urine excreted as uric acid.
b. air sacs.
c. an amniotic egg.
d. a cloaca.

____ 7. Which of the following is NOT a characteristic of mammals?
a. ability to nourish their young with milk
b. lack of pharyngeal slits during development
c. hair
d. endothermy

Name______________________ Class______________ Date ____________

Figure 1

____ **8.** The mammals in the figure above have similar characteristics because

a. they are closely related.

b. they have adapted to different environments on the same continent.

c. of divergent evolution.

d. of convergent evolution.

____ **9.** The hominoid group of primates includes

a. prosimians and anthropoids.

b. New World monkeys and Old World monkeys.

c. apes and hominids.

d. anthropoids only.

____ **10.** Endothermy is characterized by

a. requiring large amounts of food.

b. a low metabolic rate.

c. behaviors that help regulate body temperature.

d. a lack of insulating feathers, fat, or hair.

____ **11.** A single-loop circulatory system is present in all animals with

a. jaws. c. a backbone.

b. a four-chambered heart. d. gills.

____ **12.** Which of the following trends in reproduction is evident as you move through the vertebrate groups?

a. external fertilization to internal fertilization

b. endothermy to ectothermy

c. progressively fewer openings in the digestive tract

d. closed circulatory system to open circulatory system

____ **13.** When a bird learns to press a button to get food, it has learned by

a. habituation.

b. classical conditioning.

c. operant conditioning.

d. insight learning.

____ **14.** Sea turtles that travel each spring from their feeding grounds to their nesting grounds are

a. courting.

b. migrating.

c. socializing.

d. communicating.

Name________________________________ Class_____________________ Date _______________

____15. When a lion snaps, claws, and snarls at the other lions before settling down to eat, it is
a. showing aggression.
b. communicating.
c. competing.
d. all of the above

Completion

Complete each statement on the line provided.

16. A vertebrate with moist skin that contains mucous glands, and lacks scales and claws is a(an) ____________________________.

17. Unlike other reptiles, crocodiles and alligators have a(an) __________________________ like birds and mammals.

18. The first true mammals appeared during the late ________________, about 220 million years ago.

19. All chordates, both nonvertebrate and vertebrate, share a common ________________ ancestor.

20. When a dog barks at a ringing doorbell, the barking is the ________________ and the ringing doorbell is the ________________

Short Answer

In complete sentences, write the answers to the questions on the lines provided.

21. Why are tunicates and lancelets classified as chordates even though they are nonvertebrates?

____

____

____

22. Why are amphibians dependent on water?

____

____

____

____

23. What adaptations do birds have for flight?

____

____

____

Name________________________ Class________________ Date ____________

24. What characteristics do all primates share?

25. How is a behavior produced in an animal?

Using Science Skills

Use the diagram below to answer the following questions on the lines provided.

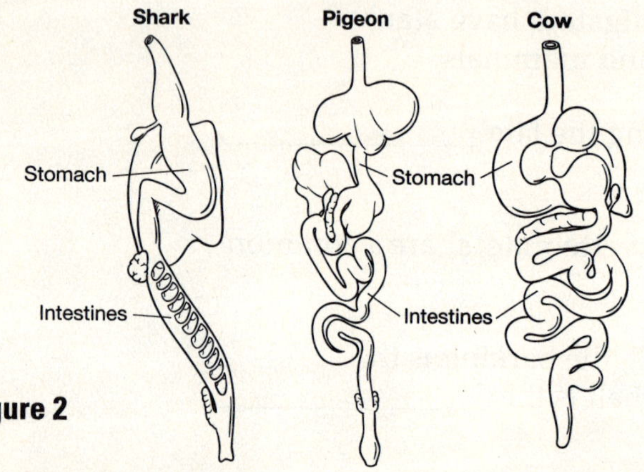

Figure 2

26. **Interpreting Graphics** Which vertebrate digestive system in Figure 2 has the largest stomach?

27. **Interpreting Graphics** Which of the vertebrates in Figure 2 are carnivores? Which are herbivores? Explain.

28. **Applying Concepts** Why are the digestive systems of birds and mammals more complex than the digestive system of fishes?

29. **Applying Concepts** What structures are found only in the pigeon's digestive system? What digestive functions do these structures have?

30. **Inferring** Describe the structure of the cow's teeth. Relate teeth structure to their function.

____

____

Essay

Write the answer to each question in the space provided.

31. Compare hominid evolution with the evolution of other chordate groups.

32. Relate the complexity of human communication and the structure of the human nervous system and sense organs.

33. Compare and contrast the structures of the respiratory systems in all vertebrate groups.

Name________________________________ Class__________________ Date ______________

34. Explain how the control of body temperature is important for maintaining homeostasis.

35. How does an animal's behavior make it better adapted to its environment?

Name________________________ Class________________ Date ____________

Unit 9 Chordates Unit Test B

Multiple Choice

Write the letter that best answers the question or completes the statement on the line provided.

____ 1.. Which of the following is NOT a characteristic of all chordates?
a. dorsal, hollow nerve cord
b. tail that extends past the anus
c. vertebrae
d. pharyngeal pouches

____ 2. All of the following are examples of cartilaginous fishes EXCEPT
a. skates. c. sharks.
b. hagfishes. d. rays.

____ 3. The presence of lungs enables adult amphibians to
a. breathe under water.
b. take in carbon dioxide.
c. reproduce more effectively.
d. live on land.

____ 4. One way in which reptiles are adapted to a fully terrestrial life is
a. external fertilization.
b. an amniotic egg.
c. endothermy.
d. viviparous development.

____ 5. Compared with that of a reptile, a bird's body temperature is
a. higher and more constant.
b. higher and more variable.
c. lower and more constant.
d. lower and more variable.

____ 6. The single most important characteristic that separates birds from other chordates is
a. a four-chambered heart.
b. wings.
c. feathers.
d. the amniotic egg.

____ 7. Characteristics of mammals include all of the following EXCEPT
a. ectothermy.
b. hair.
c. the ability to nourish their young with milk.
d. lungs.

Name______________________ Class______________ Date ___________

____ **8.** Mammals that lay eggs are called
a. monotremes. c. placentals.
b. marsupials. d. reptiles.

____ **9.** The two main groups of primates are
a. Old World monkeys and New World monkeys.
b. apes and humans.
c. hominids and hominoids.
d. prosimians and anthropoids.

____ **10.** Which of the following animals are ectotherms?
a. shark and ostrich
b. salamander and rattlesnake
c. duckbill platypus and crocodile
d. sea turtle and whale

____ **11.** The vertebrate group with the most efficient respiratory system is
a. mammals. c. birds.
b. reptiles. d. amphibians.

____ **12.** A double-loop circulatory system is present in all animals with
a. jaws. c. a two-chambered heart.
b. vertebrae. d. lungs.

____ **13.** An innate behavior
a. appears in fully functional form the first time it is performed.
b. requires habituation.
c. occurs with trial and error.
d. requires reasoning.

____ **14.** When an animal associates a stimulus with a reward or a punishment, it has learned by
a. habituation. c. operant conditioning.
b. classical conditioning. d. insight.

____ **15.** An animal that communicates visually will have
a. good hearing. c. good eyesight.
b. a good sense of smell. d. a well-developed cerebellum.

Completion

Complete each statement on the line provided.

16. The evolution of ______________ and ______________ were important developments during the rise of fishes.

Name______________________________ Class__________________ Date ______________

17. Reptiles belonging to the order Testudines are characterized by a ________________ built into the skeleton, which is used for protection.

18. The evolution of a(an) ________________ enabled hominids to grasp objects and use tools.

19. A(An) ________________ is an animal whose body temperature is controlled from within.

20. When you get up to answer the telephone, the ________________ is the ringing telephone and the ________________ is getting up to answer it.

Short Answer

In complete sentences, write the answers to the questions on the lines provided.

Figure 1

21. Is the chordate in Figure 1 adapted to an aquatic or a terrestrial lifestyle? How do you know?

____

____

____

22. What similarities between reptiles and birds lead scientists to think that birds evolved from extinct reptiles?

____

____

____

23. How do mammalian kidneys help maintain homeostasis?

____

____

____

24. Briefly describe two evolutionary trends scientists use to describe the evolution of chordates.

____

____

____

25. How do social behaviors increase an animal's evolutionary fitness?

____

____

____

Name________________ Class________________ Date ________________

Using Science Skills

Use the diagram below to answer the following questions on the lines provided.

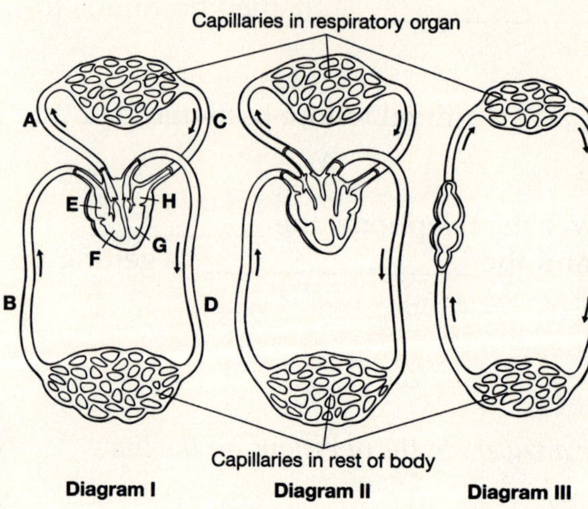

Figure 2

26. Interpreting Graphics Which of the diagrams in Figure 2 illustrate(s) a single-loop circulatory system?

____

____

27. Applying Concepts Which of the circulatory systems in Figure 2 support(s) gills and which support(s) lungs?

____

____

28. Interpreting Graphics In Diagram I, which of the vessels labeled A, B, C, and D contain(s) oxygen-rich blood and which contain(s) oxygen-poor blood?

____

____

____

29. Interpreting Graphics Which of the circulatory systems in Figure 2 has a heart with a ventricle that is partially divided? What vertebrate group is characterized by a heart with this structure?

____

____

____

30. Inferring Which of the circulatory systems in Figure 2 is most efficient at providing body tissues with oxygen? Explain.

____

____

____

Name______________________________ Class__________________ Date ______________

Chapter 35 Nervous System — Chapter Test A

Multiple Choice

Write the letter that best answers the question or completes the statement on the line provided.

____ **1.** The four basic types of tissue in the human body are
a. cell, organ, and organ system.
b. sight, smell, and hearing.
c. thyroid, trachea, adenoid, and bronchus.
d. muscle, nervous, connective, and epithelial.

____ **2.** The levels of organization in the body include
a. endocrine, respiratory, digestive, and nervous.
b. cells, tissues, organs, and organ systems.
c. cells, tissues, and organs.
d. lymphatic, respiratory, and circulatory.

____ **3.** Which of the following general categories of sensory receptors are located everywhere in the body except the brain?
a. thermoreceptors
b. mechanoreceptors
c. photoreceptors
d. pain receptors

____ **4.** What is the function of neurotransmitters?
a. to transmit nerve impulses through dendrites
b. to stimulate the production of epinephrine
c. to transmit nerve impulses across synapses
d. none of the above

____ **5.** Which of the following is a function of the cerebrum?
a. control conscious activities of the body
b. control heart rate
c. control blood pressure
d. control breathing

____ **6.** What types of drugs slow down the activity of the central nervous system?
a. stimulants
b. depressants
c. opiates
d. cocaine

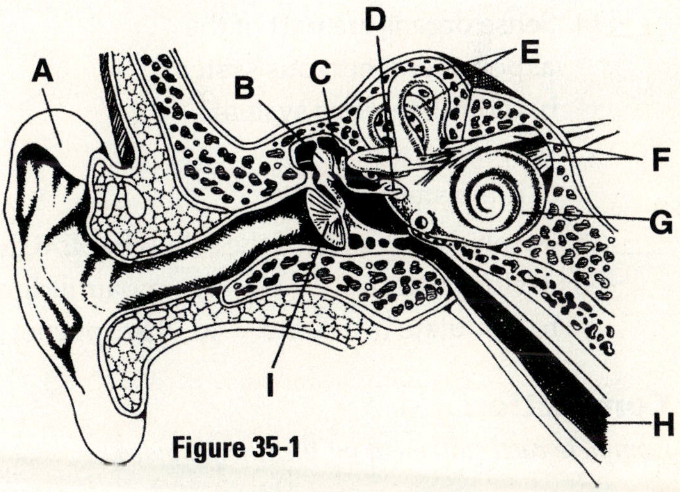

Figure 35-1

____ **7.** In Figure 35-1, which labeled structure is filled with fluid and lined with tiny hair cells?
a. structure D
b. structure E
c. structure F
d. structure G

Name______________________ Class________________ Date ____________

____ 8. The region of the brain that recognizes hunger is the
a. brain stem. c. hypothalamus.
b. medulla oblongata. d. thalamus.

____ 9. When an action potential reaches the end of a neuron, it triggers the release of
a. neurotransmitters. c. dendrites.
b. sodium ions. d. receptors.

____10. Neurons are categorized by the
a. direction in which they carry impulses.
b. amount of metabolic activity that takes place.
c. number of dendrites that branch out.
d. number of impulses that they carry.

____11. Which process enables the body to maintain a stable temperature?
a. heating c. feedback inhibition
b. circulation d. cellular activity

____12. What is the smallest structural and functional unit of the nervous system?
a. receptor c. organ
b. neuron d. tissue

____13. The division of the nervous system that regulates activities that are under conscious control is the
a. somatic nervous system.
b. sensory nervous system.
c. autonomic nervous system.
d. sympathetic nervous system.

____14. Sense organs are part of the
a. peripheral nervous system.
b. central nervous system.
c. spinal cord.
d. brain stem.

____15. Alcohol does each of the following EXCEPT
a. slow reflexes. c. impair judgment.
b. stimulate heart rate. d. disrupt coordination.

Completion

Complete each statement on the line provided.

16. The difference in electrical charge across the cell membrane of a resting neuron is its resting ________________.

17. In most animals, axons and dendrites are clustered into bundles of fibers called ________________.

Name____________________ Class______________ Date ___________

18. Neurons are classified into three types according to the ______________ in which an impulse travels.

19. ______________ is the process by which organisms maintain a relatively constant internal environment.

20. The myelin sheath that surrounds a single long axon leaves many gaps, called ______________, where the axon membrane is exposed.

Short Answer

In complete sentences, write the answers to the questions on the lines provided.

21. Compare and contrast tissues and organs.

22. What are the cause and effect of fetal alcohol syndrome?

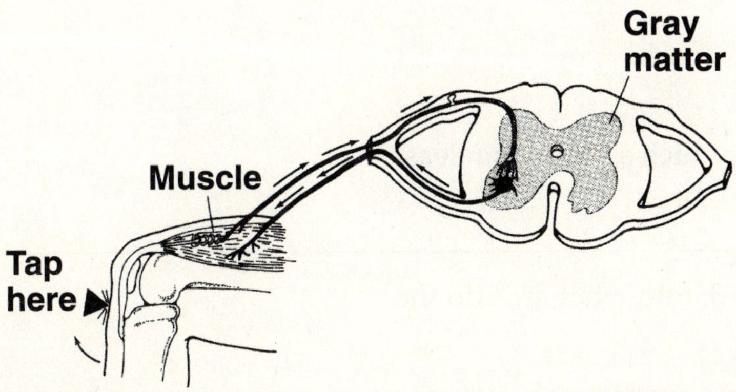

Figure 35-2

23. Describe the reflex arc shown in Figure 35-2.

24. Compare and contrast the functions of rods and cones.

Name________________________ Class________________ Date ____________

25. How is the spinal cord like a major telephone line?

____

____

____

____

Using Science Skills

Use the diagram below to answer the following questions on the lines provided.

This diagram shows the structure of a synapse between the axon of one neuron and the dendrite of a neighboring neuron.

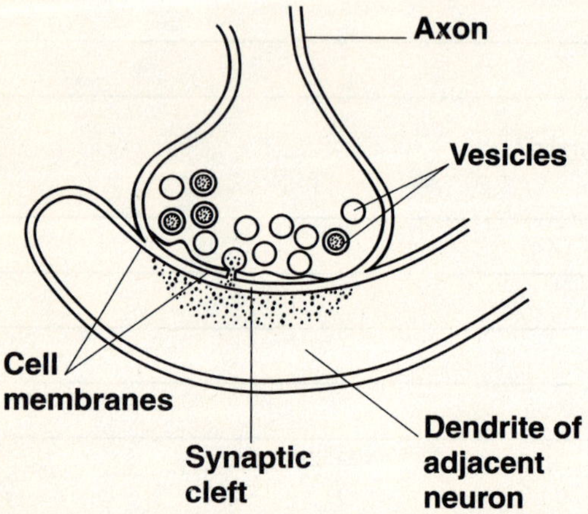

Figure 35-3

26. **Applying Concepts** In Figure 35-3, which structures release neurotransmitters?

____

27. **Interpreting Graphics** In Figure 35-3, into what area do the neurotransmitters diffuse?

____

28. **Predicting** Referring to Figure 35-3, predict the direction of the impulse.

____

29. **Applying Concepts** Referring to Figure 35-3, after the neurotransmitters are released from the cell surface, what happens to the neurotransmitters?

____

____

____

30. **Applying Concepts** If the axon in Figure 35-3 is part of a motor neuron, to what cell are the impulses being passed?

____

Name______________________________ Class_________________ Date ______________

Essay

Write the answer to each question in the space provided.

31. Compare the effects of the sympathetic and the parasympathetic divisions of the automatic nervous system.

32. What is the function of each of the main regions of the brain?

33. What are the five general categories of sensory receptors? Where are they located in the body?

Name______________________________ Class__________________ Date ______________

34. People who must drive long distances sometimes take stimulants. How might this practice be dangerous?

35. Compare resting potential and action potential in a neuron.

Name________________________________ Class__________________ Date ______________

Multiple Choice

Write the letter that best answers the question or completes the statement on the line provided.

____ **1.** The process by which organ systems maintain a relatively constant internal environment is called

a. circulation. c. homeostasis.
b. organization. d. teamwork.

____ **2.** Which system coordinates the body's response to changes in its internal and external environments?

a. lymphatic system c. excretory system
b. nervous system d. reproductive system

____ **3.** What begins when a neuron is stimulated by another neuron in its environment?

a. a threshold c. an impulse
b. a resting potential d. a dendrite

____ **4.** What is the function of the central nervous system?

a. to relay messages
b. to process information
c. to analyze information
d. all of the above

____ **5.** Which division(s) of the peripheral nervous system transmit(s) impulses from sense organs to the central nervous system?

a. sensory division
b. motor division
c. sensory and motor division
d. spinal cord division

____ **6.** Which general category of sensory receptors detects variations in temperature?

a. thermoreceptors c. photoreceptors
b. mechanoreceptors d. pain receptors

____ **7.** Which of the five senses contains two types of photoreceptors called rods and cones?

a. vision c. smell
b. hearing d. taste

____ **8.** Drugs that increase heart rate, blood pressure, and breathing rate are called

a. stimulants. c. opiates.
b. depressants. d. alcohol.

Name________________________________ Class____________________ Date ______________

____ **9.** Which illegal drug is most widely abused?

a. marijuana
c. amphetamines
b. tranquilizers
d. alcohol

____ **10.** How many organ systems make up the human body?

a. 4
c. 8
b. 5
d. 11

____ **11.** A group of similar cells that perform a single function is called a(an)

a. nerve.
c. tissue.
b. organ.
d. organ system.

____ **12.** Which type of tissue lines your internal organs?

a. epithelial
c. nerve
b. connective
d. muscle

____ **13.** Which type of tissue provides support for the body?

a. epithelial
c. nerve
b. connective
d. muscle

____ **14.** What are the two divisions of the peripheral nervous system?

a. brain and spinal cord
b. thalamus and hypothalamus
c. somatic and autonomic
d. sensory and motor

____ **15.** What system does alcohol immediately affect?

a. digestive
c. nervous
b. circulatory
d. endocrine

Completion

Complete each statement on the line provided.

16. The two major divisions of the human nervous system are the central and the ____________________ nervous systems.

17. The process by which the product of a system shuts down the system or limits its operation is called ____________________ inhibition.

18. Sensory receptors called ____________________ are found in the skin, skeletal muscles, and inner ears and are sensitive to touch, pressure, stretching of muscles, sound, and motion.

19. Sensory neurons and motor neurons are connected by neurons called ________________, which carry impulses between them.

20. Cirrhosis of the liver is a possible result of the long-term use of ____________________.

Name________________________________ Class__________________ Date ______________

Short Answer

In complete sentences, write the answers to the questions on the lines provided.

21. Distinguish between the functions of dendrites and axons.

22. How do opiates help people overcome sensations of pain?

23. What are the major regions of the brain?

24. Identify the five main senses.

25. What is homeostasis?

Name________________________ Class________________ Date ____________

Using Science Skills

Use the diagram below to answer the following questions on the lines provided.

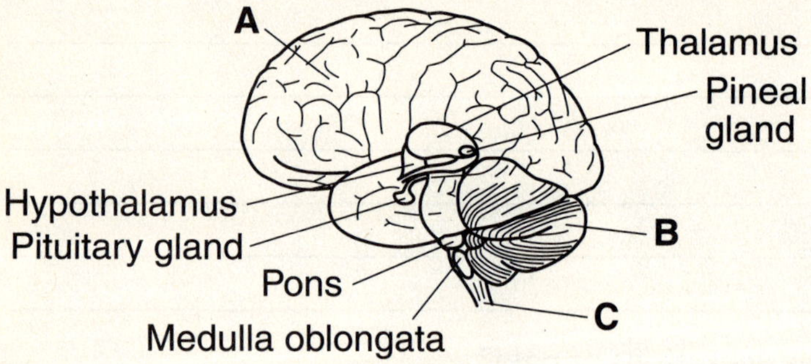

Figure 35-1

26. Applying Concepts What is the name of structure C in Figure 35-1?

____

____

27. Applying Concepts Referring to Figure 35-1, which structure is responsible for the voluntary, or conscious, activities of the body?

____

____

28. Applying Concepts Referring to Figure 35-1, which structure is the control center for recognition and analysis of hunger, thirst, fatigue, anger, and body temperature?

____

____

29. Applying Concepts In Figure 35-1, which structure is the cerebellum?

____

____

30. Applying Concepts Referring to Figure 35-1, what two regions of the brain stem act as neural "switchboards," regulating the flow of information between the brain and the rest of the body?

____

____

Name______________________ Class______________ Date__________

Multiple Choice

Write the letter that best answers the question or completes the statement on the line provided.

____ **1.** Which of the following is NOT part of the axial skeleton?
a. skull
b. vertebral column
c. pelvis
d. rib cage

____ **2.** How many bones are found in an adult human skeleton?
a. 150
b. 200
c. 206
d. 212

____ **3.** Which of the following contains nerves and blood vessels and runs through the compact bone?
a. osteocytes
b. Haversian canals
c. bone marrow
d. periosteum

____ **4.** Which of the following is true about red marrow?
a. It makes blood cells.
b. It stores fat.
c. It produces new bone.
d. It is found in compact bones of adults.

____ **5.** What is the function of cells called osteoclasts?
a. build and maintain bone
b. break down bone
c. support the skin
d. produce blood cells

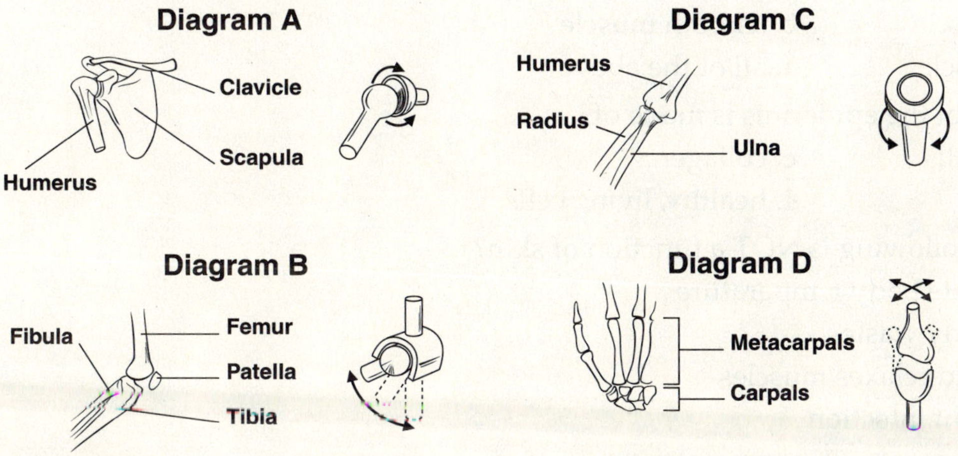

Figure 36-1

____ **6.** Which diagram in Figure 36-1 is an example of a ball-and-socket joint?
a. Diagram A
b. Diagram B
c. Diagram C
d. Diagram D

Name_____________________________ Class___________________ Date _____________

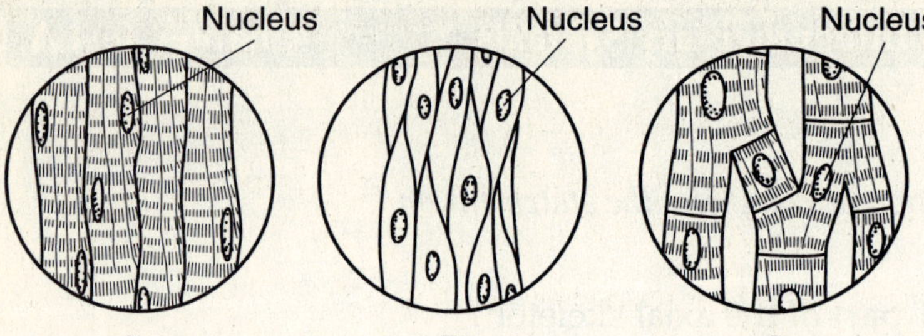

Figure 36-2

_____ **7.** Which diagram(s) in Figure 36-2 show(s) muscles that are striated?

a. A and B c. C
b. B d. A and C

_____ **8.** About what percent of the mass of the human body is made up of muscle?

a. under 10 percent c. 30 percent
b. 20 percent d. over 40 percent

_____ **9.** Where is the protein called actin located in the body?

a. ATP molecules c. myosin filaments
b. thin filaments d. acetylcholine neurotransmitters

_____**10.** In addition to myosin, what other protein is involved in skeletal muscle contraction?

a. collagen c. keratin
b. actin d. melanin

_____**11.** Which of the following types of muscle generally remain in a state of partial contraction when you are keeping your legs straight?

a. skeletal muscle c. smooth muscle
b. cardiac muscle d. all of the above

_____**12.** The top layer of the epidermis is made of

a. flat, dead cells. c. collagen.
b. melanin. d. healthy, living cells.

_____**13.** Which of the following is NOT a function of skin?

a. helps regulate body temperature
b. removes body wastes
c. contracts and relaxes muscles
d. helps prevent infection

_____**14.** Which of the following is NOT part of the integumentary system?

a. skin c. nails
b. cartilage d. hair

Name______________________ Class________________ Date ____________

____15. From which of the following does the basic structure of hair and nails form?

a. sweat
b. melanin
c. collagen
d. keratin

Completion

Complete each statement on the line provided.

16. A less dense tissue known as ________________ is found in the ends of long bones and in the middle part of short, flat bones.

17. The joints between adjacent vertebrae are slightly ________________ joints.

18. The concept used to explain the action of filaments in muscle contraction is the ________________.

19. The point of contact between a motor neuron and a skeletal muscular cell is called a(an) ________________.

20. A slight scratch of the skin does not cause bleeding because the epidermis lacks ________________.

Short Answer

In complete sentences, write the answers to the questions on the lines provided.

21. Name and describe the two major parts of the skeletal system.

____

____

____

____

22. Where is cartilage located in an adult human? Why is cartilage necessary?

____

____

____

____

____

____

23. In Figure 36-3, identify and explain the function of the structure labeled A.

____

____

____

____

____

____

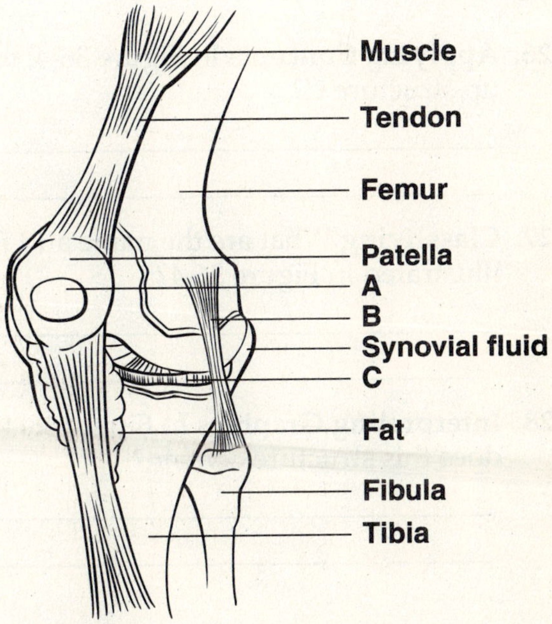

Figure 36-3

Name________________________ Class________________ Date____________

24. In a left-handed person, which hand would probably have more strength? Why?

____

____

____

____

25. Compare and contrast keratin and melanin.

____

____

____

____

Using Science Skills

Use the diagram below to answer the following questions on the lines provided.

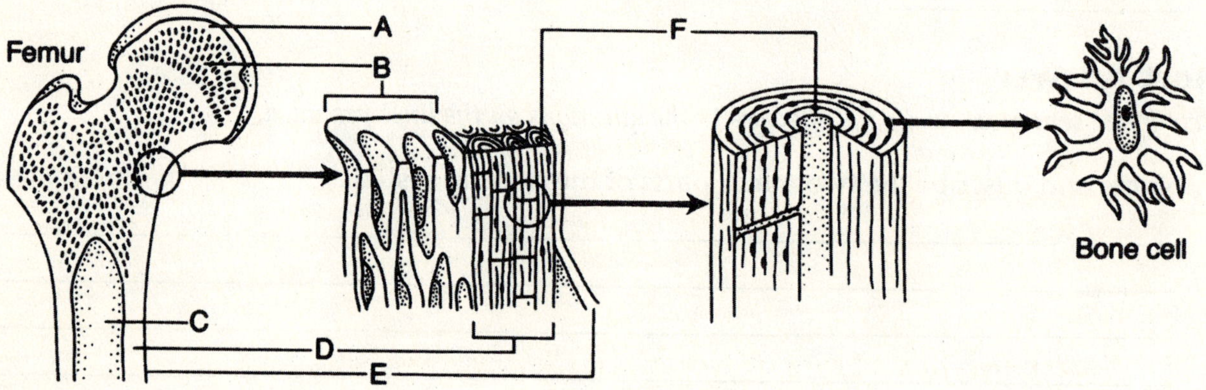

Figure 36-4

26. **Applying Concepts** In Figure 36-4, what two tissues might make up structure C?

____

____

27. **Classifying** What are the name and the function of the bone cell illustrated in Figure 36-4?

____

____

28. **Interpreting Graphics** In Figure 36-4, identify structure F. What does this structure contain?

____

____

Name________________________ Class________________ Date ____________

29. **Applying Concepts** What are the name and the function of structure E in Figure 36-4?

____

____

30. **Interpreting Graphics** What structures are found inside the compact bone of the femur illustrated in Figure 36-4?

____

____

Essay

Write the answer to each question in the space provided.

31. Compare and contrast spongy bone and compact bone.

32. Name the three types of joints. Give an example of each type.

Name________________________________ Class____________________ Date ______________

33. Name and compare the three types of muscle tissue.

34. Why is regular exercise important for the body?

35. How is hair produced, and how does it grow?

Name________________ Class________________ Date________________

Multiple Choice

Write the letter that best answers the question or completes the statement on the line provided.

____ 1. Which of the following provides support for the body, attachment sites for muscles, and protection for internal organs?

a. skin c. skeleton

b. spinal cord d. joints

____ 2. Which of the following is NOT a function of bones in the human skeletal system?

a. storing minerals

b. regulating body temperature

c. moving body parts

d. protecting organs

____ 3. The main function of the skull is to

a. produce blood cells.

b. protect the heart and lungs.

c. protect the brain.

d. act like a lever for muscle attachment.

____ 4. The periosteum is a

a. place for storing fat cells.

b. location for red blood cell production.

c. tough layer of connective tissue surrounding a bone.

d. latticework structure that adds strength to the bone.

____ 5. Of what is the skeleton of a newborn baby mainly composed?

a. bone c. bone marrow

b. cartilage d. growth plates

____ 6. A saddle joint is represented in which diagram in Figure 36-1?

a. Diagram A c. Diagram C

b. Diagram B d. Diagram D

____ 7. Ligaments connect

a. cartilage to bone. c. bone to muscle.

b. muscle to muscle. d. bone to bone.

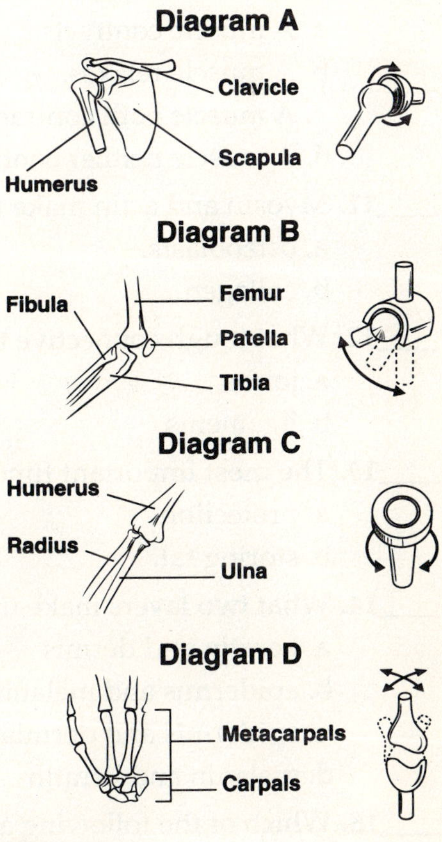

Figure 36-1

Name________________________________ Class____________________ Date ______________

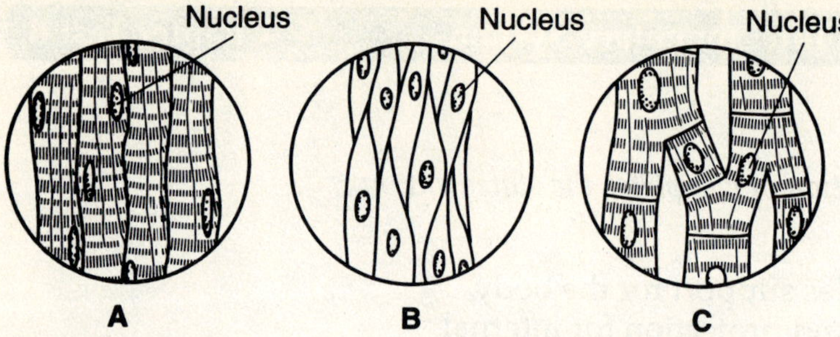

Figure 36-2

____ 8. In Figure 36-2, diagram B is an example of
a. cardiac muscle.
b. skeletal muscle.
c. smooth muscle.
d. heart muscle.

____ 9. Where is cardiac muscle tissue located in the body?
a. heart
b. ribs
c. skull
d. bones

____10. What happens when the thin filaments in a muscle fiber slide over the thick filaments?
a. A muscle contracts.
b. A muscle relaxes.
c. A muscle both contracts and relaxes.
d. A muscle neither contracts nor relaxes.

____11. Myosin and actin make up
a. osteoblasts.
b. collagen.
c. myofibrils.
d. red marrow.

____12. What tough connective tissue joins skeletal muscles to bones?
a. joints
b. ligaments
c. periosteum
d. tendons

____13. The most important function of the skin is
a. protection.
b. storing fat.
c. sweating.
d. insulation.

____14. What two layers make up skin?
a. keratin and dermis
b. epidermis and melanin
c. epidermis and dermis
d. melanin and keratin

____15. Which of the following are tubelike pockets of epidermal cells that extend into the dermis?
a. epidermal layers
b. keratin layers
c. melanin cells
d. hair follicles

Name______________________________ Class________________ Date ____________

Completion

Complete each statement on the line provided.

16. The vertebral column and skull are part of the ________________ skeleton.

17. The ________________ form a cage that protects the heart and lungs.

18. The process by which cartilage is replaced by bone is ________________.

19. The dark pigment in the skin is called ________________.

20. Hair and nails are part of the ________________ system.

Short Answer

In complete sentences, write the answers to the questions on the lines provided.

21. What are three functions of the human skeleton?

____

____

____

____

22. Of what are bones made?

____

____

____

____

____

23. What type of freely movable joint is being illustrated in Figure 36-3?

____

____

24. What are the three different types of muscles?

____

____

____

____

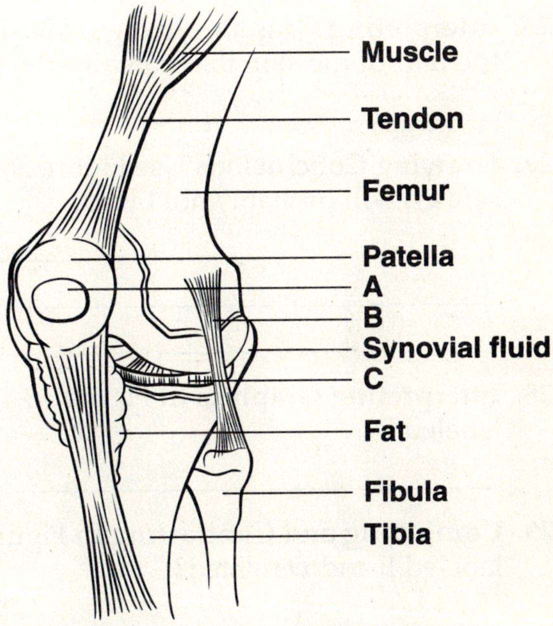

Figure 36-3

25. What are the four functions of the integumentary system?

____

____

____

____

____

Name________________________________ Class__________________ Date ______________

Using Science Skills

Use the diagram below to answer the following questions on the lines provided.

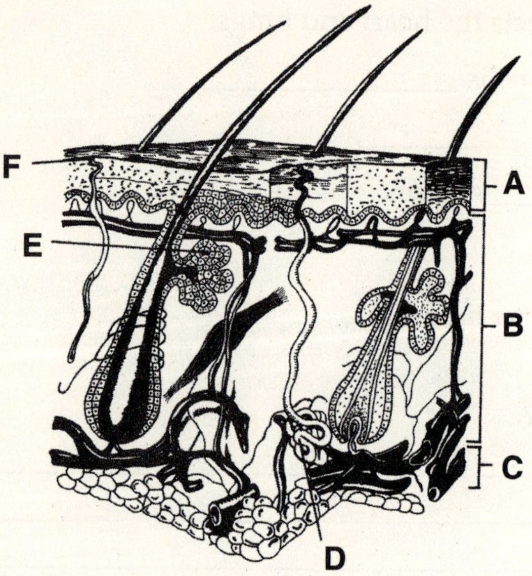

Figure 36-4

26. **Interpreting Graphics** In Figure 36-4, what letter is used to label the part of the skin that contains the melanin?

____

27. **Drawing Conclusions** Use Figure 36-4 to explain why a slight scratch will probably not bleed.

____

____

____

28. **Interpreting Graphics** In Figure 36-4, what is the part of the skin labeled B?

____

29. **Comparing and Contrasting** In Figure 36-4, how are the structures labeled F and D related?

____

____

30. **Interpreting Graphics** What is the function of the structure labeled E in Figure 36-4?

____

____

Name______________________________ Class________________ Date ____________

Multiple Choice

Write the letter that best answers the question or completes the statement on the line provided.

____ **1.** Which of the following is NOT a part of the circulatory system?
a. heart c. blood vessels
b. air passageway d. blood

____ **2.** Which of the following pathways is the largest of the circulatory system?
a. systemic circulation
b. pulmonary circulation
c. lymphatic circulation
d. coronary circulation

____ **3.** Where are the cells that make up the sinoatrial node, or pacemaker, located?
a. right atrium c. right ventricle
b. left atrium d. left ventricle

____ **4.** Compared with the walls of arteries, the walls of veins
a. are thicker. c. lack valves.
b. are thinner. d. have more resistance.

____ **5.** The function of valves in the human circulatory system is to
a. stimulate the heartbeat.
b. accelerate the flow of blood.
c. prevent the backward flow of blood.
d. serve as a cushion to prevent friction.

____ **6.** Which helps regulate blood pressure?
a. spleen c. bronchus
b. kidney d. alveolus

____ **7.** Which of the following activities is(are) key(s) to avoiding cardiovascular disease?
a. exercise c. weight control
b. low-fat diet d. all of the above

____ **8.** How much blood does the human body contain?
a. 1–2 liters c. 8–10 liters
b. 4–6 liters d. 12–14 liters

____ **9.** In Figure 37-1, what does step B show?
a. clot forming
b. capillary wall breaking
c. clumping of platelets
d. conversion of fibrinogen into fibrin

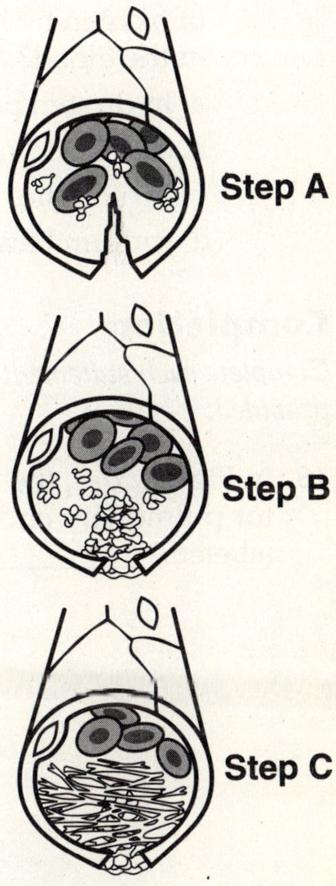

Figure 37-1

Name_________________________ Class_________________ Date _____________

_____10. Into what substance is fibrinogen converted?

a. thrombin
b. fibrin
c. platelets
d. thromboplastin

_____11. Which of the following is NOT a function of the lymphatic system?

a. collects and returns lost fluid to the circulatory system
b. traps bacteria
c. absorbs protein
d. carries fat-soluble vitamins to the blood

_____12. Which of the following are signs that a person is breathing?

a. hearing or feeling air being exhaled from the nose
b. chest rising and falling
c. hearing or feeling air being exhaled from the mouth
d. all of the above

_____13. Air is filtered, warmed, and moistened in the

a. nose.
b. alveoli.
c. lungs.
d. bronchioles.

_____14. Air is forced into the lungs by the contraction of the

a. alveoli.
b. bronchioles.
c. diaphragm.
d. heart.

_____15. What gas, found in cigarette smoke, blocks the transport of oxygen by hemoglobin in the blood?

a. hydrogen peroxide
b. carbon monoxide
c. carbon dioxide
d. sodium bicarbonate

Completion

Complete each statement on the line provided.

16. In Figure 37-2, the pathway for pulmonary circulation is labeled _________________.

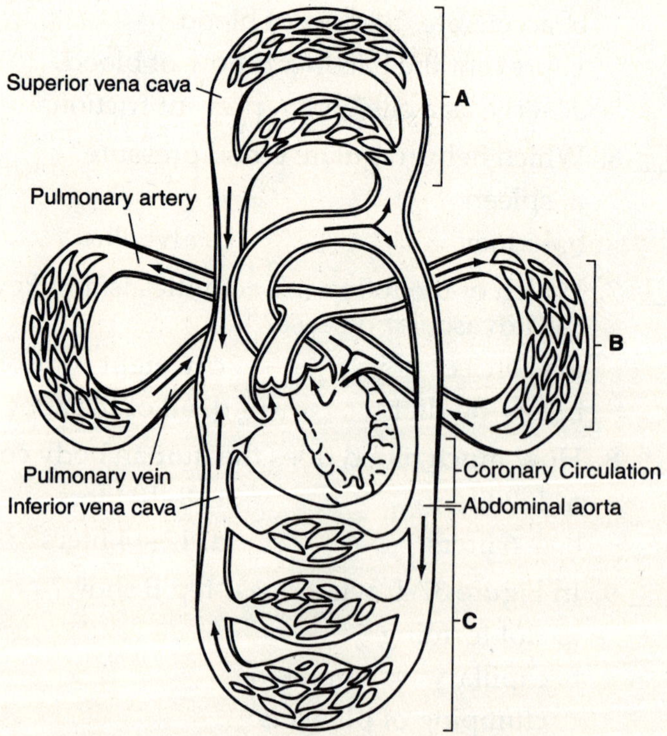

Figure 37-2

Name ______________________ Class ______________ Date ____________

17. Medical workers use a sphygmomanometer to measure ______________.

18. Fatty deposits called plaque build up on the inner walls of arteries, causing a condition known as ______________.

19. Plasma consists mostly of ______________.

20. A genetic disorder called ______________ results from a defective protein in the clotting pathway.

Short Answer

In complete sentences, write the answers to the questions on the lines provided.

21. Are the ribs raised or lowered after you exhale?

____

22. List the main functions of the three types of blood cells.

____

____

____

____

23. List in order the structures of the respiratory system through which air passes from the outside environment to the alveoli.

____

____

____

24. In what stage of breathing is the total volume of the chest cavity largest?

____

____

____

____

____

____

____

____

25. Name and describe three respiratory diseases that can be caused by smoking.

____

____

____

____

Name______________________ Class__________________ Date ____________

Using Science Skills

Use the diagram below to answer the following questions on the lines provided.

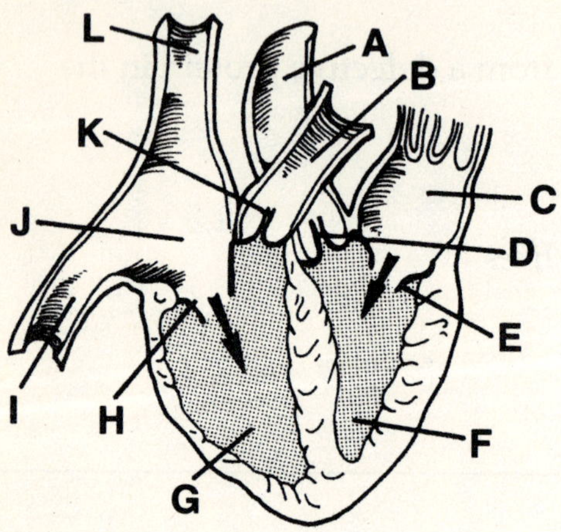

Figure 37-3

26. Interpreting Graphics Name and describe the function of the structure labeled I in Figure 37-3.

____

____

____

27. Drawing Conclusions What is the function of the structures labeled D, E, H, and K in Figure 37-3?

____

____

____

____

____

28. Interpreting Graphics In Figure 37-3, what label shows the vessel in which oxygen-rich blood leaves the heart and goes to the rest of the body?

____

____

____

____

____

29. Drawing Conclusions In Figure 37-3, what happens to valves E and H when the ventricles contract?

____

Name________________________ Class________________ Date ____________

30. **Interpreting Graphics** In Figure 37-3, what is the function of the structure labeled B?

____

____

Essay

Write the answer to each question in the space provided.

31. Describe the function of each of the following: (a) the muscular walls of the large arteries, (b) the valves in the veins, and (c) the valves in the heart.

32. How does the body regulate blood pressure?

33. How do white blood cells guard against infection, fight parasites, and attack bacteria?

34. What is the lymphatic system, and what are its functions?

35. Compare the two types of biological respiration.

Name______________________ Class______________ Date __________

Multiple Choice

Write the letter that best answers the question or completes the statement on the line provided.

____ 1. Which body system acts as a transportation system?
a. circulatory c. nervous
b. respiratory d. excretory

____ 2. In the walls of the heart, there are two thin layers that form a sandwich around a thick layer of muscle called the
a. epithelial tissue layer.
b. pericardium.
c. connective tissue layer.
d. myocardium.

____ 3. Which of the following are the smallest of the blood vessels?
a. arteries c. lymphatic cells
b. veins d. capillaries

____ 4. Which of the following is NOT a type of blood vessel?
a. artery c. lymphatic cell
b. vein d. capillary

____ 5. When the heart contracts, it produces a wave of fluid pressure in the
a. veins. c. capillaries.
b. arteries. d. pharynx.

____ 6. When the pressure of blood pumping through the blood vessels is constantly too high, the condition is known as
a. a heart attack. c. atherosclerosis.
b. hypertension. d. a stroke.

____ 7. Which of the following is correct about the composition of plasma?
a. 90 percent water c. 10 percent water
b. 50 percent water d. no water

____ 8. Which plasma proteins help to regulate osmotic pressure and blood volume?
a. albumins c. fibrinogens
b. globulins d. platelets

____ 9. Which of the following blood cells contain hemoglobin?
a. red blood cells c. platelets
b. white blood cells d. all of the above

____10. Which body system collects the fluid that is lost by the blood and returns it to the body's transport system?

a. circulatory c. respiratory
b. lymphatic d. nervous

____11. What is the term used to describe a swelling of the tissues due to the accumulation of excess fluid, which can occur when lymphatic vessels are blocked because of injury or disease?

a. hemophilia c. edema
b. stroke d. hypertension

____12. The process by which oxygen and carbon dioxide are exchanged between cells, the blood, and air in the lungs is one type of

a. systemic circulation.
b. respiration.
c. emphysema.
d. cellular respiration.

____13. What structure serves as a passageway for both air and food?

a. pharynx c. larynx
b. trachea d. bronchus

____14. Generally speaking, what controls breathing?

a. the brain c. the diaphragm
b. the lungs d. the heart

____15. Which of the following is NOT contained in tobacco smoke?

a. carbon monoxide c. nicotine
b. caffeine d. tar

Completion

Complete each statement on the line provided.

16. A circulating fluid called ____________________ is pumped through a system of vessels in the body.

17. In Figure 37-1 on the next page, the area labeled B represents the capillaries of a(an) ____________________.

18. The iron-containing protein called ____________________ binds to oxygen in the lungs and transports it to tissues throughout the body, where the oxygen is released.

19. Inhaled air passes from the trachea to one of the two ____________________.

20. Breathing is such an important function that your ____________________ system will not let you have complete control over it.

Name______________________ Class______________ Date__________

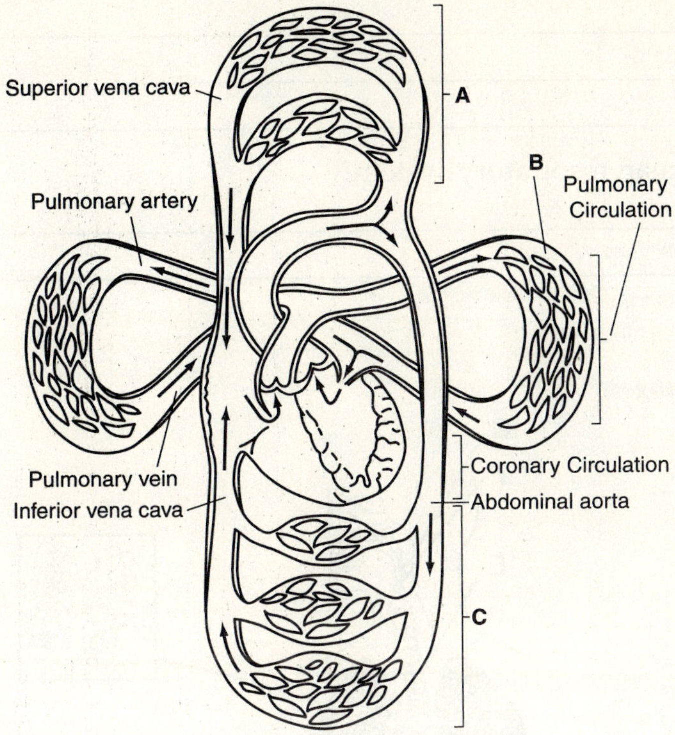

Figure 37-1

Short Answer

In complete sentences, write the answers to the questions on the lines provided.

21. What are the three structures of the circulatory system?

22. Identify and compare the functions of the three types of blood vessels of the circulatory system.

23. What is blood pressure?

Name______________________________ Class____________________ Date ______________

24. Describe blood plasma.

25. What is the basic function of the human respiratory system?

Using Science Skills

Use the diagram below to answer the following questions on the lines provided.

26. **Interpreting Graphics** In Figure 37-2, which labeled structure is the diaphragm?

27. **Interpreting Graphics** In Figure 37-2, what are two names for the structure labeled E?

28. **Interpreting Graphics** In Figure 37-2, what is the label and name of the structures in which the exchange of gases takes place?

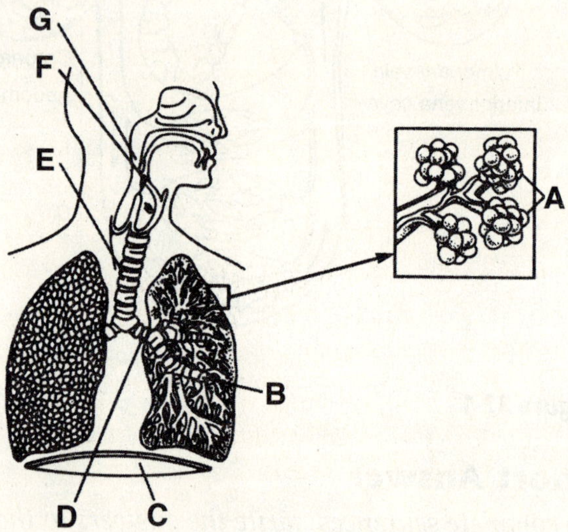

Figure 37-2

29. **Interpreting Graphics** What is the label and name of the structure in Figure 37-2 that prevents the passage of food from the pharynx into the lungs?

30. **Interpreting Graphics** In Figure 37-2, what is the label and name of the structure that carries air from the trachea to a lung?

Name______________________________ Class_________________ Date ______________

Multiple Choice

Write the letter that best answers the question or completes the statement on the line provided.

____ **1.** The energy available in foods can be measured by
a. first determining which nutrients the body needs.
b. burning the foods.
c. tracing chemical pathways.
d. all of the above

____ **2.** Inorganic molecules that perform vital functions in the body are called
a. lipids.
b. proteins.
c. vitamins.
d. minerals.

____ **3.** The American Heart Association recommends a diet with a maximum of
a. 40 percent of Calories from fat, of which only 20 percent should be from unsaturated fats.
b. 30 percent of Calories from saturated fat, in addition to 10 percent from unsaturated fats.
c. 50 percent of Calories from saturated fats.
d. 30 percent of Calories from fat, of which only 10 percent should be from saturated fats.

____ **4.** What is the minimum amount of water you should drink every day?
a. 0.5 liter
b. 1 liter
c. 2 liters
d. 3 liters

____ **5.** Water is lost from the body by each of the following EXCEPT
a. sweating.
b. urinating.
c. respiration.
d. digestion.

____ **6.** Which of the following statements is NOT true about the process of sweating?
a. Sweating is more likely to occur on hot days than on cool days.
b. Sweating is likely to occur after strenuous exercise.
c. Evaporation of sweat warms the body.
d. Excessive sweating can lead to dehydration.

____ **7.** According to the Food Guide Pyramid, the majority of food in your diet should be from the
a. Fruit Group.
b. Bread, Cereal, Rice, and Pasta Group.
c. Vegetable Group.
d. Milk, Yogurt, and Cheese Group.

Name_______________________ Class_______________ Date ___________

____ 8. What is one of the roles of the pancreas in nutrition?
a. absorbs nutrients
b. churns food
c. dissolves food
d. neutralizes acids

____ 9. Where does the process of chemical digestion begin?
a. stomach
b. esophagus
c. small intestine
d. mouth

____10. Through which structure do wastes pass into the rectum?
a. duodenum
b. large intestine
c. small intestine
d. villus

____11. Water is extracted from digested food in the body primarily by the
a. bladder.
b. gallbladder.
c. large intestine.
d. pancreas.

____12. The function of the excretory system is to help maintain homeostasis by
a. breaking down nutrients.
b. removing wastes.
c. absorbing nutrients.
d. preventing infection.

____13. Which of the following is NOT part of a nephron?
a. urethra
b. renal vein
c. glomerulus
d. Bowman's capsule

____14. What percentage of the filtrate's water that enters Bowman's capsule is reabsorbed into the blood?
a. 100 percent
b. almost 99 percent
c. 50 percent
d. less than 25 percent

____15. What materials are removed from the blood through dialysis?
a. urea and excess salts
b. chyme
c. amylase and pepsin
d. none of the above

Completion

Complete each statement on the line provided.

16. Calcium, iron, and magnesium are all examples of the group of nutrients called ________________.

17. Most of the blood, lymph, and other bodily fluids consist of ________________.

18. The teeth are extremely important in ________________ digestion.

19. Gastric glands produce mucus, hydrochloric acid, and the enzyme ________________.

20. To a large extent, the activity of the kidneys is controlled by the composition of ________________.

Name______________________________ Class__________________ Date ______________

Short Answer

In complete sentences, write the answers to the questions on the lines provided.

21. Name two functions of fats.

____

____

____

____

22. How does a calorie differ from a Calorie?

____

____

____

____

____

23. What is the Food Guide Pyramid?

____

____

____

____

____

24. What is the function of the ureter?

____

____

____

____

____

____

____

25. What is the function of the urinary bladder?

____

____

____

____

____

____

____

Name________________________ Class________________ Date ____________

Using Science Skills

Use the diagram below to answer the following questions on the lines provided.

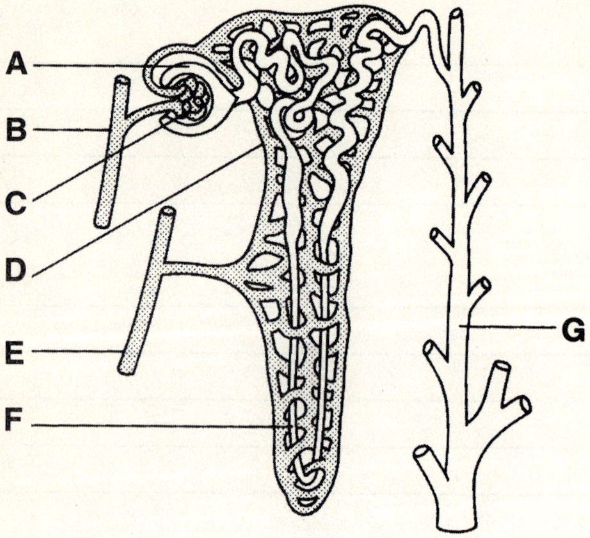

Figure 38-1

26. **Interpreting Graphics** In Figure 38-1, what is the name and what is the function of structure F?

____

27. **Interpreting Graphics** In Figure 38-1, in which structure are useful substances reabsorbed into the blood?

____

28. **Drawing Conclusions** In Figure 38-1, what is the name and what is the purpose of structure G?

____

____

____

____

29. **Comparing and Contrasting** What two processes take place in the structures illustrated in Figure 38-1?

____

____

30. **Inferring** In what organ are the structures illustrated in Figure 38-1 located?

____

____

Name________________________________ Class____________________ Date ______________

Essay

Write the answer to each question in the space provided.

31. How does the body lose water?

32. Discuss the functions of the following structures of the excretory system: kidneys, ureter, urinary bladder, and urethra.

33. What organs and major accessory structures make up the digestive system?

34. How do the kidneys respond to an increased intake of water? Of salt?

35. Describe the process of dialysis.

Name______________________________ Class__________________ Date ____________

Multiple Choice

Write the letter that best answers the question or completes the statement on the line provided.

____ **1.** The energy to perform actions, as well as the materials from which body cells and tissues are made, comes from

a. cellulose. c. food.
b. oxygen. d. blood.

____ **2.** A calorie is the amount of energy needed to

a. raise the temperature of the body by 1° Celsius.
b. raise the temperature of 1 g of fat by 1° Celsius.
c. raise the temperature of 1 g of water by 1° Celsius.
d. none of the above

____ **3.** Substances that are needed by the body for growth, repair, and maintenance are called

a. enzymes. c. ATP.
b. nutrients. d. Calories.

____ **4.** The raw materials that the body needs for growth and repair come from

a. proteins. c. carbohydrates.
b. unsaturated fats. d. water.

____ **5.** Proteins are made of polymers called

a. lipids. c. amino acids.
b. unsaturated fats. d. vitamins.

____ **6.** Which of the following structures is made up mostly of water?

a. blood c. body fluids
b. lymph d. all of the above

____ **7.** How many food groups make up the Food Guide Pyramid?

a. four c. six
b. five d. seven

____ **8.** What passage carries food between the pharynx and the stomach?

a. small intestine c. esophagus
b. epiglottis d. large intestine

____ **9.** What enzyme found in saliva breaks chemical bonds between the sugar monomers in starches?

a. amylase c. pepsin
b. chyme d. hydrochloric acid

____**10.** The stomach breaks down food into a soft, partially digested mixture called

a. pepsin. c. bolus.
b. chyme. d. bile.

____**11.** In what structure in Figure 38-1 do filtration and reabsorption occur?

a. kidney
b. renal vein
c. urinary bladder
d. ureter

____**12.** The main organs of the excretory system are the

a. kidneys. c. intestines.
b. lungs. d. ureters.

____**13.** In the kidneys, both useful substances and wastes are removed from the blood by

a. reabsorption. c. dialysis.
b. excretion. d. filtration.

____**14.** The activity of the kidneys is controlled by hormones and by the

a. volume of nutrients.
b. volume of filtrate.
c. composition of the blood.
d. composition of the urine.

____**15.** Dialysis performs the function of the

a. liver. c. kidneys.
b. ureter. d. urinary bladder.

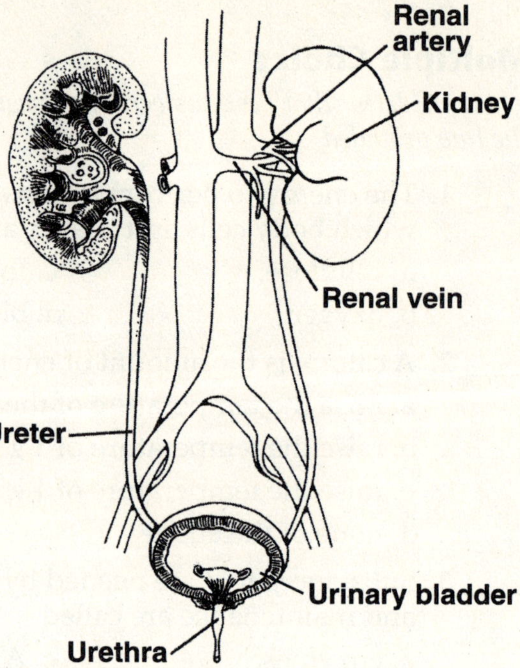

Figure 38-1

Completion

Complete each statement on the line provided.

16. The science of ____________________ is the study of food and its effects on the body.

17. Contractions, known as ____________________, squeeze food through the length of the esophagus into the stomach.

18. If a part of the stomach wall digests itself, a(an) ____________________ develops.

19. The functional units of the kidneys are the ____________________.

20. The kidneys play an important role in maintaining ____________________ in the body.

Name________________________ Class________________ Date ____________

Short Answer

In complete sentences, write the answers to the questions on the lines provided.

21. List the six nutrients needed by the body.

22. How is chyme produced?

23. What is the function of the digestive system?

24. How do kidneys maintain homeostasis?

25. What are a person's two options for survival if both of his or her kidneys are unable to function?

Name______________________________ Class__________________ Date ______________

Using Science Skills

Use the diagram below to answer the following questions on the lines provided.

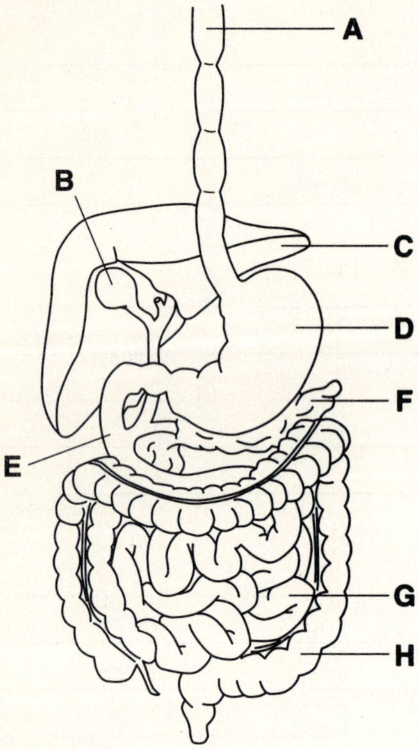

Figure 38-2

26. **Interpreting Graphics** Which structure shown in Figure 38-2 is the esophagus?

____

27. **Interpreting Graphics** Which structure shown in Figure 38-2 is the stomach?

____

28. **Interpreting Graphics** Which structure shown in Figure 38-2 produces bile? What is the name of this structure?

____

29. **Interpreting Graphics** Hormones that regulate blood sugar are produced in which structure shown in Figure 38-2? What is the name of this structure?

____

30. **Interpreting Graphics** Which structure shown in Figure 38-2 contains villi, which aid in absorbing nutrients? What is the name of this structure?

____

Name______________________________ Class__________________ Date ______________

Multiple Choice

Write the letter that best answers the question or completes the statement on the line provided.

____ **1.** The endocrine system
a. affects only the reproductive system.
b. releases hormones into the bloodstream.
c. competes with the nervous system.
d. is made up primarily of glands with ducts.

____ **2.** Unlike endocrine glands, exocrine glands
a. have ducts.
b. release hormones.
c. release secretions directly into the bloodstream.
d. are found throughout the body.

____ **3.** A thermostat is a good example of a(an)
a. hormone-receptor complex.
b. feedback system.
c. prostaglandin.
d. exocrine gland.

____ **4.** Which structure in Figure 39-1 regulates the level of calcium in the blood?
a. C c. E
b. D d. F

____ **5.** Which gland produces epinephrine and norepinephrine?
a. parathyroid c. pituitary
b. hypothalamus d. adrenal

____ **6.** Which of the following secondary sex characteristics is caused by testosterone?
a. deepening of the voice
b. widening of the hips
c. development of breasts
d. all of the above

____ **7.** Which structure produces sperm?
a. scrotum
b. epididymis
c. seminiferous tubules
d. vas deferens

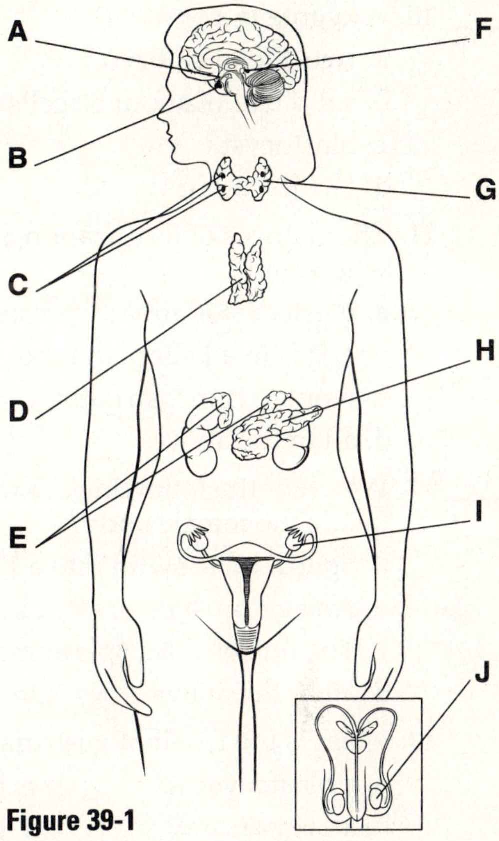

Figure 39-1

Name______________________________ Class__________________ Date ______________

____ 8. When during the menstrual cycle does an egg have the best chance of being fertilized?
a. during the follicular phase
b. just before menstruation
c. at the beginning of the luteal phase
d. on the day of ovulation

____ 9. Menstruation does not occur if the
a. uterine lining thickens.
b. estrogen level falls.
c. progesterone level falls.
d. egg is fertilized.

____10. A zygote is a
a. two-celled embryo.
b. solid ball of about 50 cells.
c. blastocyst.
d. fertilized egg.

____11. The chances of fertilization are very good if sperm are present when a(an)
a. blastocyst is already present.
b. egg is in a Fallopian tube.
c. woman is menstruating.
d. all of the above

____12. Which of the following are required for fertilization to occur inside the female body?
a. Sperm must swim into a Fallopian tube.
b. An egg must be present in the Fallopian tube.
c. The nucleus of a sperm must enter an egg cell.
d. all of the above

____13. What is the result of gastrulation?
a. a blastocyst
b. a zygote
c. the amnion
d. germ layers

____14. Which of the following is a function of the placenta?
a. mixing the blood of the mother and the fetus
b. protecting the fetus from any drugs or alcohol in the mother's body
c. providing nutrients to the fetus
d. cushioning and protecting the fetus

____15. Which of the following is characteristic of infancy?
a. Teeth appear.
b. Puberty begins.
c. The first signs of aging appear.
d. An individual reaches 70 percent of his or her height.

Name__________________ Class____________ Date __________

Completion

Complete each statement on the line provided.

16. The ________________ system is made up of glands that release their products into the ________________ .

17. The shortest phase of the menstrual cycle is ________________ .

18. The testes are contained in an external sac called the ________________ .

19. During ________________ , three germ layers form. They are called ________________ , ectoderm, and mesoderm.

20. A problem with the ________________ can disrupt an embryo's respiration, nourishment, and excretion.

Short Answer

In complete sentences, write the answers to the questions on the lines provided.

21. Compare endocrine glands and exocrine glands. Give an example of each kind of gland.

22. What are prostaglandins? How are they different from true hormones?

23. Give an example of how the endocrine system maintains homeostasis.

24. List the structures that sperm pass through from the time they are produced until they are expelled from the male body.

Name________________________ Class____________________ Date ____________

25. Many sperm will take the same path toward fertilizing an egg. What prevents more than one sperm from fertilizing an egg?

____

____

____

Using Science Skills

Use the diagram below to answer the following questions on the lines provided.

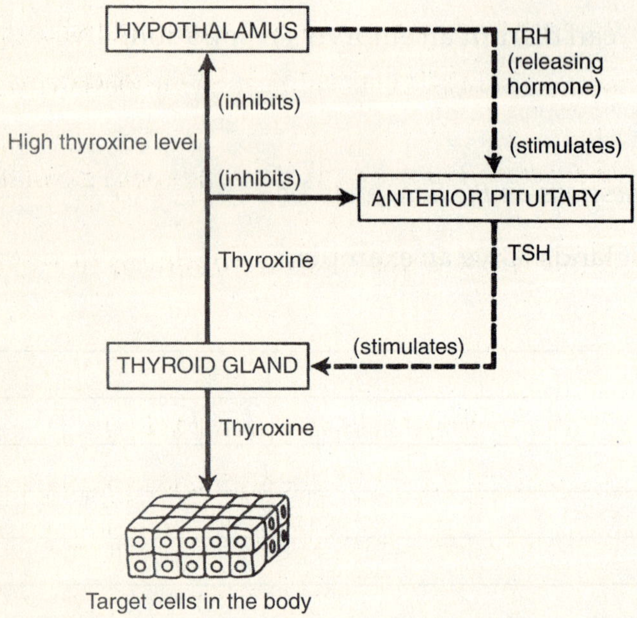

Figure 39-2

26. **Interpreting Graphics** What does Figure 39-2 show?

____

____

27. **Interpreting Graphics** Based on Figure 39-2, what happens as the level of thyroxine increases in the blood?

____

____

____

____

____

28. **Inferring** Based on Figure 39-2, which hormone directly stimulates target cells in the body?

____

Name______________________ Class______________ Date______________

29. **Predicting** Give an example of how Figure 39-2 would change to show a person who has hyperthyroidism. What are some symptoms of this condition?

30. **Applying Concepts** Look at Figure 39-2. How will the three endocrine glands shown behave on a cold day when the body needs a higher level of metabolic activity to stay warm?

Essay

Write the answer to each question in the space provided.

31. Describe complementary hormone action and give an example.

32. Identify the functions of the hormones released by the pituitary gland. What controls the release of these hormones?

33. Describe the action of nonsteroid hormones.

34. What is puberty?

35. What characteristics are produced in males and females by their sex hormones?

Name______________________________ Class__________________ Date ______________

Multiple Choice

Write the letter that best answers the question or completes the statement on the line provided.

____ **1.** The endocrine system is made up of
a. hormones. c. gonads.
b. glands. d. prostaglandins.

____ **2.** Which of the following is a gland of the endocrine system?
a. sweat gland c. pituitary gland
b. tear gland d. all of the above

____ **3.** Figure 39-1 shows the body's
a. hormones.
b. target cells.
c. endocrine glands.
d. exocrine glands.

____ **4.** In the action of steroid hormones, the hormone-receptor complex
a. activates genes.
b. involves a second messenger.
c. acts directly on the blood.
d. acts directly on the liver.

____ **5.** Which endocrine gland secretes sex hormones?
a. adrenal medulla c. hypothalamus
b. testis d. pituitary

____ **6.** Which labeled structure in Figure 39-1 releases nine hormones and regulates many of the other endocrine glands?
a. B c. H
b. D d. J

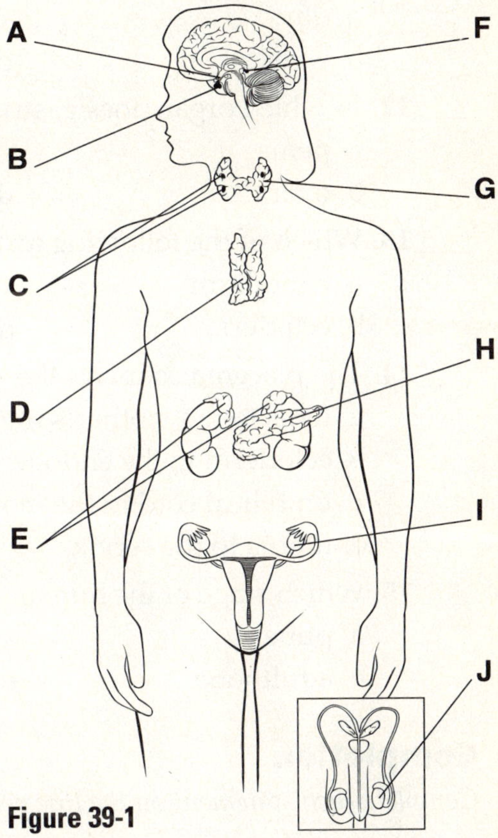

Figure 39-1

____ **7.** Which gland fails to produce enough of its hormone in the disease diabetes mellitus?
a. adrenal c. pancreas
b. hypothalamus d. parathyroid

____ **8.** Puberty usually begins between the ages of
a. 5 and 8. c. 16 and 19.
b. 9 and 15. d. 20 and 25.

Name________________________ Class________________ Date ____________

____ **9.** The testes and the ovaries do not begin making active reproductive cells until

a. birth. c. gastrulation.

b. fertilization. d. puberty.

____ **10.** Which organ system is responsible for making and delivering sperm?

a. female reproductive system

b. endocrine system

c. nervous system

d. male reproductive system

____ **11.** One menstrual cycle usually lasts about a

a. day. c. month.

b. week. d. year.

____ **12.** In which organ does gastrulation occur?

a. penis c. uterus

b. ovary d. vagina

____ **13.** Which of the following forms during gastrulation?

a. endoderm c. mesoderm

b. ectoderm d. all of the above

____ **14.** The placenta connects the

a. fetus to the mother's uterus.

b. ectoderm to the endoderm.

c. umbilical cord to the mother's vagina.

d. uterus to the cervix.

____ **15.** Which stage of the human life cycle occurs first?

a. puberty c. adolescence

b. adulthood d. childhood

Completion

Complete each statement on the line provided.

16. ________________ are chemicals that travel through the bloodstream and affect the activities of other cells.

17. Hormonelike substances that generally affect only nearby cells and tissues are called ________________.

18. ________________ is a period of rapid growth and sexual maturation during which the reproductive system becomes fully functional.

19. In the female body, each egg is surrounded by a(an) ________________, in which the egg matures.

20. Almost everything that the mother takes into her body passes through the ________________ to the embryo.

Name____________________ Class______________ Date ____________

Short Answer

In complete sentences, write the answers to the questions on the lines provided.

21. What is the function of the endocrine system?

____

____

____

____

22. Which glands are labeled C in Figure 39-2? What is the main function of these glands?

____

____

____

____

____

____

____

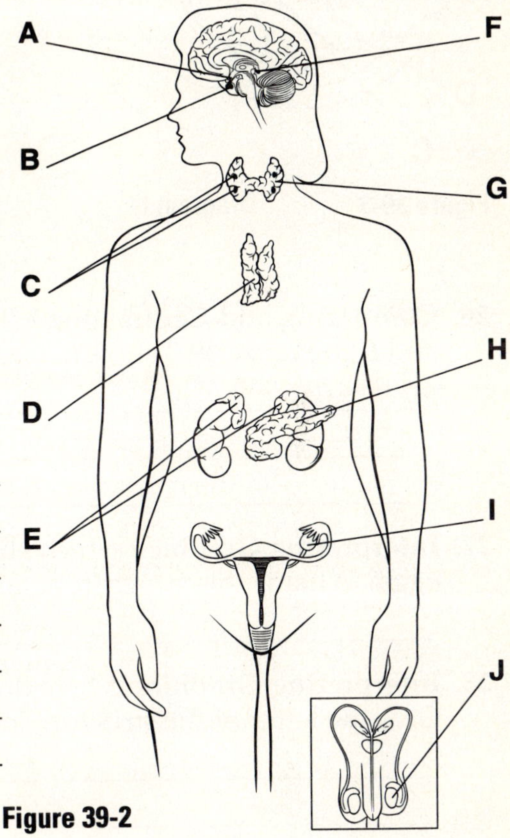

Figure 39-2

23. What is each of the four phases of the menstrual cycle called?

____

____

____

____

____

____

____

24. List the three germ layers that result from gastrulation.

____

____

____

____

____

25. Name and describe one of the four stages of the human life cycle.

____

____

____

____

____

Name_______________ Class_______________ Date_______________

Using Science Skills

Use the diagram below to answer the following questions on the lines provided.

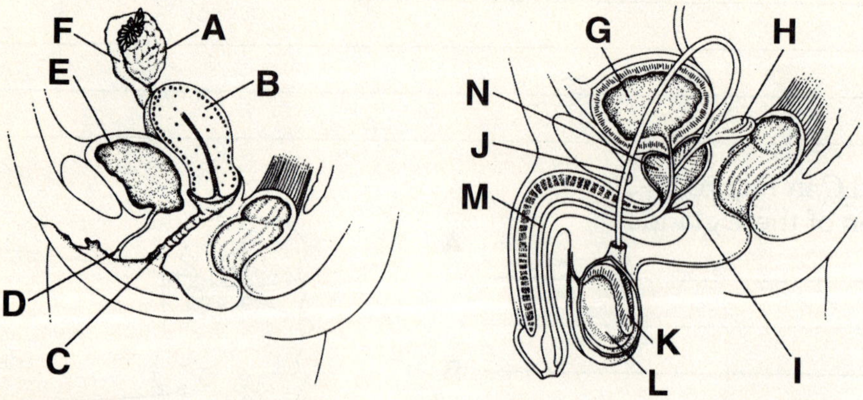

Figure 39-3 Diagram I Diagram II

26. Comparing and Contrasting Which body systems do Diagrams I and II in Figure 39-3 show?

____

____

____

27. Interpreting Graphics Which structure in Figure 39-3 releases eggs? What is this structure called?

____

28. Interpreting Graphics Where does fertilization typically occur? Give the letter of the structure in Figure 39-3.

____

____

____

29. Interpreting Graphics Which structure in Figure 39-3 produces sperm? Give both the name and the letter of this structure.

____

____

30. Interpreting Graphics Through which structure do sperm exit the male body? Give both the name and the letter of the structure in Figure 39-3.

____

____

Name______________________________ Class__________________ Date ______________

Multiple Choice

Write the letter that best answers the question or completes the statement on the line provided.

____ **1.** Diseases are caused by
a. pathogens. c. fungi.
b. cigarette smoke. d. all of the above

____ **2.** An example of an infectious disease that is spread by viruses in the air is
a. athlete's foot. c. flu.
b. tuberculosis. d. tetanus.

____ **3.** How can you prevent the spread of Lyme disease?
a. Wash your hands frequently.
b. Avoid sexual contact.
c. Avoid areas where ticks may be present.
d. Cover your mouth with a tissue when you cough.

____ **4.** Antibiotics fight infections by
a. preventing viruses from replicating.
b. killing bacteria.
c. killing infected cells.
d. growing green mold that inhibits bacterial growth.

____ **5.** The inflammatory response can cause
a. permanent immunity.
b. pain, swelling, and fever.
c. antibodies to bind to antigens.
d. killer T cells to attack infected cells.

____ **6.** If the skin is cut or broken, an infection can result from microorganisms
a already inside the body. c. in the blood.
b. on the skin. d. in the mucus.

____ **7.** Unlike in passive immunity, in active immunity antibodies are produced by
a. the mother of an infant.
b. your own body.
c. other animals.
d. an autoimmune disease.

____ **8.** Humoral immunity is carried out by
a. killer T cells. c. antibodies.
b. lymphocytes. d. macrophages.

Name______________________________ Class__________________ Date ______________

____ **9.** Asthma is an example of

a. the immune system's attacking its own body cells.

b. the immune system's overreacting to an antigen.

c. an autoimmune disease.

d. cancer.

____ **10.** The sneezing, runny nose, and itchy eyes associated with allergies are caused when

a. smooth muscles reduce the size of air passageways in the lungs.

b. the immune system attacks the body's own cells.

c. mast cells release histamines.

d. infected cells produce interferon.

____ **11.** Autoimmune diseases result when the immune system

a. fails to distinguish self from nonself.

b. overreacts to certain antigens.

c. is weakened by asthma.

d. all of the above

____ **12.** All of the following are ways that HIV can be spread EXCEPT

a. sharing needles for intravenous drug use.

b. tending to another person's bleeding wound when you have a cut on your hand.

c. using another person's hairbrush.

d. engaging in sexual activity.

____ **13.** HIV spreads through the body by

a. replicating inside the cells of the immune system.

b. preventing the body from producing antibodies against HIV.

c. causing the body to have asthma attacks.

d. strengthening the immune system.

____ **14.** Cancer cells affect other cells in the body by

a. taking in nutrients needed by other cells.

b. increasing nerve connections.

c. forming benign tumors throughout the body.

d. all of the above

____ **15.** Transplanted organs may be rejected in the type of immune response called

a. cell-mediated immunity.

b. inflammatory response.

c. humoral immunity.

d. antibody response.

Name______________________________ Class__________________ Date ______________

Completion

Complete each statement on the line provided.

16. Smallpox, botulism, and malaria are all examples of ____________________ diseases.

17. Insecticides are used to kill mosquitoes, which are ____________________ for West Nile virus and the protists that cause malaria.

18. ____________________ are effective at treating strep throat but not at treating colds.

19. Any opening in the skin is a potential entrance for ____________________ .

20. A person who has ____________________ is likely to suffer from a number of other rare infections because a virus attacks the immune system.

Short Answer

In complete sentences, write the answers to the questions on the lines provided.

21. What role does interferon play in the immune system? Is it effective against all pathogens?

____

____

____

____

____

22. Describe the mechanism that causes allergies.

____

____

____

____

____

23. What causes autoimmune diseases?

____

____

____

____

____

24. What effects does HIV have on the immune system?

____

____

____

____

____

Name________________ Class________________ Date________________

25. Describe the relationship between antigens and antibodies.

____

____

____

____

____

Using Science Skills

Use the diagram below to answer the following questions on the lines provided.

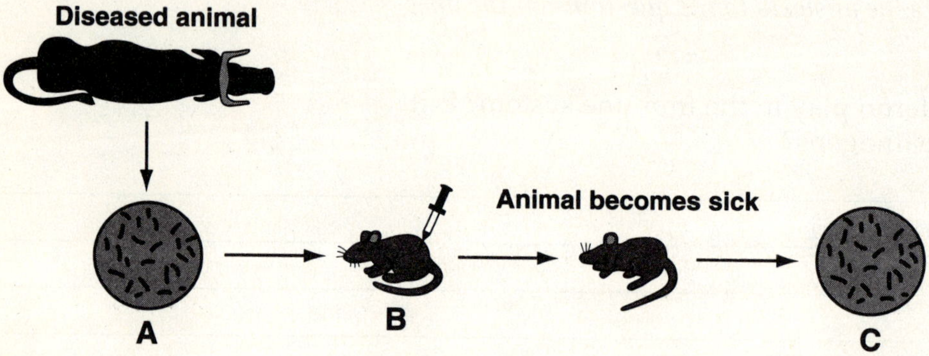

Figure 40-1

26. **Inferring** What is the purpose of the culture labeled A in Figure 40-1?

____

____

____

27. **Interpreting Graphics** What is happening in step B of Figure 40-1?

____

____

____

28. **Inferring** What is the purpose of the culture labeled C in Figure 40-1?

____

____

____

____

____

Name______________________________ Class_________________ Date _____________

29. **Predicting** Look at Figure 40-1. Predict what will happen if the microorganisms grown in the petri dish labeled C are injected into a healthy mouse. Support your answer using the figure.

____

____

____

____

30. **Drawing Conclusions** Assume that the cultures in A and C in Figure 40-1 are identical. What can you conclude from this experiment based on this information?

____

____

____

____

Essay

Write the answer to each question in the space provided.

31. Identify three main causes of disease. Give an example of a disease caused by each factor you identify.

32. Describe active and passive immunity and explain how an individual develops each type.

Name______________________________ Class____________________ Date ______________

33. Describe the inflammatory response in terms of a wound.

34. What are three main causes of cancer? Explain how each factor causes cancer.

35. Compare and contrast cell-mediated immunity and humoral immunity.

Name______________________ Class______________ Date ___________

Multiple Choice

Write the letter that best answers the question or completes the statement on the line provided.

____ **1.** The germ theory of disease states that infectious diseases are caused by
a. toxins.
b. microorganisms.
c. heredity.
d. materials in the environment.

____ **2.** An infectious disease is one that is caused by
a. heredity.
b. materials in the environment.
c. pathogens.
d. hemophilia.

____ **3.** How are infectious diseases spread?
a. through coughing, sneezing, or physical contact
b. through contaminated water and food
c. by infected animals
d. all of the above

____ **4.** Compounds that kill bacterial cells without harming the cells of humans or other animals are called
a. antiviral drugs.
b. insecticides.
c. antibiotics.
d. carcinogens.

____ **5.** The body's nonspecific defenses against invading pathogens include
a. antibiotics.
b. mucus, sweat, and tears.
c. antibodies.
d. killer T cells.

____ **6.** The body's most important nonspecific defense is
a. the skin.
b. cell-mediated immunity.
c. the inflammatory response.
d. permanent immunity.

____ **7.** An immune response is triggered by a(an)
a. antibiotic.
b. antibody.
c. antigen.
d. histamine.

____ **8.** The symptoms of allergies include
a. fever and rare types of cancer.
b. sneezing and watery eyes.
c. unusual infections of the lungs, mouth, throat, and skin.
d. formation of a tumor.

____ 9. An example of an autoimmune disease is
a. asthma.
b. allergies.
c. multiple sclerosis.
d. strep throat.

____10. Doctors first suspected that AIDS was weakening the immune system of infected patients when their patients developed
a. asthma.
b. allergies.
c. rare infections.
d. strep throat.

____11. HIV weakens the immune system by killing
a. antibodies.
b. B cells.
c. helper T cells.
d. killer T cells.

____12. HIV is spread by
a. intravenous drug use.
b. sexual contact.
c. contact with infected blood.
d. all of the above

____13. Cancer can be caused by
a. bacteria.
b. asthma.
c. radiation.
d. allergies.

____14. A benign tumor is one that
a. is cancerous.
b. is malignant.
c. does not spread to other parts of the body.
d. does not divide uncontrollably.

____15. Regular exercise helps maintain health by
a. moving blood through the body.
b. maintaining cardiovascular fitness.
c. helping to maintain weight.
d. all of the above

Completion

Complete each statement on the line provided.

16. A(An) ____________________ is any disease-causing organism.

17. Antibiotics are used to treat infectious diseases caused by ____________________ .

18. Killer T cells attack pathogens in ____________________ immunity.

19. A weakened or mild form of a pathogen injected to produce immunity is a ____________________ .

20. ____________________ immunity involves antibodies in the bloodstream.

Name_____________________ Class_____________ Date__________

Short Answer

In complete sentences, write the answers to the questions on the lines provided.

21. What are three general causes of disease?

22. Describe one way that antibiotics fight infection.

23. How does the skin act as a nonspecific defense against pathogens?

24. What are allergies?

25. What is the difference between a benign tumor and a malignant tumor?

Name________________ Class________________ Date________________

Using Science Skills

Use the diagram to answer the following questions on the lines provided.

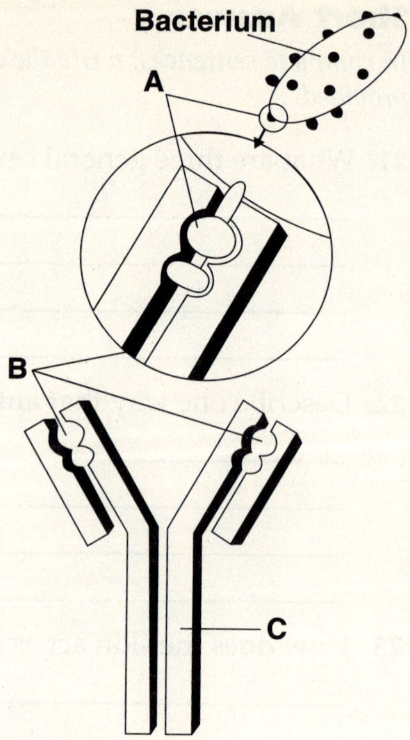

Figure 40-1

26. **Interpreting Graphics** What does Figure 40-1 represent?

27. **Interpreting Graphics** What does letter B in Figure 40-1 represent?

28. **Interpreting Graphics** Which structure or structures trigger the production of antibodies?

29. **Interpreting Graphics** What kind of immunity is represented in Figure 40-1?

30. **Interpreting Graphics** What role do antigen-binding sites play in an infection?

Name______________________________ Class__________________ Date ______________

Multiple Choice

Write the letter that best answers the question or completes the statement on the line provided.

____ **1.** The human organ system that helps fight infection and regulate body temperature is the

a. nervous system.

b. circulatory system.

c. endocrine system.

d. immune system.

____ **2.** Neurons that carry messages from the brain to glands are called

a. sensory neurons.

b. interneurons.

c. motor neurons.

d. secretor neurons.

____ **3.** The function of cerebrospinal fluid is to

a. transmit messages between neurons.

b. carry messages from the cerebrum to the spinal cord.

c. prevent the exchange of wastes between blood and nervous tissue.

d. protect the central nervous system.

____ **4.** If a bone is modeled by a lever, what does the fulcrum represent?

a. tendon c. muscle

b. joint d. ligament

____ **5.** What is the difference between a weak and a strong muscle contraction?

a. number of muscle cells contracting

b. distance the actin filaments move

c. distance between the Z lines

d. height of the cross-bridges between filaments

____ **6.** The chamber of the heart that receives blood from the body is the

a. left atrium.

b. left ventricle.

c. right atrium.

d. right ventricle.

Name________________________ Class________________ Date ____________

____ 7. Cells in the breathing center monitor the blood level of
a. carbon monoxide
b. hemoglobin
c. oxygen
d. carbon dioxide.

____ 8. Which of the following is NOT needed for blood clotting?
a. vitamin A
b. vitamin K
c. calcium
d. thrombin

____ 9. The enzyme amylase is found in the
a. mouth and stomach.
b. mouth and small intestine.
c. stomach and small intestine.
d. liver and mouth.

____10. The volume of urine is minimized in the
a. glomerulus.
b. urethra.
c. ureter.
d. loop of Henle.

____11. Where does chemical digestion takes place?
a. mouth, stomach, and large intestine
b. mouth, pancreas, and small intestine
c. mouth, stomach, and small intestine
d. stomach, small intestine, large intestine

____12. An example of complementary hormone action is the complementary action of insulin and
a. diabetes mellitus.
b. carbohydrates.
c. glucagon.
d. glucose.

____13. The ectoderm germ layer develops into
a. the skin and the nervous system.
b. many of the body's internal tissues and organs.
c. the lining of the digestive system and many of the digestive organs.
d. all of the above

Name________________________________ Class____________________ Date ______________

____14. All forms of cancer are ultimately caused by
a. blocked nerves.
b. harmful mutations.
c. pathogens.
d. radiation.

____15. The researcher most directly responsible for the development of vaccines was
a. Louis Pasteur.
b. Robert Koch.
c. Allen Steere.
d. Edward Jenner.

Completion

Complete each statement on the line provided.

16. Any stimulus that is stronger than the ____________________ will produce a nervous impulse.

17. Beneath the periosteum is a thick layer of ____________________ bone.

18. The ____________________ connects the trachea and bronchioles.

19. The most important nutrient is ____________________.

20. Humoral immunity involves lymphocytes called ____________________.

Short Answer

In complete sentences, write the answers to the questions on the lines provided.

21. What structures are found in the dermis?

____

____

____

22. What is the role of the pancreas in digestion?

____

____

____

23. Define and give an example of a carcinogen.

____

____

____

Name______________________________ Class__________________ Date ______________

24. What role do platelets play in blood clotting?

____

____

____

25. What is the significance of villi in the small intestine?

____

____

____

Using Science Skills

Use the diagram to answer the following questions on the lines provided.

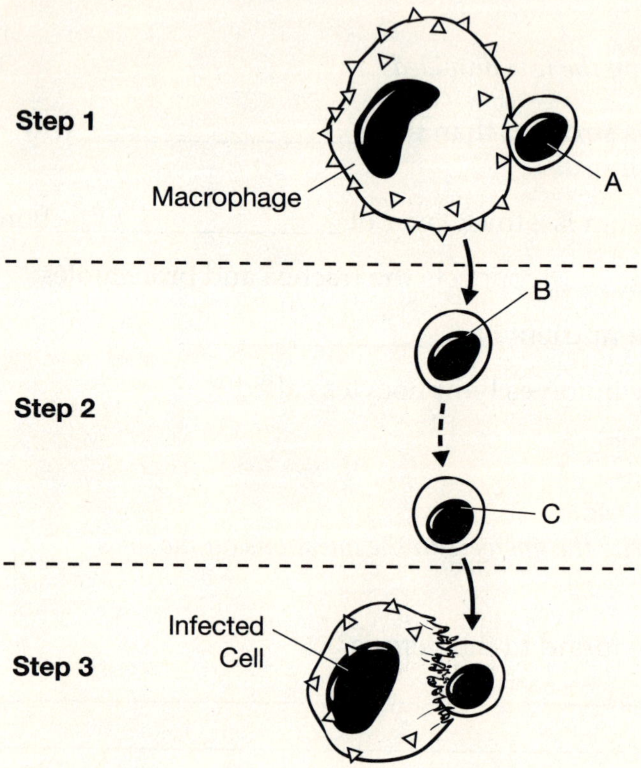

Figure 1

26. What process is shown in Figure 1?

____

____

27. In Figure 1, identify structure A, and describe what occurs in Step 1.

____

____

Name________________________ Class________________ Date ____________

28. What is the name of structure B in Figure 1, and what role does it play in the process?

____

____

29. Identify structure C in Figure 1, and describe what occurs in Step 3.

____

____

30. How is the process shown in Figure 1 different from humoral immunity? How is it similar?

____

____

Essay

Write the answer to each question in the space provided.

31. Compare and contrast the endocrine and nervous systems as communication systems.

32. List the levels of organization of the body, from smallest and simplest to largest and most complex, and give an example of each level.

Name______________________________ Class_________________ Date ____________

33. Describe how bones develop from seven months before birth until adulthood.

34. Explain how the endocrine system regulates metabolism.

35. Identify and describe the functions of the body's nonspecific defenses.

Name______________________ Class______________ Date__________

Multiple Choice

Write the letter that best answers the question or completes the statement on the line provided.

____ **1.** A group of tissues that work together to perform a complex function is a(an)

a. cell.

b. organ.

c. organ system.

d. hormone.

____ **2.** An insulating membrane that surrounds the axon in some neurons is called a

a. dendrite.
c. myelin sheath.

b. node.
d. synapse.

____ **3.** Maintaining a relatively constant internal body temperature is an example of

a. complementary hormone action.

b. resting potential.

c. threshold effect.

d. homeostasis.

____ **4.** Ossification is the process in which bone replaces

a. cartilage.

b. periosteum.

c. bone marrow.

d. calcium salts.

____ **5.** Which type of muscle is under conscious control?

a. cardiac muscle

b. smooth muscle

c. skeletal muscle

d. none of the above

____ **6.** Melanin helps protect skin from

a. pathogens.
c. keratin.

b. radiation.
d. water.

____ **7.** Which chamber of the heart receives oxygen-rich blood from the lungs?

a. right atrium

b. left atrium

c. right ventricle

d. left ventricle

____ **8.** The protein in blood that binds to oxygen is
a. albumin.
b. fibrin.
c. hemoglobin.
d. globulin.

____ **9.** Bronchioles subdivide until they reach the
a. alveoli.
b. bronchus.
c. trachea.
d. pharynx.

____**10.** The body needs fatty acids to
a. produce cell membranes.
b. make myelin sheaths.
c. absorb certain vitamins.
d. all of the above

____**11.** Urea is removed from the blood by the
a. urinary bladder.
b. pancreas.
c. kidneys.
d. large intestine.

____**12.** Steroid hormones are
a. proteins.
b. carbohydrates.
c. lipids.
d. fatty acids.

____**13.** All tumors are
a. cancerous.
b. masses of growing tissue.
c. able to invade surrounding tissues.
d. benign.

____**14.** Antibiotics fight infections caused by
a. viruses.
b. worms.
c. fungi.
d. bacteria.

____**15.** Passive immunity occurs after a person
a. recovers from a disease.
b. receives a vaccination.
c. receives an injection of antibodies.
d. is infected by pathogens.

Name______________________________ Class__________________ Date ______________

Completion

Complete each statement on the line provided.

16. A quick, automatic response to a stimulus is known as a(n) __________________.

17. The tough layer of connective tissue that surrounds bone is called the __________________.

18. The smallest of the blood vessels are the __________________.

19. The liver produces a fluid called __________________ that dissolves fats in the small intestine.

20. Multiple sclerosis is an example of a(n) __________________ disease.

Short Answer

In complete sentences, write the answers to the questions on the lines provided.

21. How are stimulant and depressant drugs different?

22. What are the functions of ligaments and tendons?

23. What is atherosclerosis and how is it related to cardiovascular disease?

24. What role does the mouth play in digestion?

25. What are three ways infectious diseases can be spread?

Name________________________ Class________________ Date ____________

Using Science Skills

Use the diagram to answer the following questions on the lines provided.

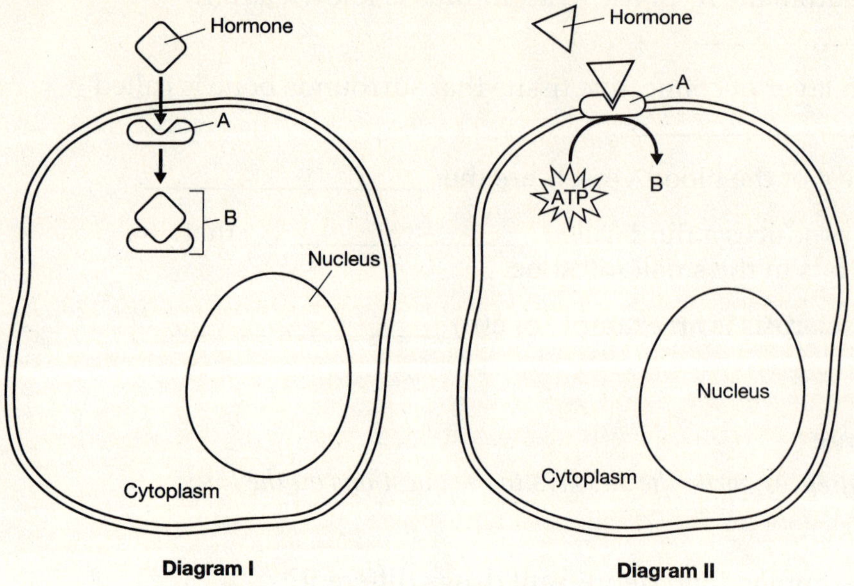

Figure 1

26. In Figure 1, which diagram, I or II, shows the action of a steroid hormone? Of a nonsteroid hormone?

____

____

27. In Figure 1, Diagram I, what are the structures labeled A and B, and what are their functions?

____

____

28. How does the hormone shown in Diagram I of Figure 1 affect the function of the target cell?

____

____

29. In Figure 1, Diagram II, what are the structures labeled A and B, and what role do they play in hormone action?

____

____

30. Based on Figure 1, what basic difference between the hormone in Diagram I and the hormone in Diagram II explains their different ways of acting?

____

____

Name________________________ Class________________ Date ____________

Final Exam Level A

Multiple Choice

Write the letter that best answers the question or completes the statement on the line provided.

____ **1.** Which of the following is NOT science?

a. deciding that human cloning will benefit society
b. observing the courtship behavior of geese
c. collecting and cataloging plant specimens from the rain forest
d. inferring that water from a reservoir is safe to drink

____ **2.** Secondary succession occurs when

a. pioneer species move in to populate an area.
b. species migrate to find abundant food sources.
c. plants begin to grow in areas where no soil existed.
d. land cleared for farming is abandoned.

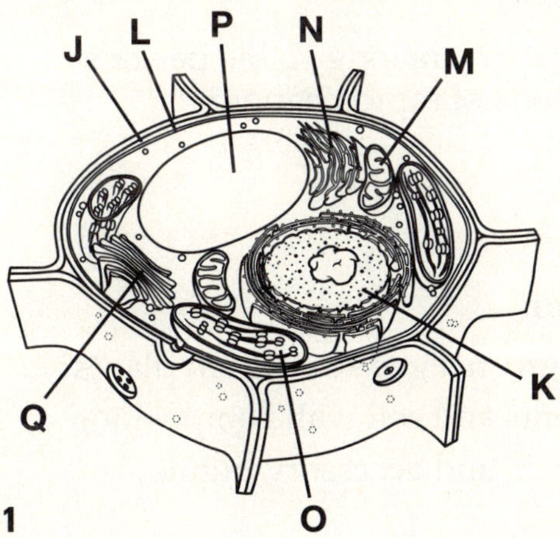

Figure 1

____ **3.** The cell illustrated in Figure 1 is a

a. eukaryotic animal cell.
b. prokaryotic animal cell.
c. eukaryotic plant cell.
d. prokaryotic plant cell.

____ **4.** Photosynthesis occurs in

a. consumers.
b. autotrophs.
c. mitochondria.
d. plant roots.

____ **5.** During cellular respiration,

a. glucose is produced with the sun's energy.
b. oxygen is exchanged with carbon dioxide.
c. energy is released when glucose is broken down.
d. glucose is broken down in the absence of oxygen.

____ 6. Mitosis differs from meiosis in that mitosis results in
a. two genetically identical diploid cells.
b. four genetically identical haploid cells.
c. two genetically different diploid cells.
d. four genetically different haploid cells.

____ 7. If a cross between two black guinea pigs produces three black and one white offspring, then
a. the genes for coat color are linked to different chromosomes.
b. a mutation occurred.
c. the alleles for coat color did not assort independently.
d. the parents were heterozygous for coat color.

____ 8. Which of the following is NOT evidence of evolution?
a. fossil record
b. homologous body structures
c. genetic variation within a population
d. similarities in embryology

____ 9. In which pattern of evolution are long stable periods interrupted by brief periods of rapid change?
a. adaptive radiation
b. convergent evolution
c. coevolution
d. punctuated equilibrium

____ 10. What characteristics do green algae share with plants?
a. photosynthetic pigments and cell wall composition
b. photosynthetic pigments and accessory pigment composition
c. accessory pigments and cell wall composition
d. accessory pigments and cell membrane composition

____ 11. An important trend in plant evolution is
a. reduction of the gametophyte and increased size of the sporophyte.
b. increased reliance on water for reproduction.
c. appearance of a life cycle with only a diploid generation.
d. use of spores for reproduction.

____ 12. Which of the following describes the most complex invertebrate?
a. radial symmetry, cephalization, pseudocoelom
b. bilateral symmetry, cephalization, true coelom
c. radial symmetry, no cephalization, true coelom
d. bilateral symmetry, no cephalization, pseudocoelom

Name______________________________ Class__________________ Date ______________

____**13.** All vertebrates

a. are chordates.

b. have bilateral symmetry.

c. have tails.

d. all of the above

____**14.** How does the endocrine system maintain homeostasis in the human body?

a. helps protect the body from disease

b. eliminates waste products from the body

c. coordinates the body's response to changes in the environment

d. produces hormones that control growth and development

____**15.** Which of the following is NOT a healthful behavior?

a. wearing sunscreen

b. exercising regularly

c. eating a high-fat diet

d. abstaining from drugs and alcohol

Completion

Complete each statement on the line provided.

16. As resources become less available, the population growth rate ____________________ .

17. During the process of ____________________ , RNA polymerase uses DNA to produce a strand of RNA.

18. Eukaryotic heterotrophs that have cell walls made of chitin are classified as ____________________ .

19. As angiosperm seeds mature, the ovary walls thicken to form a(an) ____________________ that encloses the developing seeds.

20. A soft-bodied invertebrate that usually has an internal or external shell is a(an) ____________________ .

Short Answer

In complete sentences, write the answers to the questions on the lines provided.

21. Design an experiment that tests whether plants grow faster with fertilizer. Identify the variables in the experiment.

Name________________________ Class________________ Date ____________

22. Describe the role of plants in the water cycle.

____

____

____

23. Give two reasons why cells divide.

____

____

____

24. Why is genetic diversity important to plant and animal breeders?

____

____

____

25. Explain how scientists currently think human evolution occurred.

____

____

____

Using Science Skills

Use the diagram below to answer the following questions on the lines provided.

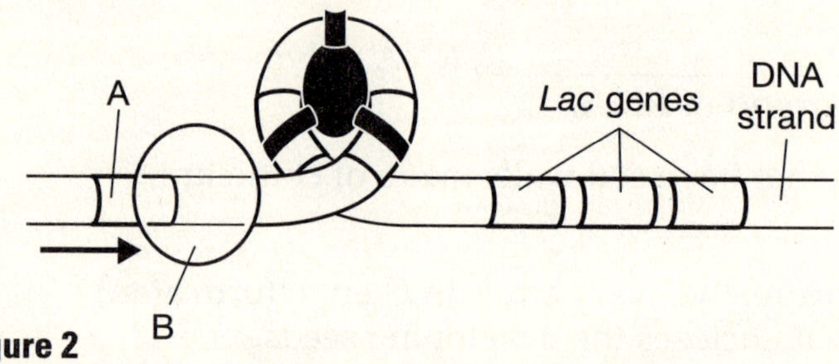

Figure 2

26. **Interpreting Graphics** In the operon in Figure 2, what is the function of the DNA sequence labeled A?

____

____

____

27. **Interpreting Graphics** In the operon in Figure 2, why can't the *lac* genes be transcribed when the repressor is bound to the operator?

____

____

____

Name______________________________ Class__________________ Date ______________

28. **Applying Concepts** How does the shape of the repressor protein affect its function?

____

____

____

29. **Inferring** In the operon illustrated in Figure 2, when are the *lac* genes turned on?

____

____

____

30. **Applying Concepts** Why do *E. coli* cells have a *lac* operon? How is it useful to the cell?

____

____

____

Essay

Write the answer to each question in the space provided.

31. How would a climate change affect the biotic and abiotic factors in an ecosystem?

32. What would happen to life on Earth if plants suddenly became extinct? Explain.

33. Explain how to construct a bacterial cell that produces human insulin.

34. Explain how changes in DNA relate to the evolution of species.

35. Describe the levels of organization involved in the human stomach.

Name________________________ Class________________ Date ______________

Final Exam — Level B

Multiple Choice

Write the letter that best answers the question or completes the statement on the line provided.

____ **1.** An enzyme speeds up a reaction by
- a. raising the activation energy.
- b. lowering the activation energy.
- c. absorbing energy.
- d. releasing energy.

____ **2.** In an ecosystem, energy flows from
- a. consumers to producers.
- b. carnivores to heterotrophs.
- c. the sun to autotrophs.
- d. omnivores to consumers.

____ **3.** A cedar waxwing eats the berries from a bush. The following spring, new berry bushes sprout from the seeds deposited in the bird droppings far from the parent plant. This is an example of
- a. predation.
- b. pollination.
- c. symbiosis.
- d. germination.

____ **4.** According to the cell theory, cells are
- a. the basic units of all living things.
- b. abiotic factors.
- c. made up of organs and tissues.
- d. observed only with microscopes.

____ **5.** Mitosis differs from meiosis in that mitosis results in
- a. two genetically identical diploid cells.
- b. four genetically identical haploid cells.
- c. two genetically different diploid cells.
- d. four genetically different haploid cells.

____ **6.** What is the inheritance pattern illustrated by the genetic cross in the Punnett square in Figure 1?
- a. simple dominance
- b. incomplete dominance
- c. codominance
- d. multiple alleles

____ **7.** Unlike DNA, RNA
- a. never leaves the nucleus.
- b. has thymine.
- c. is double-stranded.
- d. has a ribose sugar.

TT \ Tt	*T*	*t*
T	*TT*	*Tt*
T	*TT*	*Tt*

T = tall
t = short

Figure 1

Name________________________ Class________________ Date ____________

____ 8. Which of the following procedures would you use to separate fragments of DNA?

a. polymerase chain reaction

b. DNA fingerprinting

c. transformation

d. gel electrophoresis

____ 9. In order for a new species to evolve,

a. new mutations must be induced.

b. abiotic factors in the environment must change.

c. reproductive isolation must occur.

d. resources in the environment must be abundant.

____ 10. In a drop of pond water under a light microscope, you observe a single-celled organism completely covered with cilia. This organism is probably a(an)

a. bacteria.

c. plantlike protist.

b. animallike protist.

d. fungi.

____ 11. Which of the following plant adaptations does NOT allow a plant to reproduce in areas without water?

a. sporangia

c. seeds

b. pollination

d. flowers

____ 12. Phototropism is the response of plants to

a. touch.

c. light.

b. gravity.

d. periods of light and dark.

____ 13. An invertebrate with a segmented body, tough exoskeleton, and jointed appendages is a(an)

a. annelid.

c. echinoderm.

b. mollusk.

d. arthropod.

____ 14. An animal with smooth, shiny skin; a tail; and four legs sticking straight out from its body just ran past. This animal is probably a(an)

a. arthropod.

c. reptile.

b. amphibian.

d. mammal.

____ 15. How does the endocrine system help maintain homeostasis in the human body?

a. helps protect the body from disease

b. eliminates waste products from the body

c. coordinates the body's response to changes in the environment

d. produces hormones that control growth and development

Name______________________________ Class__________________ Date ______________

Completion

Complete each statement on the line provided.

16. Examples of ____________________ factors in an ecosystem are climate, wind, and soil type.

17. In the cell cycle, DNA replication occurs during ____________________.

18. Over time, ____________________, or the survival of the fittest, results in changes in the inherited characteristics of a population.

19. Invertebrates with ____________________ symmetry have body parts that extend from the center of the body.

20. Vertebrates that are ____________________ can generate and retain heat inside their bodies.

Short Answer

In complete sentences, write the answers to the questions on the lines provided.

21. Do you agree or disagree that plants are not living things because they are not mobile? Explain.

____

____

____

____

22. Explain how the decreasing genetic diversity of a species could lead to extinction.

____

____

____

23. Summarize the process of photosynthesis, using the reactants and the products.

____

____

____

24. What are the two main sources of genetic variation?

____

____

____

25. Why is behavior important for survival and reproduction?

____

____

____

Name______________________ Class______________ Date____________

Using Science Skills

Use the diagram below to answer the following questions on the lines provided.

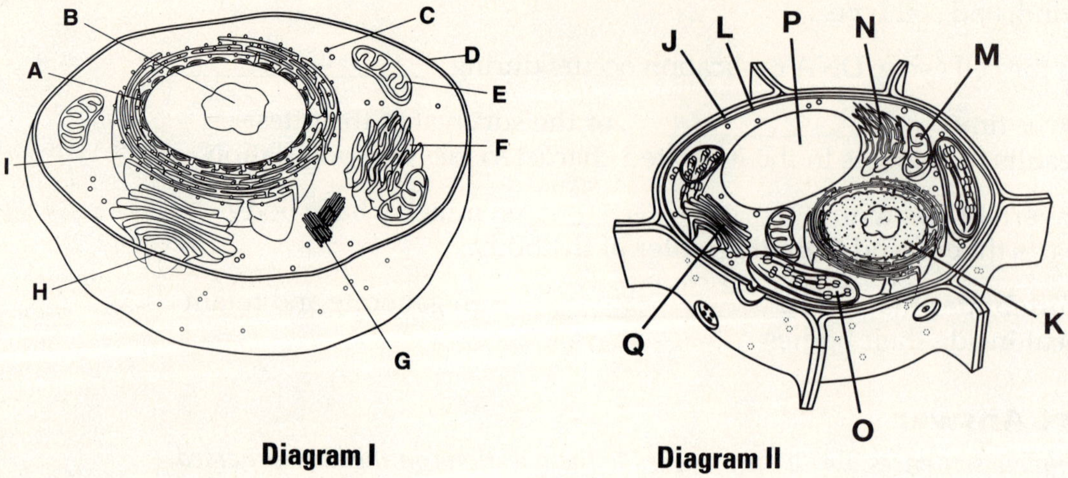

Figure 2

26. Interpreting Graphics Which of the cells in Figure 2 is a plant cell? How do you know?

____

____

27. Interpreting Graphics In Figure 2, are these cells eukaryotic cells or prokaryotic cells? Explain.

____

____

28. Applying Concepts In Diagram I of Figure 2, what is the structure labeled A? What is its function?

____

____

____

29. Applying Concepts In Diagram I of Figure 2, how does the structure labeled B control most cell processes?

____

____

____

30. Inferring In Figure 2, if the cell in Diagram II were placed in a hypertonic solution, what would happen to it? Explain.

____

____

____

Chapter 1—Test A

Multiple Choice **1.** C **2.** D **3.** D **4.** C **5.** D **6.** D **7.** C **8.** B **9.** C **10.** A **11.** D **12.** C **13.** D **14.** B **15.** C **Completion** **16.** data **17.** hypothesis **18.** spontaneous generation **19.** metabolism **20.** manipulated **Short Answer** **21.** A theory applies to a well-tested explanation that unifies a broad range of observations; a hypothesis is a possible explanation for a set of observations. **22.** In sexual reproduction, two cells from different parents unite to produce the first cell of the new organism, whereas in asexual reproduction, the new organism has a single parent. **23.** Tissues, organs, and organ systems are included in this level of organization. **24.** The metric system is a decimal system of measurement whose units are based on certain physical standards and are scaled on multiples of 10. **25.** Light microscopes produce images by focusing visible light rays, whereas electron microscopes produce images by focusing beams of electrons. **Using Science Skills** **26.** There are four populations in this community. **27.** Nonliving elements include air, rocks, and a stream. **28.** This ecosystem is part of the biosphere. **29.** Each population is made up of individual organisms. **30.** No, the rabbits live in a different area. A community is defined as the populations that live together in a defined area. **Essay** **31.** An inference is a logical interpretation based on prior knowledge or experience. An observation involves using the senses to gather information. Inferences can be supported by observations. **32.** A theory can change over time. No theory is considered absolute truth. As new evidence is uncovered, a theory may be revised or replaced by a more useful explanation. **33.** Anything lacking one or more of the following characteristics of life is nonliving: made up of cells, reproduces, based on a universal genetic code, grows and develops, obtains and uses materials and energy, responds to its environment, maintains a stable internal environment, and taken as a group, changes over time. **34.** A compound light microscope is used to view living cells and organisms. A transmission electron microscope (TEM) is used to view dead cells or organisms when greater magnifications are needed. A scanning electron microscope (SEM) is used for three-dimensional, enlarged views of the surfaces of dead organisms. **35.** Cell culturing is used to obtain material to study by placing a single cell into a nutrient solution. A group of cells then develops from the single original cell. Cell fractionation separates the different parts of a cell, which allows the study of a particular cell part.

Chapter 1—Test B

Multiple Choice **1.** B **2.** D **3.** D **4.** D **5.** B **6.** C **7.** B **8.** A **9.** C **10.** D **11.** B **12.** A **13.** B **14.** C **15.** D **Completion** **16.** data **17.** cells **18.** metric **19.** compound light **20.** centrifuge **Short Answer** **21.** An inference is a logical interpretation based on prior knowledge and experience. **22.** A hypothesis is a proposed scientific explanation for a set of observations. **23.** Answers can include any five of the following: molecules, cells, groups of cells, organism, population, community, ecosystem, biosphere. **24.** Cell fractionation allows a scientist to study a particular part of a cell. **25.** Wash your hands thoroughly. **Using Science Skills** **26.** meter, centimeter, millimeter, kilometer **27.** The metric system is based on multiples of 10, whereas traditional English units have many different conversion factors. **28.** The boiling point of water in degrees Celsius is 100°C. **29.** The prefix *kilo-* means 1000. **30.** Two liters (L) of water are equal to 2000 mL.

Chapter 2—Test A

Multiple Choice **1.** B **2.** D **3.** D **4.** A **5.** A **6.** B **7.** D **8.** B **9.** D **10.** A **11.** A **12.** C **13.** A **14.** B **15.** B **Completion** **16.** electrons **17.** van der Waals **18.** acid **19.** H^+ ions **20.** absorb **Short Answer** **21.** Mass number is the sum of the protons and neutrons in the nucleus of an atom. **22.** The ability of water to form multiple hydrogen bonds accounts for water's properties of adhesion and cohesion. **23.** Add some salt—the solute—to a container of water—the solvent—to produce a salt solution. **24.** RNA contains the sugar ribose and DNA contains the sugar deoxyribose. **25.** Answers should include two of the following: regulating chemical pathways, making materials that cells need, releasing energy, and transferring information. **Using Science Skills** **26.** hydrochloric acid **27.** 6.2 **28.** 13 **29.** greater **30.** 100 times (10 × 10) **Essay** **31.** Protons and neutrons are located in the nucleus, whereas electrons are located outside the nucleus. Electrons have a negative charge, protons have a positive charge, and neutrons have no charge. Protons and neutrons have approximately the same mass, which is 1840 times that of an electron. **32.** Isotopes of an element have the same number of protons but different numbers of neutrons. Because the mass number is the sum of the atomic number (number of protons) and the number of neutrons, isotopes of the same element have different mass numbers. **33.** In an ionic compound, valence electrons of atoms of one element have been transferred to the atoms of another element. Thus, ionic bonding involves the gain and loss of electrons. In covalently bonded compounds, atoms of different elements share electrons. **34.** Both adhesion and cohesion are attractions. They are different in that adhesion is the force of attraction between the molecules of two different substances, whereas cohesion is the force of attraction between molecules of the same substance. Possible examples include capillary action, in which the adhesion between water and glass causes the water to rise in a narrow tube against the force of gravity. Because of hydrogen bonding, water is extremely cohesive. This causes molecules on the surface of water to be drawn inward, which is why drops of water form beads on smooth surfaces. **35.** In a comparison of the activity of enzymes to a lock and key, the lock symbolizes the active site of the enzyme and the key symbolizes the substrate. A specific key can fit into only a specific lock; so, too, does a specific substrate fit into the active site of a specific enzyme. Without a perfect match, the reaction cannot occur. Likewise, without the right key, the lock cannot open.

Chapter 2—Test B

Multiple Choice **1.** D **2.** C **3.** C **4.** A **5.** B **6.** D **7.** C **8.** A **9.** C **10.** A **11.** A **12.** C **13.** B **14.** D **15.** A **Completion** **16.** electrons **17.** chemical **18.** ionic **19.** hydrogen **20.** products **Short Answer** **21.** A molecule is the smallest unit of most chemical compounds. **22.** The main types of chemical bonds are covalent bonds and ionic bonds. **23.** A mixture is a material composed of two or more elements or compounds that are physically mixed together but not chemically combined. **24.** The four groups are carbohydrates, lipids, nucleic acids, and proteins. **25.** energy changes **Using Science Skills** **26.** 12 **27.** 8 **28.** 13 **29.** helium **30.** oxygen

Unit 1—Test A

Multiple Choice **1.** B **2.** B **3.** D **4.** C **5.** C **6.** A **7.** C **8.** A **9.** B **10.** B **11.** A **12.** D **13.** B **14.** A **15.** D **Completion** **16.** hypothesis **17.** asexual **18.** protons **19.** carbohydrates **20.** bonds **Short Answer** **21.** Whenever possible, a hypothesis should be tested by an experiment in which only one variable is changed at a time. **22.** Cell cultures can be used to test cell responses under controlled conditions, to study interactions between cells, and to select specific cells for further study. **23.** Adhesion is the attraction between molecules of different substances, whereas cohesion is the attraction between molecules of the same substance. **24.** In a salt-water solution, the sodium and chloride ions that make up the salt gradually become dispersed in the water, becoming evenly distributed throughout the solution. **25.** An enzyme speeds up a chemical reaction by lowering the activation energy. **Using Science Skills** **26.** coffee **27.** 12.0 **28.** Blood is a base. **29.** Lemon juice has more H^+ ions than OH^- ions. **30.** 100 times **Essay** **31.** A hypothesis is a proposed scientific explanation for a set of observations. A particular hypothesis may become so well supported that scientists consider it a theory. A theory applies to a well-tested explanation that unifies a broad range of observations. **32.** As multicellular organisms develop, cells form that look different from one another and perform different functions. This is the process of differentiation. During development, a single fertilized egg cell divides again and again to produce the many cells of a mature organism. As those cells divide, they change shape and structure to form cells such as liver cells, brain cells, and muscle cells. **33.** A compound light microscope produces a magnified image as light passes through a specimen and then through the lenses of the microscope. A TEM produces a magnified image of slices of a specimen as it shines a beam of electrons through a thin specimen. An SEM produces a magnified three-dimensional image of a specimen as it scans a narrow beam of electrons back and forth across the surface of the specimen. **34.** An ionic bond is formed when one or more electrons are transferred from one atom to another. A covalent bond forms when electrons are shared between atoms. **35.** In an enzyme-catalyzed reaction, the substrates bind to the site on the enzyme called the active site. The active site and the substrates have complementary shapes. The enzyme and the substrates form an enzyme-substrate complex, and they remain bound until the reaction is done.

Unit 1—Test B

Multiple Choice **1.** D **2.** C **3.** B **4.** C **5.** D **6.** A **7.** B **8.** C **9.** A **10.** B **11.** A **12.** C **13.** D **14.** A **15.** A **Completion** **16.** hypothesis **17.** manipulated **18.** sexual **19.** compound **20.** enzyme **Short Answer** **21.** A theory is a well-tested explanation that unifies a broad range of observations. **22.** Students should list five of these eight characteristics: made up of units called cells, ability to reproduce, based on a universal genetic code, ability to grow and develop, ability to obtain and use materials and energy, ability to respond to the environment, ability to maintain a stable internal environment, and tendency as a group to change over time. **23.** Isotopes of an element all have the same chemical properties because they have the same number of electrons. **24.** An acidic solution contains a higher concentration of H^+ ions than pure water and has a pH value above 7. A basic solution contains a lower concentration of H^+ ions than pure water and has a pH value above 7. **25.** A chemical reaction is a process that changes one set of chemicals into another set of chemicals. **Using Science Skills** **26.** milligram, gram, kilogram, metric ton **27.** 0°C **28.** one tenth **29.** 5000 **30.** Because metric units are scaled on multiples of 10, converting meters to centimeters entails multiplying by 1000, a multiple of 10. Converting yards to inches, however, would entail multiplying the number of yards by 36, which is the number of inches in each yard.

Chapter 3—Test A

Multiple Choice **1.** D **2.** B **3.** B **4.** C **5.** D **6.** A **7.** D **8.** B **9.** D **10.** C **11.** D **12.** A **13.** B **14.** A **15.** A **Completion** **16.** chemicals **17.** decomposers (or detritivores) **18.** sharks **19.** transpiration **20.** amino acids **Short Answer** **21.** The algae are autotrophs, the producers at the base of the food chain. **22.** Energy flows from the sun to the autotrophs, and from the autotrophs to the heterotrophs. Energy also flows from the autotrophs and the heterotrophs to the decomposers. **23.** The insects would provide 10 percent (one tenth) of 1500 calories, or 150 calories, of energy to the shrews. The shrews would provide one tenth of 150 calories, or 15 calories to the owl. **24.** The tree at the base has much more available energy than all the insects that live in it. An energy diagram would be broad at the base like a typical pyramid. **25.** Water seeps through soil and into the ground water. Roots of trees and plants take the water up through their roots. Plants release the water through the process of transpiration. Both seepage and transpiration are parts of the water cycle. **Using Science Skills** **26.** Possible answer: A large omnivore would consume energy at every trophic level, reducing the amount of energy available to each level above the producers. **27.** The producers are the bluegrass. Adding a limiting nutrient to the grass would increase the amount of primary productivity. The result would be an increased amount of biomass at the producer level, and increasing biomass and available energy at each successive level. The number of birds could increase, because there would be more for them to eat. **28.** The bass is the top carnivore. It has 100 kcal available to it from the trophic level below it. The bass will use part of that energy for its life processes, lose some as heat, and be able to store only about 10 percent, or 10 kcal. **29.** observing, experimenting, and modeling **30.** Certain bacteria are chemosynthetic autotrophs and live in tidal marshes along the coast. They are producers that use inorganic molecules to produce food. The energy they produce would add to the total amount produced by plants to supply energy to the ecosystem. **Essay** **31.** Sunlight is the main energy source for life on Earth. Less than 3 percent of all the sun's energy that reaches Earth is used by living things. Inorganic chemical compounds also provide energy for life on Earth. Plants, some algae, and certain bacteria can capture energy from sunlight or chemicals and use that energy to produce food. **32.** A food chain is a series of steps in which organisms transfer energy by eating and being eaten. A food web is a feeding relationship among the various organisms in an ecosystem that forms a network of complex interactions. A food web links together all the food chains in an ecosystem. **33.** An energy pyramid is a diagram that shows the relative amounts of energy contained within each trophic level in a food chain or web. A biomass pyramid represents the amount of living tissue within each trophic level. A pyramid of numbers shows the relative number of organisms at each trophic level. **34.** Certain types of bacteria are able to use nitrogen gas directly. These bacteria, which live in the soil and on the roots of legumes, convert nitrogen gas to ammonia during the process of nitrogen fixation. Other bacteria in the soil convert ammonia into nitrates and nitrites, which are also taken up directly by producers. Still other soil bacteria convert nitrates into nitrogen gas during the process of denitrification, returning the nitrogen to the atmosphere. **35.** The growth of producers in a water environment like a lake is slowed by lack of a limiting nutrient. Phosphorus is a limiting nutrient in most freshwater environments. If phosphorus is added in large amounts, such as from runoff from heavily fertilized fields, an immediate rise in the amount of algae and other producers can occur. This result is called an algal bloom. If an algal bloom in a lake gets too extensive, it may cover the surface of the water.

Chapter 3—Test B

Multiple Choice **1.** D **2.** D **3.** B **4.** B **5.** A **6.** B **7.** A **8.** A **9.** B **10.** D **11.** C **12.** A **13.** C **14.** B **15.** D **Completion** **16.** biome **17.** biogeochemical **18.** herbivores (or consumers) **19.** energy **20.** nutrients **Short Answer** **21.** The biosphere is the combined portions of the planet in which all of life exists, including land, air, and water. It extends from about 8 kilometers above Earth's surface to about 11 kilometers below the surface of the ocean. **22.** Sharks are the top carnivore in a food chain that begins with algae. Algae are producers that get their energy from the sun. **23.** Energy flows in one direction from the sun to producers and consumers. Matter moves in biogeochemical cycles through living systems, the Earth, the atmosphere, and the oceans. **24.** Water evaporates from the lake, condenses in the atmosphere, and falls as precipitation. Some rain may fall directly on the lake, and some water may return to the lake as runoff. **25.** A large input of a limiting nutrient, such as nitrogen or phosphorus, can cause the algae to grow and reproduce more quickly. If there are not enough consumers to eat the excess algae, the algae will grow to cover the surface of the water. **Using Science Skills** **26.** The diagram is a model based on observations and experiments. Ecologists use models to explain phenomena and make predictions. **27.** The wolf is on the fourth trophic level, so it is four steps away from the sun. **28.** An ecosystem is a collection of all the organisms that live in a particular place, together with their nonliving, or physical, environment. A food web represents relationships among all the organisms and their environment. **29.** There is one first-level consumer for corn, three for carrots, four for flowering shrubs, and one for trees. **30.** The rabbit is on the second trophic level and has more energy available to the wolf than the shrew on the third trophic level.

Chapter 4—Test A

Multiple Choice **1.** C **2.** B **3.** B **4.** D **5.** C **6.** C **7.** D **8.** A **9.** B **10.** D **11.** B **12.** C **13.** A **14.** B **15.** C **Completion** **16.** climate **17.** 66.5°N **18.** 23.5°N, 23.5°S **19.** chemistry **20.** wetland **Short Answer** **21.** Five factors are: the trapping of heat by the atmosphere, latitude, the transport of heat by winds and ocean currents, amount of precipitation, the shape and elevation of landmasses. **22.** The sun's rays strike Earth at increasingly lower angles with increasing distance from the equator, thus providing less and less direct sunlight. **23.** A niche includes all of the physical and biological conditions in which an organism lives and how the organism uses those conditions. Food is part of the biological conditions that a bullfrog uses. **24.** In primary succession, pioneer plants help rocks to break up in the process of soil formation. They also contribute organic material to the forming soil in which plants can grow. **25.** Whale-fall succession takes place on the deep, permanently dark ocean floor where there is no light for photosynthesis and no plants can grow. **Using Science Skills** **26.** Point A is the origin of this flowing-water ecosystem. The water is most likely turbulent and cold. Region B would have swiftly moving water, which would be turbulent where it goes over the waterfall. There would be little plant life in the water due to the turbulence, and the animals in the water would have adaptations to keep themselves from being swept away. **27.** The hills show a growth of shrubs and a few small trees. This indicates a later stage of secondary succession. Eventually, an ecologically characteristic boreal forest will grow if there are no more disturbances. **28.** A lake would form above the dam, creating a standing-water ecosystem. Without a flow of water, the river below the dam would dry up. The dry land would then undergo secondary succession, and because of the climate and latitude, would eventually have the ecological characteristics of a land biome, a boreal forest. **29.** Regions D and E would contain more nutrients because the water is moving very slowly and more plants can grow. It can support algae and bacteria and many more organisms than farther up the mountain where they would be washed away in the current. The river also carries nutrient-rich sediments down to regions D and E from higher up the mountain. **30.** Sample answer: In region D, the water is moving slowly and meandering across the land. In region E, the sediments carried by the water have formed a marsh where the flow is moving gradually toward a lake. The regions are different in that the water is moving faster in region D than in region E, and region E likely has more numbers and kinds of organisms. They are alike in that they are both part of the same aquatic ecosystem. **Essay** **31.** Cold water near the poles sinks and then flows parallel to the ocean bottom, eventually rising again in warmer regions through a process called upwelling. Meanwhile, surface water is moved by winds. In both cases, the water flow creates ocean currents. Like air currents, ocean currents transport heat energy within the biosphere. Surface ocean currents warm or cool the air above them, thus affecting the weather and climate of nearby landmasses. **32.** Solar energy has an important effect on the temperature of the atmosphere. Atmospheric gases, such as carbon dioxide, methane, and water vapor, do not allow heat energy to pass out of the atmosphere. This natural situation in which heat is retained by these greenhouse gases is called the greenhouse effect. This effect provides a natural insulating blanket around Earth and maintains its temperature range. **33.** The three main classes of symbiotic relationships are mutualism, commensalism, and parasitism. In mutualism, both species benefit from the relationship. For example, flowers depend on certain insects to pollinate them. The flowers provide the insects with food in the form of nectar, pollen, or other substances. In commensalism, one member of the association benefits and the other is neither helped nor harmed. Barnacles attached to the skin of whales benefit from food particles in the water moving past the swimming whale, but the whale is not affected. In parasitism, one member benefits by obtaining all its nutritional needs from the host. The host can be damaged, but is usually not killed. Fleas, ticks, and lice are examples of parasites that live on the bodies of mammals. **34.** The climate within a small area often differs significantly from the climate of the surrounding area and is called a microclimate. An example of a microclimate is the redwood forests on parts of California's northern coast. Some areas a short distance inland from these forests are desertlike, inhabited by cacti and other drought-resistant plants. **35.** The main zones of the marine biome are the photic zone, the aphotic zone, the intertidal zone, the coastal ocean zone, the open ocean zone, and the benthic zone. The photic zone is the well-lit upper layer where photosynthesis takes place. Below the photic zone is the aphotic zone, which is permanently dark. The benthic zone extends along the ocean floor from the coastal ocean through the ocean zones. The intertidal zone is the area most affected by tides. The coastal ocean zone extends from the low-tide mark to the end of the continental shelf. The open ocean, or oceanic zone, begins at the edge of the continental shelf and extends outward. This is the largest marine zone, covering over 90 percent of the surface area of the world's oceans.

Chapter 4—Test B

Multiple Choice **1.** D **2.** D **3.** B **4.** A **5.** A **6.** A **7.** D **8.** B **9.** B **10.** B **11.** B **12.** C **13.** B **14.** C **15.** B **Completion** **16.** sunlight **17.** greenhouse **18.** biotic **19.** microclimate **20.** phytoplankton **Short Answer** **21.** Glass traps heat energy inside the greenhouse. Atmospheric gases trap heat energy inside Earth's atmosphere. **22.** Greenhouse gases are atmospheric gases, such as carbon dioxide, methane, and water vapor that trap heat energy. **23.** The climate is most likely a hot, rainy climate in the tropical zone. **24.** An ecological resource is any necessity of life for an organism, such as water, nutrients, light, food, or space. **25.** The amount of light received by an aquatic ecosystem is determined by the depth of the overlying water. **Using Science Skills** **26.** The tropical rain forest has an average daytime temperature of 34°C. The tundra has an average summer temperature of 12°C. The difference is that the tropical rain forest receives almost direct sunlight all year long. The tundra receives sunlight at a much lower angle, resulting in a colder climate than the tropical rain forest has. **27.** highest temperature = 38°C; lowest temperature = –26°C; range = 38°C above zero + 26°C below zero = 64 degrees. The greenhouse effect is responsible for maintaining Earth's range of temperatures. **28.** temperate zone **29.** Biomes that have less than 75 cm of yearly rainfall include: boreal forest, desert, temperate grassland, temperate woodland and shrubland, and tundra. Biomes that have more than 200 cm of yearly rainfall include: tropical rain forest and tropical dry forest. **30.** The tropical dry forest and the tropical savanna each have a wet season and a dry season.

Chapter 5—Test A

Multiple Choice **1.** D **2.** B **3.** A **4.** D **5.** C **6.** C **7.** C **8.** A **9.** A **10.** A **11.** D **12.** B **13.** C **14.** B **15.** A **Completion** **16.** immigration **17.** logistic **18.** carrying capacity **19.** dependent **20.** economic **Short Answer** **21.** Emigration is the movement of individuals out of a population and reduces its size if all other factors are equal. **22.** A density-independent limiting factor is a limiting factor that affects all populations regardless of size, such as a rainstorm, drought, fire, and certain human activities. **23.** Limiting factors, such as food scarcity and diseases, kept the human population growth rate low. **24.** A demographer studies the social and economic factors that affect a population and the age structure in order to predict how it will change. **25.** Population growth has slowed in the United States, but China and India are only in the beginning stage of demographic transition where the population is still growing rapidly. **Using Science Skills** **26.** *P. caudatum* **27.** Overall, *P. aurelia* had the greater growth rate. *P. caudatum* had the greater initial growth rate, but was eventually excluded by *P. aurelia*. *P. caudatum* eventually died out. **28.** Both populations show a logistic growth curve. **29.** *P. aurelia* is a better competitor for limited resources, such as food. **30.** The limiting factors are density-dependent. In Graphs I and II, the population growth rate decreased and leveled out when the population size reached carrying capacity, as did *P. aurelia* in Graph III. **Essay** **31.** A population will increase or decrease in size depending on how many individuals are added to it or removed from it. There are two ways individuals can be added to a population. Individuals can be born into the population, or they can move into it from outside the population, or immigrate. There are two ways individuals can be removed from a population. They can die, or they can move out of the population, or emigrate. **32.** Exponential growth occurs when the individuals in a population reproduce at a constant rate. At first, the size of the population increases slowly, then it increases more and more rapidly until it approaches an infinitely large size. Under ideal conditions with unlimited resources, a population will grow exponentially. Exponential growth does not continue in natural populations for very long. Resources become less available as the population grows, and the rate of population growth slows down. Logistic growth occurs when a population's growth slows or stops following a period of exponential growth. **33.** Under ideal conditions, populations will continue to grow if there are not factors that limit their growth. Some factors in a population's environment that can limit growth are predation, competition, parasitism, and disease. Other limiting factors are climate extremes and human activities. **34.** Density-dependent limiting factors become limiting only when population density—the number of organisms in a given area—reaches a certain level. These factors include competition, predation, parasitism, and disease. Density-independent limiting factors affect all populations in similar ways, regardless of population size. These limiting factors include weather, natural disasters, and certain human activities, such as damming rivers and clear-cutting forests. **35.** During most of human history, population growth was slow. Food was hard to find, and there was no cure or treatment for diseases. Death rates and birthrates were high. With the beginning of agriculture, life became easier. The food supply became more reliable. With the Industrial Revolution, other advances, such as medicine and improved healthcare and sanitation, reduced the death rate and made it possible to live longer. This is when the human population began to grow exponentially.

Chapter 5—Test B

Multiple Choice **1.** D **2.** B **3.** D **4.** A **5.** A **6.** C **7.** A **8.** C **9.** A **10.** B **11.** A **12.** B **13.** C **14.** B **15.** B **Completion** **16.** density **17.** decrease **18.** exponentially **19.** predator-prey **20.** age-structure diagram **Short Answer** **21.** geographic distribution, density, growth rate, and age structure **22.** The pattern of growth is a J-shaped curve. **23.** A limiting factor is a factor that causes population growth to decrease, such as competition, drought, predation, or disease. **24.** The food supply improved; sanitation and medical and health care reduced the death rate; and the birthrate was high. **25.** Some economists believe that science, technology, and positive changes in society will control the negative impacts of population growth. **Using Science Skills** **26.** 1900 **27.** The deaths per thousand did not decrease from 1960 to 1970. **28.** 1990 (6.9) **29.** The number of births over deaths since 1950 steadily decreases from 14.5 in 1950 to 6.9 in 1990. **30.** Life expectancy at birth is increasing. People are living longer and the birthrate is larger than the death rate, so our population is increasing.

Chapter 6—Test A

Multiple Choice **1.** D **2.** B **3.** D **4.** B **5.** C **6.** C **7.** D **8.** C **9.** C **10.** D **11.** C **12.** D **13.** B **14.** D **15.** C **Completion** **16.** fisheries **17.** species diversity **18.** biological magnification **19.** captive breeding **20.** phytoplankton **Short Answer** **21.** Nearly all modern subsistence hunters use some form of technology, such as guns, snowmobiles, or manufactured tools. Ancient hunter-gatherers hunted birds and animals and fished in rivers. **22.** The resources common to everyone, such as air or parts of the oceans, will eventually be destroyed because no one is responsible for preserving them. **23.** Sustainable-development practices are aimed at using natural resources at a rate that does not deplete them. Sustainable-development practices are based on principles of ecology and economics. **24.** A biodiversity hot spot is an area where endangered species are most threatened, such as the California Floristic Province. **25.** Possible answers include: solar energy, production of oxygen, temperature control, air and water purification, food production, habitats for wildlife, storage and recycling of nutrients, and other services listed in Figure 6-22 of the student text. **Using Science Skills** **26.** 1980 **27.** The use of chemical fertilizers increased steadily from the beginning of the Green Revolution until 1990. During those years, global food production nearly doubled, which was the goal of the Green Revolution. Other circumstances, however, caused fertilizer use to decrease after 1990. **28.** A strategy of the Green Revolution was to use a method called monoculture, in which farmers planted large fields with a single, highly productive crop year after year. To support these fields, farmers relied on chemical fertilizers to boost plant growth. **29.** Possible answers include: increased awareness of the damage being done to the biosphere, which led to the reduced use of chemical fertilizers; development of better land-use techniques, such as contour plowing and alternating crops, which led to a reduced dependence on chemical fertilizers. **30.** The kind of increase in the use of chemical fertilizers from 1965 to 1990 is not sustainable. The application of fertilizers can interfere with food webs and biogeochemical cycles. **Essay** **31.** Earth is like an island in space. All the living things on Earth share a limited resource base. We all depend on the natural ecological processes that sustain these resources. The human population is still growing, but the planet we live on is not. Along with this growing population come increasing demands on Earth's air, water, land, resources, and biodiversity. **32.** Many combustion processes, such as the burning of fossil fuels, release acidic gases containing nitrogen and sulfur compounds into the air. When these gases combine with water vapor, they form drops of nitric and sulfuric acids. These drops then fall as acid rain. Acid rain can kill plants by damaging their leaves and changing the chemistry of soils and standing-water ecosystems. Acid rain may also dissolve and release toxic elements, such as mercury, from the soil, allowing them to enter the biosphere. **33.** Biodiversity is one of Earth's greatest natural resources. Species of many kinds have provided us with foods, industrial products, and medicines—including painkillers, antibiotics, heart drugs, antidepressants, and anticancer drugs. For example, the rosy periwinkle plant is the source of substances used to treat certain cancers. When biodiversity is lost, potential sources of material with significant value to the biosphere and to humankind may be lost with it. **34.** One of the most important threats to biodiversity today comes from apparently harmless plants and animals that humans transport around the world either accidentally or intentionally. Introduced into new habitats, these organisms often become invasive species that reproduce rapidly and crowd out native species. Invasive species increase their populations rapidly because their new habitat lacks the parasites and predators that control their populations in their native environments. **35.** Sometimes, the need to protect biodiversity is greatest in countries that are least able to do so. The destruction of tropical rain forests, for example, is the single greatest threat to biodiversity on land. Most rain forests are located in developing countries, where conservation goals must be weighed against the survival needs of the human population.

Chapter 6—Test B

Multiple Choice **1.** A **2.** A **3.** D **4.** B **5.** D **6.** B **7.** A **8.** D **9.** C **10.** C **11.** B **12.** D **13.** C **14.** D **15.** C **Completion** **16.** renewable **17.** particulates **18.** economics **19.** nonbiodegradable **20.** ozone layer **Short Answer** **21.** Suburban sprawl is a pattern of growth that occurs as cities become crowded and people move to the suburbs. **22.** The spread of agriculture and a dependable supply of food enabled humans to remain in one place and live together in larger settlements. **23.** Species of many kinds provide us with food, industrial products, and medicines. **24.** Animals, such as the passenger pigeon, have often been hunted to extinction. Populations of rare animals are hunted for fur, hides, meat, and other body parts such as horns, which makes them vulnerable to extinction. **25.** Ozone is a gas made of molecules containing three oxygen atoms that, in the upper atmosphere, absorbs harmful UV radiation. **Using Science Skills** **26.** Students should total all the numbers in the column labeled Number Identified to obtain 1,584,000 species. **27.** The most diverse group of species is insects, with 950,000 identified species. **28.** The group flowering plants has the highest number of species on the list, with 565 endangered species. **29.** Table II **30.** The list will likely have a smaller number of endangered species.

Unit 2—Test A

Multiple Choice **1.** B **2.** C **3.** D **4.** A **5.** B **6.** D **7.** B **8.** C **9.** D **10.** A **11.** A **12.** B **13.** D **14.** D **15.** C **Completion** **16.** ten **17.** denitrification **18.** tundra **19.** exponential **20.** monoculture **Short Answer** **21.** Photosynthesis depends on light energy; chemosynthesis relies on the energy within inorganic molecules **22.** Energy flows in one direction: first from sunlight to producers and then from producers to consumers. **23.** Foreign species in new habitats often become invasive because they lack parasites and predators in the new habitats. **24.** Possible answer: A sustainable strategy for protecting land resources is contour plowing; for protecting forest resources, replanting trees; for protecting fishery resources, regulating the number and size of fish caught. **25.** Biological magnification is the process in which increasing concentrations of toxic substances accumulate at higher trophic levels of food chains and food webs. **Using Science Skills** **26.** From 1880 to 1980, the acidity increased from a pH of about 6.8 to a pH of about 4.8. The likely cause of the change is acid rain. **27.** Mussels are least tolerant of increased acidity. Lake trout are most tolerant. **28.** The survival of smallmouth bass is first affected by increasing acidity at a pH of about 5.8. The survival of brown trout is first affected at a pH of about 5.5. **29.** Mussels are most likely to be threatened by extinction if the above trend continues. **30.** The change in acidity is a density-independent factor, because it is caused by human activities and not by the size of mussel and fish populations. **Essay** **31.** A predator-prey relationship is one of the best-known mechanisms of population control. Increases in the prey population are followed by increases in the predator population. As the predators kill the prey, the prey population falls. The decline in the prey population is followed by a decline in the predator population because there are fewer prey for the predators to feed upon. This cycle of predator and prey populations can be repeated indefinitely. **32.** The unequal heating of Earth's surface drives winds and ocean currents, which transport heat throughout the biosphere. Winds form because warm air tends to rise and cool air tends to sink. The upward movement of warm air near the equator and the downward movement of cold air over the poles create winds. The winds move heat throughout the atmosphere, from regions of sinking air to regions of rising air. Similar patterns of heating and cooling occur in Earth's oceans and create ocean currents. **33.** For most of human existence, the population grew slowly. Life was harsh, and limiting factors, such as diseases and lack of food, kept population sizes low. About 500 years ago, the human population began growing more rapidly. Agriculture and industry made life easier and safer. The food supply increased, and improved sanitation and medicine reduced the death rate. At the same time, birthrates in most places remained high, causing the human population to experience exponential growth. Over the past century, population growth in the United States and some other countries has slowed dramatically, as both birth and death rates have fallen with increasing education and standards of living. **34.** Habitat fragmentation refers to the splitting of ecosystems into pieces. As a result, remaining pieces of habitat become biological "islands." For example, in suburbs, patches of forest can be surrounded by farms, houses, and shopping malls. The smaller the "island," the fewer species can live there, the smaller their populations can be, and the more vulnerable they are to further disturbance or climate change. **35.** Population growth depends, in part, on how many people of different ages make up a given population. In a population with similar numbers of people in each age group, the population growth rate is likely to be low. There are relatively few people of reproductive age to produce children, and there are relatively great numbers of people in the older ages at high risk of dying. In populations in which there are many more young people than middle-aged and elderly people, the population growth rate is likely to be high. Many people are of reproductive age, and there are relatively few people in the older ages at high risk of dying.

Unit 2—Test B

Multiple Choice **1.** C **2.** B **3.** D **4.** A **5.** C **6.** C **7.** D **8.** B **9.** A **10.** B **11.** C **12.** D **13.** D **14.** C **15.** A **Completion** **16.** biosphere **17.** sunlight **18.** niche **19.** logistic **20.** biological magnification **Short Answer** **21.** Aquatic ecosystems are determined primarily by the depth, flow, temperature, and chemistry of the overlying water. **22.** Latitude determines the angle at which sunlight strikes Earth's surface. This, in turn, determines how much the surface is heated and, thus, climate. **23.** Four important characteristics of populations are geographic distribution, density, growth rate, and age structure. **24.** The green revolution refers to the effort to increase the world's food supply with the introduction of modern agricultural techniques, such as monoculture and the use of chemical fertilizers. **25.** The three different types of ecological pyramids are energy pyramids, biomass pyramids, and numbers pyramids. **Using Science Skills** **26.** The graph represents the interaction between a predator population and a prey population. **27.** The lynx population generally rises whenever the hare population increases, because there are more hares for the lynxes to feed upon. **28.** An increase in the number of lynxes usually is followed by a decline in the hare population, because the large number of lynxes feeds upon hares until they are reduced in number. **29.** There are almost always fewer lynxes than hares, because lynxes are at a higher trophic level than hares, and fewer organisms generally can be supported at a higher trophic level. **30.** The graph shows the action of density-dependent limiting factors, because the size of one population is limited by the size of the other population.

Chapter 7—Test A

Multiple Choice **1.** D **2.** B **3.** C **4.** D **5.** A **6.** C **7.** B **8.** D **9.** D **10.** D **11.** A **12.** A **13.** C **14.** A **15.** C **Completion** **16.** cells **17.** cytoplasm **18.** chromosomes **19.** ribosomes **20.** cell membrane **Short Answer** **21.** Prokaryotes are generally simpler and smaller than eukaryotes, whereas eukaryotes have a nucleus and other specialized organelles. **22.** Structure Q in drawing II corresponds to structure H in drawing I, the Golgi apparatus. The function of the Golgi apparatus is to modify, sort, and package proteins and other materials from the endoplasmic reticulum for storage in the cell or secretion outside the cell. **23.** Structures J, O, and P are found in drawing II but not in drawing I. J is the cell wall, which provides support and protection for the cell. O is a chloroplast, an organelle that captures the energy of sunlight and converts it into chemical energy. P is a central vacuole, which stores water, salts, protein, and carbohydrates and also provides support. **24.** Diffusion is a process in which particles tend to move from an area where they are more concentrated to an area where they are less concentrated. **25.** When a blood cell is placed in a hypertonic solution, the water flows from the blood cell, through the cell membrane, into the solution. As a result of losing water, the cell shrinks. **Using Science Skills** **26.** The experimental setup shows two solutions with different concentrations of two different solutes separated by a selectively permeable membrane. Over time, the solutes should reach an equilibrium. **27.** Some of the C molecules will move to the B side of the container. **28.** Some of the D molecules will move from side A to side B. The net movement of D molecules will be from side B to side A. **29.** At equilibrium, the concentrations of both C and D molecules will be equal on either side of the selectively permeable membrane. **30.** Yes, both kinds of molecules will continue to move across the membrane; however, there will be no net movement of either molecule. **Essay** **31.** The cell theory states that all living things are composed of cells. It also says that cells are the basic units of life and new cells come from preexisting cells. The cell theory is significant to biology because all living thing are made of cells. Differences in the structure and function of different life forms are reflected in differences in their cell structures. **32.** Like a piece of mosaic art, which is made of different pieces, the cell membrane is made of many different molecules. The background is a lipid bilayer. Within this bilayer are proteins that form channels and pumps that help move materials from one side of the membrane to the other. Carbohydrates on the outer surface of the membrane act like chemical identification cards and allow cells to identify one another. **33.** Facilitated diffusion is the movement through a protein channel of molecules that could not otherwise cross the membrane. Facilitated diffusion only occurs with a concentration difference and does not require additional energy. Active transport is the movement of materials across a cell membrane against a concentration difference and does require the addition of energy. **34.** The cell from the unicellular organism carries out all the life processes of the organism. The cell from the multicellular organism is specialized and carries out only certain tasks in the organism, while relying on other cells in the multicellular organism to complete other life processes. **35.** The levels of organization in a multicellular organism include cells, tissues, organs, and organ systems. Similar cells are grouped into tissues; groups of tissues that work together form organs; a group of organs that work together make up an organ system. Unicellular organisms cannot have cell specialization. Instead, all the activities of the organism must be carried out by the single cell.

Chapter 7—Test B

Multiple Choice **1.** B **2.** A **3.** C **4.** C **5.** A **6.** D **7.** A **8.** B **9.** C **10.** B **11.** B **12.** B **13.** B **14.** A **15.** C **Completion** **16.** cytoplasm **17.** organelles **18.** diffusion **19.** specialization **20.** cells, organs **Short Answer** **21.** The cell theory states that all living things are composed of cells, that cells are the basic units of structure and function in living things, and that new cells are produced from existing cells. **22.** The nucleus contains the cell's genetic material in the form of DNA. The nucleus also controls many of the cell's activities. **23.** The cytoskeleton helps the cell maintain its shape. It is also involved in movement. **24.** Because the concentration of water in the cup is greater than the concentration of water in the raisin, water will flow from the cup into the raisin. The water is a hypotonic solution in this example. **25.** cell, tissue, organ, organ system **Using Science Skills** **26.** Drawing II, a plant cell, contains such a structure, a chloroplast, which is labeled O. **27.** Structure E in drawing I corresponds to structure M in drawing II. This structure is a mitochondrion. **28.** Structure D in drawing I corresponds to structure L in drawing II. This structure is the cell membrane. **29.** Structure P is a central vacuole. It stores materials such as water, salts, proteins, and carbohydrates. The pressure in the vacuole helps the cell support heavy structures. **30.** Both drawings are of eukaryotes, because each cell has a nucleus.

Chapter 8—Test A

Multiple Choice **1.** D **2.** C **3.** D **4.** D **5.** C **6.** B **7.** D **8.** A **9.** B **10.** B **11.** D **12.** B **13.** A **14.** B **15.** C **Completion** **16.** chlorophyll **17.** green **18.** green **19.** B **20.** decreases **Short Answer** **21.** Heterotrophs obtain energy from the food they consume. **22.** ATP, or adenosine triphosphate, is one of the principal compounds that cells use to store and release energy. Energy is released from ATP when the chemical bond between the second and third phosphates is broken and the third phosphate group is released.

23.

$$6CO_2 + 6H_2O \xrightarrow{\text{light}} C_6H_{12}O_6 + 6O_2$$

carbon dioxide + water $\xrightarrow{\text{light}}$ sugar + oxygen

24. Sample answer: By being close together, electrons and energy can easily move through the photosystems powering each successive step of the light-dependent reactions. **25.** The Calvin cycle uses ATP and NADPH from the light-dependent reactions to produce high-energy sugars. **Using Science Skills** **26.** The beaker she placed in the shade is the control. **27.** The air bubbles are probably oxygen gas. Jan Ingenhousz showed that plants produce oxygen bubbles in the light but not in the dark. **28.** 5 cm **29.** The student's data show that as the light gets closer to the water plant, the water plant gives off more bubbles. **30.** Probably not; light intensity increases the rate of photosynthesis up to a certain point and then levels off. **Essay** **31.** Heterotrophs depend on autotrophs to obtain energy from the sun. This energy is then passed on to heterotrophs in the form of food. Without autotrophs, the sun's energy would not be available to heterotrophs and heterotrophs would eventually die out or find a new way of obtaining energy. **32.** A glucose molecule can store more than 90 times the chemical energy than an ATP molecule. Glucose is used by cells to store large amounts of energy for long periods of time. In contrast, ATP is used to store smaller amounts of energy that will be used in the next few seconds. Cells can regenerate ATP from ADP as needed by using the energy in foods like glucose. **33.** Light is a form of energy; thus, when a pigment absorbs light it also absorbs the energy from that light. When the pigment chlorophyll absorbs light, much of the energy is transferred directly to electrons in the chlorophyll molecule, raising the energy levels of these electrons. This energy is then passed on in the reactions of photosynthesis. **34.** Three of the factors that affect the rate of photosynthesis are light intensity, temperature, and water. The rate of photosynthesis increases with light intensity up to a certain point, then levels off. Photosynthesis slows at extreme temperatures and usually has an optimal temperature for each kind of plant. Lack of water slows down photosynthesis. **35.** Electrons in the pigments in photosystem II absorb energy from light. These electrons then move through the electron transport chain to photosystem I. H^+ ions move from the stroma into the inner thylakoid membrane. $NADP^+$ becomes NADPH. The charge difference on either side of the thylakoid membrane provides the energy to make ATP via ATP synthase.

Chapter 8—Test B

Multiple Choice **1.** A **2.** B **3.** C **4.** C **5.** A **6.** C **7.** D **8.** A **9.** D **10.** A **11.** C **12.** B **13.** A **14.** C **15.** D **Completion** **16.** heterotrophs **17.** light **18.** high-energy sugars **19.** the stroma **20.** ATP **Short Answer** **21.** Autotrophs are organisms, such as plants, that can make their own food. Heterotrophs, such as animals, must use food made by other organisms to get energy. **22.** Heterotrophs get energy by eating autotrophs and/or other heterotrophs. **23.** When a phosphate group is removed from ATP, energy is released and ADP results. **24.** The structures are thylakoids. They contain photosystems that are able to capture the energy of sunlight. **25.** Any two of the following: water, light intensity, temperature **Using Science Skills** **26.** photosynthesis **27.** a chloroplast **28.** Oxygen gas, ATP, and NADPH are the products of the light-dependent reactions. **29.** sugars **30.** CO_2

Chapter 9—Test A

Multiple Choice **1.** A **2.** C **3.** D **4.** C **5.** C **6.** D **7.** C **8.** C **9.** A **10.** C **11.** B **12.** B **13.** B **14.** C **15.** B **Completion** **16.** oxygen **17.** 2 **18.** alcoholic **19.** C **20.** lactic acid **Short Answer** **21.** The three stages are as follows: glycolysis (which occurs in the cytoplasm), the Krebs cycle (which occurs in the mitochondria), and electron transport (which occurs in the mitochondria). **22.** Alcoholic fermentation produces carbon dioxide, alcohol, and NAD^+; lactic acid fermentation produces lactic acid and NAD^+. **23.** Sample answer: Lactic acid fermentation does not give off carbon dioxide as shown by pathway A in the figure. **24.** Oxygen is the final electron acceptor in the electron transport chain, which means that it is needed to get rid of low-energy electrons and H^+ ions. **25.** The body uses ATP that is already present in the muscle cells, ATP made from lactic acid fermentation, and ATP made from cellular respiration. **Using Science Skills** **26.** Sample answer: The equation for cellular respiration is $6O_2 + C_6H_{12}O_6 \rightarrow 6CO_2 + 6H_2O$ + Energy. The mouse should give off CO_2. **27.** Sample answer: The mouse requires oxygen. Fresh air containing oxygen flows from outside the flasks into flasks B, C, and D. Air mixed with whatever the mouse gives off flows from flask B into flask A. The mouse receives fresh air and should be able to survive in the chamber for the duration of the experiment. **28.** Sample answer: If the mouse is carrying out cellular respiration, it will give off CO_2, the CO_2 will flow into flask A, and the phenolphthalein in flask A will change from pink to clear. **29.** Sample answer: The cricket, like all living organisms, is carrying out cellular respiration. However, the mouse is larger than the cricket and will give off more CO_2 than the cricket. After one hour, the cricket probably has not given off enough CO_2 to measure. If the scientist allows the experiment to continue for several hours, she will see that more CO_2 is given off by the cricket over time. **30.** Sample answer: The mouse that had been exercising should give off more CO_2 because this mouse will be breathing more heavily. This mouse might even have an oxygen debt to repay. **Essay** **31.** Sample answer: Glycolysis is the breakdown of glucose into two molecules of pyruvic acid, producing 4 ATP molecules. An initial input of 2 ATP molecules is required to start glycolysis; thus, there is a net gain of 2 ATP molecules. This process produces high-energy electrons, which are passed to NAD^+ forming 2 NADH. If oxygen is present, glycolysis leads to the Krebs cycle and the electron transport chain. If oxygen is not present, glycolysis is followed by the rest of fermentation. **32.** Both main types of fermentation release energy from food molecules by producing ATP in the absence of oxygen. Alcoholic fermentation occurs in yeasts and a few other microorganisms, and its products are alcohol, CO_2, and NAD^+. Lactic acid fermentation occurs in muscle cells and some unicellular organisms, and its products are lactic acid and NAD^+. **33.** Sample answer: The electron carriers of cellular respiration are NAD^+ and FAD. These molecules accept high-energy electrons and move to the electron transport chain. The electron transport chain produces ATP molecules. **34.** The electron transport chain uses the high-energy electrons from the Krebs cycle to convert ADP into ATP. High-energy electrons are passed from one carrier protein to the next. Every 2 high-energy electrons that move down the chain result in enough energy to convert 3 ADP molecules to 3 ATP molecules, on average. **35.** Sample answer: The first few seconds of a race, for example, are powered by the ATP that the muscle cells already have on hand. Then, the muscle cells begin producing ATP by lactic acid fermentation—this would be around the 50-meter mark of the race. After about 90 seconds, however, cellular respiration is the only way to generate a continuing supply of ATP.

Chapter 9—Test B

Multiple Choice **1.** C **2.** A **3.** B **4.** A **5.** A **6.** C **7.** C **8.** D **9.** C **10.** C **11.** A **12.** D **13.** C **14.** D **15.** B **Completion** **16.** pyruvic acid **17.** 2 **18.** electron transport chain **19.** oxygen **20.** 36 **Short Answer** **21.** Cellular respiration is the process that releases energy by breaking down food molecules in the presence of oxygen. **22.** During glycolysis, one molecule of glucose is broken in half, producing two molecules of pyruvic acid. **23.** Citric acid is the first compound formed in the process. **24.** The electron transport chain uses high-energy electrons from the Krebs cycle to convert ADP into ATP. **25.** Photosynthesis releases oxygen into the atmosphere, whereas cellular respiration uses oxygen to release energy from food. **Using Science Skills** **26.** The figure shows the electron transport chain. **27.** The electrons come from NADH and $FADH_2$, electron carriers that come from the Krebs cycle. **28.** They join with 4 H^+ and O_2 to form water molecules. **29.** The H^+ ions are mostly moving into the intermembrane space. This movement causes the intermembrane space to become positively charged. **30.** Sample answer: ATP synthase catalyzes the reaction that produces a high-energy ATP from a low-energy ADP and a phosphate.

Chapter 10—Test A

Multiple Choice **1.** A **2.** B **3.** B **4.** C **5.** B **6.** C **7.** A **8.** C **9.** A **10.** A **11.** D **12.** A **13.** C **14.** A **15.** C **Completion** **16.** less **17.** daughter **18.** cell plate **19.** internal **20.** cancer **Short Answer** **21.** A large cell carries out its activities less efficiently than a small cell does. For example, as a cell's size increases, it becomes more difficult for needed materials, such as oxygen and nutrients, to reach all the parts of the cell. **22.** As the size of a cell increases, its ratio of surface area to volume decreases. **23.** Interphase is the period of cell growth in between cell divisions. Interphase includes the G_1 phase, the S phase, and the G_2 phase. **24.** During cytokinesis, a cell plate forms in the cytoplasm midway between each new nucleus. The cell plate gradually develops into a separating membrane, and a cell wall begins to appear in the cell plate. **25.** Contact with other cells can stop cell growth. **Using Science Skills** **26.** Diagram A shows cancer cells because it shows cells that are not growing in an organized way. **27.** Cancer cells do not respond to the signals that regulate growth in most cells. Diagram B shows cells that are normal and have responded to such signals. Diagram A shows cells that have divided until they have formed a mass of cells, or a tumor. The cancer cells are dividing uncontrollably and are disrupting nearby normal cells. **28.** They can break loose from the tumor they are now a part of and spread throughout the body, disrupting normal activities and causing serious medical problems. **29.** Diagram A—these cells might have a defect in the p53 gene, which has allowed the cells to multiply more quickly than the normal cells. Diagram B—these cells probably have a healthy copy of the p53 gene, which has stopped the cell cycle until the genetic material in these cells has been properly copied. **30.** Students may suggest removing the cancerous cells in hopes of preventing their continued division, growth, and spread throughout the body, or treating them with radiation or chemicals that will destroy the cells. **Essay** **31.** As a cell grows, it functions less efficiently because it places more demands on its DNA and it is less able to move materials to their proper destinations quickly. Cell division results in two daughter cells. Each cell has its own copy of the parent cell's DNA and has a size that allows it to efficiently exchange materials. **32.** During the G_1 phase, the cell grows; during the S phase, the DNA replicates; during the G_2 phase, the cell prepares for mitosis. The M phase or cell division includes mitosis and cytokinesis. Figures should approximate Figure 10-4 on page 245 of the student text. **33.** Prophase—the chromatin condenses into chromosomes, the centrioles separate (in animal cells), and the nuclear envelope breaks down; Metaphase—the chromosomes line up across the midline of the cell and each chromosome is attached to a spindle fiber and centromere; Anaphase—sister chromatids separate into individual chromsomes; Telophase—chromosomes move to opposite sides of the dividing cell and two new nuclear envelopes form. **34.** Cells at the edges of the cut are stimulated to divide rapidly, producing new cells, and healing the wound. When the cut is almost completely healed, the rate of cell division slows down. Thus, when the cells on either side of the cut are no longer in contact with one another, they divide to fill in the gap. Once the cells are in contact with one another (the cut is healed), the cells resume their normal growth rate. **35.** Sample answer: Unlike normal cells, cancer cells do not respond to normal controls on growth and division. By dividing uncontrollably, cancer cells form tumors and spread throughout the body. A cure for cancer includes a way to both prevent cancer cells from dividing uncontrollably and to allow normal cells to continue dividing normally. Finding a way to stop the cell cycle in some cells, but not interfere with the cell cycle in other cells, has made it difficult to cure cancer.

Chapter 10—Test B

Multiple Choice **1.** D **2.** C **3.** A **4.** D **5.** D **6.** D **7.** A **8.** A **9.** C **10.** C **11.** A **12.** A **13.** B **14.** B **15.** B **Completion** **16.** cell division **17.** interphase **18.** M **19.** cytokinesis **20.** cyclins **Short Answer** **21.** As a cell grows larger, more demands are placed on its DNA and the cell has more trouble moving enough nutrients and wastes across the cell membrane. **22.** Chromosomes aren't visible because the DNA and protein molecules that make up the chromosomes are spread throughout the nucleus. **23.** A—G_1 phase, cell growth; B—S phase, DNA replication; C—G_2 phase, preparation for mitosis; D—M phase, cell division (mitosis and cytokinesis) **24.** Any two of the following: contact with other cells, cyclins, growth factors, and any other internal or external regulators **25.** Cancer cells do not respond to the signals that control the growth of normal cells. As a result, cancer cells form tumors and can spread throughout the body. **Using Science Skills** **26.** It shows various stages of mitosis in an animal cell. This is an animal cell because of the presence of centrioles and the shape of the cells. **27.** Four **28.** X is the centriole; Y is the spindle. **29.** D, A, C, B **30.** The final step would be cytokinesis. It would show two daughter cells forming.

Unit 3—Test A

Multiple Choice **1.** C **2.** A **3.** C **4.** B **5.** B **6.** A **7.** D **8.** D **9.** A **10.** B **11.** A **12.** C **13.** C **14.** D **15.** B **Completion** **16.** osmosis **17.** heterotrophs **18.** photosynthesis **19.** glycolysis **20.** mitosis **Short Answer** **21.** All living things are composed of cells. Cells are the basic units of structure and function in living things. New cells are produced from existing cells. **22.** ATP, or adenosine triphosphate, is one of the principal compounds that cells use to store and release energy. Energy is released when the chemical bond between the second and third phosphates is broken.

23. $6CO_2 + 6\,H_2O \xrightarrow{\text{light}} C_6H_{12}O_6 + 6O_2$

carbon dioxide + water $\xrightarrow{\text{light}}$ sugars + oxygen

24. $6O_2 + C_6H_{12}O_6 \rightarrow 6CO_2 + 6\,H_2O$ + energy

oxygen + glucose → carbon dioxide + water + energy

25. Cell division causes the ratio of surface area to volume to become greater in the daughter cells. **Using Science Skills** **26.** Both drawings are of eukaryotes because each cell has a nucleus, structure A in diagram I and structure K in diagram II. **27.** Structure L corresponds to structure D. Each is a cell membrane. The function of both is the same. In diagram I, the cell membrane is the cell's outer boundary. In diagram II, the cell membrane is surrounded by a cell wall. **28.** No structure in diagram I corresponds to structure O. The structure is a chloroplast, where photosynthesis occurs. **29.** Structure M corresponds to structure E. Both are mitochondria, where cellular respiration occurs. **30.** Sample answer: Diagram I is probably an animal cell. It has a nucleus and is therefore a eukaryotic cell, but it has no cell wall. Diagram II is probably a plant cell, because it has a nucleus, a chloroplast, and a central vacuole. **Essay** **31.** They are similar in that both involve the movement of materials across a membrane. Facilitated diffusion involves movement of particles across the membrane through protein channels. Yet, this process does not require use of the cell's energy because it is still diffusion. Active transport, by contrast, does require use of the cell's energy because it is the movement of materials against a concentration difference. **32.** When pigments in photosystem II absorb light, the light energy is absorbed by electrons, increasing their energy level. These high-energy electrons are used in the light-dependent reactions to convert ADP and $NADP^+$ into the energy carriers ATP and NADPH. They provide the energy to build high-energy sugars in the Calvin cycle. **33.** The process of releasing energy from glucose is begun with glycolysis, which yields 2 molecules of ATP. When oxygen is not available, glycolysis is followed by fermentation, which produces no more ATP molecules. When oxygen is available, cellular respiration can occur. That pathway results in the production of 34 more ATP molecules, for a total of 36 ATP molecules from a single molecule of glucose. **34.** In the process of photosynthesis, the light-dependent reactions convert the electron carrier $NADP^+$ into NADPH, which provides energy to build high-energy sugars in the Calvin cycle. In cellular respiration, the electron carriers $NADP^+$ and FAD^+ are used to make NADPH and $FADH_2$. The electrons in those compounds are used in the electron transport chain to convert ADP to ATP. **35.** During prophase, the chromatin condenses into chromosomes. The centrioles separate, and a spindle begins to form. The nuclear envelope breaks down. During metaphase, the chromosomes line up across the center of the cell. Each chromosome is connected to a spindle fiber at its centromere.

During anaphase, the sister chromosomes separate into individual chromosomes and are moved apart. During telophase, the chromosomes gather at opposite ends of the cell and lose their distinct shapes. Two new nuclear envelopes form.

Unit 3—Test B

Multiple Choice **1.** B **2.** B **3.** D **4.** A **5.** C **6.** C **7.** A **8.** D **9.** D **10.** B **11.** C **12.** B **13.** C **14.** A **15.** A **Completion** **16.** cell membrane **17.** tissue **18.** sunlight **19.** oxygen **20.** chromatids **Short Answer** **21.** All living things are composed of cells. Cells are the basic units of structure and function in living things. New cells are produced from existing cells. **22.** An autotroph makes its own food, whereas a heterotroph obtains energy from the foods it eats.

23. carbon dioxide + water $\xrightarrow{\text{light}}$ sugars + oxygen **24.** oxygen + glucose → carbon dioxide + water + energy **25.** Prophase, metaphase, anaphase, telophase **Using Science Skills** **26.** Structure K corresponds to structure A. It is a cell nucleus. **27.** Structure E corresponds to structure M. It is a mitochondrion. **28.** Structure J is a cell wall. Its function is to provide support and protection for the cell. **29.** Structure O is a chloroplast. There is no corresponding structure in drawing I because drawing I is not a plant cell or any other cell that carries out photosynthesis. **30.** They represent eukaryotic cells, because each of the two cells has a nucleus.

Chapter 11—Test A

Multiple Choice **1.** D **2.** B **3.** C **4.** C **5.** A **6.** D **7.** A **8.** C **9.** A **10.** B **11.** C **12.** A **13.** B **14.** C **15.** D **Completion** **16.** P **17.** heterozygous **18.** *Tt* **19.** half **20.** prophase I **Short Answer** **21.** An organism must inherit two recessive alleles for a trait in order to show that trait. **22.** The phenotype ratio is 9 round, yellow peas: 3 round, green peas : 3 wrinkled, yellow peas : 1 wrinkled, green pea. **23.** Mitosis produces genetically identical diploid cells, whereas meiosis produces genetically different haploid cells. **24.** A linkage group is all the genes on the same chromosome. **25.** A gene map shows the relative locations of genes on a chromosome. **Using Science Skills** **26.** The genotype of the offspring is *bbRR.* **27.** The phenotype of the offspring is white, rough hair. **28.** The phenotypes of the offspring are black, rough hair; black, smooth hair; white, rough hair; and white, smooth hair. **29.** Offspring with black, rough hair have the genotypes *BBRR, BBRr, BbRR,* and *BbRr.* **30.** One sixteenth of the offspring are expected to have white, smooth hair.

Essay **31.** When the heterozygous tall F_1 plants produced gametes, their dominant allele for tallness segregated from their recessive allele for shortness. As a result, some of their gametes had the dominant allele, while others had the recessive allele. When the F_1 plants self-pollinated, some male gametes with the recessive allele fused with female gametes with the recessive allele during fertilization. The offspring that resulted, which were part of the F_2 generation, had two alleles for shortness and were therefore short. **32.** The tall pea plant should be crossed with a short pea plant. If the tall pea plant is homozygous, all of the offspring will be tall. If the tall pea plant is heterozygous, it is likely that about half of the offspring will be tall and half will be short. **33.** In incomplete dominance, one allele is not completely dominant over another. As a result, the heterozygous phenotype is somewhere in between the two homozygous phenotypes. In codominance, both alleles contribute to the phenotype of the organism. As a result, the heterozygous phenotype shows both homozygous phenotypes. **34.** Both meiosis I and meiosis II contain a prophase, metaphase, anaphase, and telophase. However, chromosomes replicate prior to meiosis I but not prior to meiosis II. Also, during meiosis I, tetrads form and align along the center of the cell. Then, the homologous chromosomes are separated and two haploid daughter cells form. During meiosis II, sister chromatids align along the center of the cell and are then separated. Four haploid daughter cells form. **35.** Linkage is the condition in which two genes are located on the same chromosome. As a result, the genes' alleles are usually inherited together. However, if crossing-over occurs between the alleles of the linked genes, those alleles no longer are inherited together. The frequency of crossing-over between linked genes (or the frequency with which linked genes are inherited separately) is used to determine the relative locations of genes on the same chromosome, resulting in a gene map. The greater the frequency, the farther apart are the genes on the chromosome.

Chapter 11—Test B

Multiple Choice **1.** C **2.** C **3.** C **4.** D **5.** A **6.** B **7.** B **8.** A **9.** B **10.** D **11.** C **12.** B **13.** B **14.** A **15.** A **Completion** **16.** alleles **17.** Probability **18.** *TT* and *Tt* **19.** genes **20.** haploid **Short Answer** **21.** Genetics is the scientific study of heredity. **22.** A diploid cell has two sets of chromosomes. **23.** Homologous chromosomes are corresponding sets of chromosomes; one set inherited from the male parent and the other inherited from the female parent. **24.** The number of chromosomes is reduced by half. **25.**

The genes that Mendel studied were located on different chromosomes or were located far apart on the same chromosome. **Using Science Skills 26.** *R* represents the allele for red flowers. *I* represents the allele for ivory flowers. **27.** The genotype of the pink-flowered snapdragons is *RI*. **28.** The alleles show incomplete dominance because a cross between red-flowered snapdragons and ivory-flowered snapdragons produces snapdragons with an in-between trait—pink flowers. **29.** One hundred percent of the offspring are expected to be pink-flowered. **30.** Fifty percent of the offspring are expected to be pink-flowered.

Chapter 12—Test A

Multiple Choice 1. B **2.** D **3.** C **4.** C **5.** B **6.** D **7.** D **8.** A **9.** A **10.** C **11.** A **12.** C **13.** C **14.** C **15.** B **Completion 16.** nucleotide **17.** amino acids **18.** amino acid **19.** lactose **20.** enhancer **Short Answer 21.** The hydrogen bonds between the base pairs must be broken and the molecule must unwind. **22.** Molecule B is tRNA, which carries amino acids to the ribosomes. **23.** AGG, AGA, CGA, CGC, CGU, and CGG specify arginine. **24.** Lactose binds to the *lac* repressor, causing the repressor to release the operator. **25.** The hox genes descended from the genes of common ancestors. **Using Science Skills 26.** Translation (or protein synthesis) is illustrated. **27.** Structure C is a ribosome. **28.** Structure F is a codon. **29.** The codons and anticodons have complementary nitrogenous bases, allowing them to base pair. Since each tRNA carries only one kind of amino acid, the base pairing between the anticodons and codons brings a specific sequence of amino acids to the ribosomes. **30.** The bond between the methionine and its tRNA will be broken. The tRNA will move away from the ribosome, allowing the ribosome to bind with another tRNA. The ribosome will move down the mRNA to the next codon. **Essay 31.** Hershey and Chase grew bacteriophages in cultures containing radioactive isotopes of ^{32}P and ^{35}S. The ^{32}P became incorporated into the bacteriophage's DNA because DNA contains phosphorus. The ^{35}S became incorporated into the bacteriophage's protein coat because proteins contain sulfur. After the bacteriophages were allowed to infect bacteria, Hershey and Chase found that nearly all the radioactivity in the bacteria was from ^{32}P. This indicated that the bacteriophage's DNA was injected into the bacteria. The results were important because they showed that the bacteriophage's genetic material was DNA, not protein. **32.** A DNA molecule has the shape of a double helix, or that of a twisted ladder. Each strand of the helix is a chain of nucleotides. Each nucleotide is made up of a deoxyribose sugar, a phosphate group, and a nitrogenous base. The two strands are held together by hydrogen bonds between the nitrogenous bases of the nucleotides on opposite strands. The nitrogenous bases form hydrogen bonds with each other in pairs. Adenine forms hydrogen bonds with thymine, and guanine forms hydrogen bonds with cytosine. **33.** Messenger RNA carries copies of instructions for assembling proteins from DNA to the ribosomes. Ribosomal RNA is a component of the ribosomes. Transfer RNA carries amino acids to the ribosomes for assembly into proteins. **34.** Proteins have many different functions in an organism. These functions include catalyzing and regulating chemical reactions, regulating growth patterns, and providing the actual structural components of the organisms. Together, these functions play a key role in producing an organism's traits, or phenotype. **35.** Point mutations include substitutions, insertions, and deletions of single nucleotides in DNA. Insertions and deletions have a greater effect on proteins than do substitutions, because insertions and deletions change the reading frame of the genetic code, affecting every amino acid that follows the point of mutation. In contrast, a substitution affects a single amino acid. A change in more than one amino acid is more likely to alter the function of the protein than is a change in a single amino acid.

Chapter 12—Test B

Multiple Choice 1. B **2.** A **3.** A **4.** A **5.** A **6.** B **7.** A **8.** D **9.** B **10.** C **11.** C **12.** B **13.** A **14.** A **15.** A **Completion 16.** base pairing **17.** histones **18.** RNA **19.** exons **20.** hox genes **Short Answer 21.** A bacteriophage is a kind of virus that infects bacteria. **22.** A ribose molecule, a phosphate group, and a nitrogenous base are the three main parts of an RNA nucleotide. **23.** The DNA molecule must be separated into two strands. **24.** A stop codon on the mRNA causes translation to stop. **25.** A mutation is a change in the genetic material. **Using Science Skills 26.** A chromosomal mutation results from processes A, B, C, and D. **27.** Process D involves two chromosomes. **28.** Process A results in the deletion of a segment of a chromosome. Process B results in the duplication of a segment of a chromosome. **29.** A segment of a chromosome becomes oriented in the reverse direction during process C. **30.** Process D is a translocation.

Chapter 13—Test A

Multiple Choice 1. C **2.** B **3.** D **4.** A **5.** A **6.** C **7.** A **8.** D **9.** D **10.** D **11.** C **12.** B **13.** A **14.** C **15.** D **Completion 16.** hybridization **17.** offspring

18. colored **19.** chromosomes **20.** transgenic **Short Answer 21.** A mule has some characteristics of a donkey and some characteristics of a horse. It may be hardier than either parent. **22.** In hybridization, organisms with dissimilar traits are crossed. In inbreeding, organisms with similar traits are crossed. **23.** Increasing the mutation rate in a population increases the population's genetic variation. This increases the chance that an individual will have the desired trait. **24.** Structure A is the recognition sequence of a restriction enzyme, which recognizes and cuts different sequences of nucleotides on DNA molecules. **25.** Scientists inject DNA directly into plant cells. They also remove the cell walls of plant cells and then allow the cells to take up DNA on their own. **Using Science Skills 26.** The bands consist of DNA fragments. **27.** The bands in group D moved faster because they consist of smaller DNA fragments. **28.** The restriction enzyme is cutting the DNA into fragments. **29.** The bands consist of DNA, which is negatively charged. **30.** No, the DNA samples were not identical because they produced different patterns of bands on the gel. **Essay 31.** Selective breeding has allowed humans to domesticate animals, such as horses, cats, and dogs, and to produce crop plants. It also has allowed humans to maintain and improve existing varieties of organisms and to produce new varieties. **32.** Examples will vary. The advantage of inbreeding is that it maintains desired traits within a breed. The disadvantage is that it can lead to genetic defects within a breed, such as blindness in German shepherds. **33.** In genetic engineering, organisms with desired traits are produced by directly changing the DNA of the organisms. This is done by cutting out desirable genes from the DNA of certain organisms and then inserting them into the DNA of other organisms. In selective breeding, organisms with desired traits are produced by selecting organisms for their traits and then mating, or crossing, them. Selective breeding does not directly change the DNA of living organisms. **34.** If a plant cell contains a recombinant plasmid, it contains a foreign gene. However, if that gene has not been incorporated into one of the plant cell's chromosomes, the plant cell has not been successfully transformed because the unincorporated gene cannot be expressed. **35.** Answers will vary. Transgenic bacteria make human proteins cheaply and in large amounts. Transgenic animals are used to study human genes and to improve animal foods. Transgenic plants help increase crop yields and may be healthier for people to eat.

Chapter 13—Test B

Multiple Choice 1. C **2.** D **3.** A **4.** B **5.** C **6.** A **7.** B **8.** D **9.** C **10.** B **11.** D **12.** C **13.** B **14.** D **15.** B **Completion 16.** traits (or characteristics) **17.** Inbreeding **18.** mutations **19.** bacteria **20.** antibiotics **Short Answer 21.** A polyploid organism is one that has many sets of chromosomes. **22.** PCR enables scientists to make many copies of a gene. **23.** A plasmid is a small, circular DNA molecule that scientists use to transform bacteria and plants. It is found naturally in some bacteria. **24.** A transgenic organism is an organism produced by genetic engineering that contains genes from another species. **25.** Dolly and the sheep from which she was cloned have identical genes. **Using Science Skills 26.** Sheep A **27.** Sheep B **28.** The DNA was removed from the egg cell to make sure that all of the DNA in the clone was from a single sheep. **29.** The lamb is a clone. **30.** Sheep A and the lamb are genetically identical.

Chapter 14—Test A

Multiple Choice 1. D **2.** C **3.** C **4.** D **5.** B **6.** B **7.** B **8.** C **9.** C **10.** A **11.** C **12.** D **13.** B **14.** C **15.** D **Completion 16.** karyotype **17.** *i* **18.** two **19.** mother **20.** DNA sequence **Short Answer 21.** Most human traits cannot be associated with single genes because many traits are polygenic, or controlled by many genes, and the environment influences many traits. **22.** Chromosomes 21 and 22 are the smallest human autosomes, they contain hundreds of genes, and they contain long stretches of repetitive DNA that do not code for proteins. **23.** The probability that their son has hemophilia is 50%. **24.** If the cells have a Y chromosome, the person is a male regardless of how many X chromosomes are in the cells. The Y chromosome contains genes that produce male sexual development. **25.** Viruses are used in gene therapy because of their ability to enter a cell's DNA. **Using Science Skills 26.** The individual is heterozygous *(Ff),* since her daughter has attached earlobes. The daughter inherited one allele for attached earlobes from individual 2 and another from individual 1. **27.** None of the children have attached earlobes. **28.** No, the genotype of individual 5 is uncertain because his children have free earlobes. Thus, individual 5 could be homozygous *(FF)* or heterozygous *(Ff)* for free earlobes. If, however, one of his children had attached earlobes, it would be certain that individual 5 is heterozygous. **29.** Individual 14 will be homozygous for the recessive allele *(ff)* and will have attached earlobes. **30.** No, each descendant would have received a recessive allele (f) for attached earlobes from individual 1. **Essay 31.** The amino acid sequence of abnormal hemoglobin varies from normal hemoglobin by one amino acid. Abnormal hemoglobin has valine instead of glutamic acid. As a result, the abnormal hemoglobin is less soluble. When the oxygen level of the blood decreases, the abnormal hemoglobin molecules come out of solution, stick together, and form long chains and fibers that cause the red blood cells to become sickle shaped. **32.** Because the allele for colorblindness is recessive and X-linked, the girl must have inherited alleles for colorblindness on X chromosomes from both her mother and father. Since the father has a single X chromosome, it must carry the allele for colorblindness, and he must be colorblind. **33.** A male who inherits the allele always has the trait because he has a single Y chromosome. A female never has the trait because a female does not inherit a Y chromosome. **34.** The male cat inherited an extra X chromosome. One of the X chromosomes carries the allele for black spots, the other carries the allele for orange spots. The two X chromosomes are randomly inactivated in the cat's cells, producing the orange and black spots. **35.** Both Turner's syndrome and Klinefelter's syndrome result from nondisjunction of the sex chromosomes. In Turner's syndrome, the individual has too few sex chromosomes and is a female with the genotype XO. In Klinefelter's syndrome, the individual has too many sex chromosomes and is a male with the genotype XXY, XXXY, or XXXXY.

Chapter 14—Test B

Multiple Choice **1.** D **2.** C **3.** A **4.** B **5.** A **6.** B **7.** A **8.** C **9.** A **10.** D **11.** C **12.** A **13.** C **14.** B **15.** A **Completion** **16.** pedigree **17.** O **18.** PKU **19.** X **20.** DNA fingerprinting **Short Answer** **21.** The DNA sequence of the allele that causes cystic fibrosis has a deletion of three bases. **22.** Males have just one X chromosome. **23.** Most male cells do not have two X chromosomes. **24.** These sections of DNA vary widely from person to person. **25.** Scientists expected humans to have more genes because a human is a very complex organism. **Using Science Skills** **26.** The chromosomes in each group are called homologous chromosomes. **27.** The chromosomes in groups 1 through 22 are autosomes. **28.** Forty-seven chromosomes are shown. **29.** The karyotype shows two X chromosomes and one Y chromosome, which is not normal. The normal number of sex chromosomes is two. **30.** The individual is male. In the presence of the Y chromosome, the human embryo will also develop as a male, no matter how many X chromosomes are present.

Unit 4—Test A

Multiple Choice **1.** C **2.** B **3.** A **4.** A **5.** D **6.** C **7.** B **8.** C **9.** D **10.** C **11.** D **12.** B **13.** A **14.** C **15.** D **Completion** **16.** independent assortment **17.** nucleotide **18.** messenger RNA **19.** inbreeding **20.** nondisjunction **Short Answer** **21.** Mitosis produces two genetically identical diploid cells. Meiosis produces four genetically different haploid cells. **22.** DNA polymerase joins together individual nucleotides to produce a DNA molecule and it proofreads each new DNA strand. **23.** Increasing the genetic variation of a population gives breeders a larger selection of traits from which to choose, increasing the chance that a desired trait is available in the population. **24.** During cell transformation, a cell takes in DNA from outside the cell. This external DNA becomes part of the cell's DNA. **25.** A deletion of three bases from the middle of the protein sequence causes most cases of cystic fibrosis. **Using Science Skills** **26.** A bacterial cell is being transformed. **27.** Gel electrophoresis is required to separate the human gene for growth hormone from the rest of the human DNA. It is also required to separate plasmid DNA from the bacterial chromosome. **28.** The plasmid and human gene are cut with the same restriction enzyme, so their cut ends are complementary to each other and will easily join together. **29.** The recombinant DNA molecule in the diagram is made up of the bacterial plasmid and the human gene for growth hormone. **30.** Researchers can identify transformed cells because the plasmid also carries a genetic marker. The genetic marker is a gene that gives the bacterial cell resistance to a certain antibiotic. With the addition of the antibiotic to the growth medium, only those bacteria that have the genetic marker—and the human gene—will grow. **31.** To produce the recessive phenotype, both parents must be heterozygous for the dominant phenotype. Punnett squares should illustrate a two-factor cross like the one diagrammed in Figure 11-10 on textbook page 271, in which both parents are heterozygous for seed color and seed shape. **32.** DNA is a double helix in which two strands are wound around each other. Each strand is a chain of nucleotides. Each nucleotide is made up of a deoxyribose sugar, a phosphate backbone, and a nitrogenous base. The four nitrogenous bases are adenine, guanine, thymine, and cytosine. The two strands are held together by hydrogen bonds between adenine and thymine and between guanine and cytosine. The sugar and phosphate groups form the backbone of the DNA strand. **33.** All three types of RNA molecules are single-stranded molecules, have a ribose sugar, and have uracil instead of thymine. However, each type has a different shape and function. Messenger RNA carries copies of the genetic instructions for making proteins. Ribosomal RNA combines with proteins to form ribosomes. Proteins are assembled on the ribosomes. During translation, transfer RNA carries the amino acid to the ribosome that is specified by messenger RNA. **34.** Prokaryotes are single-celled organisms. Their DNA molecule only needs to control the needs for one cell. Multicellular organisms have the same DNA in every cell nucleus. However, the cells are specialized for different functions. Genes that code for liver enzymes, for example, are not expressed in nerve cells. Therefore, only a small fraction of the available genes need to be expressed in the appropriate cells of different tissues throughout the body. The complexity of gene regulation in eukaryotes makes this specificity possible. **35.** A genetic disorder caused by a nonfunctioning or abnormally functioning allele gives scientists clues to the function of the normal protein produced by the normal allele.

Unit 4—Test B

Multiple Choice **1.** C **2.** B **3.** A **4.** B **5.** D **6.** A **7.** D **8.** C **9.** D **10.** A **11.** B **12.** D **13.** B **14.** A **15.** C **Completion** **16.** independent assortment **17.** transcription **18.** hybridization **19.** restriction enzymes **20.** DNA fingerprinting **Short Answer** **21.** There is a 25 percent chance that a white guinea pig will be produced. **22.** Oswald Avery and his group concluded that DNA was the transforming factor. **23.** Nucleotides are the units that make up the DNA molecule. Each nucleotide is made up of a deoxyribose sugar, a phosphate group, and a nitrogenous base. **24.** A transgenic organism contains genes from another species. **25.** Males have only one X chromosome. **Using Science Skills** **26.** Gel electrophoresis is illustrated in the figure. **27.** The DNA must be extracted from the cell before step A can occur. **28.** The longer fragments are located at point C because longer fragments move more slowly through the gel. **29.** DNA is negatively charged because it moves toward the positive end of the gel. **30.** The DNA is being cut into fragments at point A in the diagram when it is mixed with restriction enzymes.

Chapter 15—Test A

Multiple Choice **1.** C **2.** B **3.** B **4.** B **5.** A **6.** B **7.** C **8.** C **9.** C **10.** D **11.** B **12.** B **13.** B **14.** A **15.** C **Completion** **16.** many millions (or millions) **17.** evolution **18.** natural selection **19.** common descent **20.** environment **Short Answer** **21.** Darwin was able to collect specimens from many different places where the ship anchored, and while he was at sea, he could record his observations. **22.** Darwin collected many examples of similar, but not identical, modern organisms and fossils that were previously unknown. **23.** The tortoises varied from island to island in predictable ways, such as shell shape and neck length. **24.** No; Darwin's hypotheses challenged the fundamental scientific ideas of the times. **25.** Ancestors of modern whales likely had legs and lived on land. **Using Science Skills** **26.** The tortoises have necks of different length and shells that differ in overall shape and in the size of the opening for the neck. **27.** the Hood Island tortoise **28.** No, the diagram does not provide information on the ancestral species. **29.** Ancestral tortoises with long necks and shells that permit greater neck movement obtained food more easily, survived longer, and produced more offspring than other tortoises. **30.** The Pinta Island tortoises likely would survive more successfully because they appear to be more similar to the Hood Island tortoises. **Essay** **31.** Darwin began to think that the varied tortoise species on the islands had evolved from a single tortoise species from the mainland. **32.** Darwin made a wealth of observations about the natural world and developed a hypothesis that became the theory of evolution. He explained that species change over time through natural selection: the survival and reproduction of those members of a population that best suit their environment. **33.** Lamarck was among the first to recognize that living things changed over time and that organisms were adapted to their environments. Lamarck's explanation for the process of evolution, however, was incorrect. **34.** The population's survival would depend on how well adapted its members were to the new environment. If they were not well adapted, the population would probably not survive because it would have so little variation. **35.** In their mature forms, homologous structures, such as bird wings and mammal forelimbs, appear somewhat different, but they develop from the same kind of embryonic tissues. From this evidence, scientists infer that the particular species evolved from a common ancestor population that moved into different environments, where the populations were changed through natural selection.

Chapter 15—Test B

Multiple Choice **1.** C **2.** A **3.** D **4.** B **5.** D **6.** A **7.** B **8.** A **9.** B **10.** B **11.** A **12.** B **13.** A **14.** A **15.** D **Completion** **16.** mountains, mountain ranges **17.** acquired **18.** grows (or increases) **19.** variation (different traits) **20.** species **Short Answer 21.** Darwin collected many fossils. **22.** Lamarck believed that all organisms tended toward complexity and perfection. **23.** They were all species of finches. **24.** Farmers bred only animals with desirable traits, which resulted in improved offspring. **25.** Fossils provided evidence that life had changed over time. **Using Science Skills 26.** It increased in mass and volume. (The horses became taller and heavier.) **27.** The head became larger and larger. **28.** Scientists probably used the fossil bones of ancient horses and compared them with the bodies of modern horses. **29.** The number of toes decreased from three to one. **30.** No, this diagram shows only a pattern in horse evolution. Other species have remained small.

Chapter 16—Test A

Multiple Choice 1. C **2.** A **3.** C **4.** C **5.** C **6.** B **7.** A **8.** D **9.** B **10.** B **11.** C **12.** C **13.** C **14.** A **15.** B **Completion 16.** gametes **17.** phenotypes **18.** phenotypes **19.** genetic equilibrium **20.** directional **Short Answer 21.** A trait with two distinct phenotypes likely would be a single-gene trait. Because many genes control a polygenic trait, and each gene often has two or more alleles, a polygenic trait can have many possible phenotypes. **22.** No. An allele for a trait having no effect on fitness will not be affected by natural selection because it does not affect survival or reproduction. **23.** The five conditions are random mating, extremely large population, no movement into or out of the population, no mutations, and no natural selection. **24.** A large river would keep populations of small rodents apart, but it would not necessarily isolate populations that can fly. **25.** Darwin formed the hypothesis that the Galápagos finch species evolved through natural selection. The Grants confirmed that hypothesis with a scientific study. **Using Science Skills 26.** The average beak size of the birds represented in graph A has increased through directional selection. There are now no birds with very small beaks, and some birds with very large beaks have evolved. **27.** Stabilizing selection has occurred. There are more birds with a body mass of average size. **28.** Disruptive selection has occurred. There are more birds with smaller or larger beak sizes, and no birds having the average beak size. **29.** Graph C in Figure 16-1 shows a population with two very different beak sizes, which indicates that the birds could be eating different foods. **30.** Sample: A change in the birds' environment, such as the introduction of a larger kind of food, could have caused the directional selection in favor of larger beaks. **Essay 31.** An individual would be more likely to have an allele with a high relative frequency. A high frequency indicates that the allele occurs more often than do alleles with a low relative frequency. **32.** If each card in the deck represents an allele in the population, a "hand" represents a particular combination of alleles. The exchange of genes during gene shuffling is similar to shuffling the deck. Shuffling leads to different hands, but it does not change the relative numbers of specific cards in the deck. Similarly, sexual reproduction produces many different combinations of genes, but it does not alter the relative frequency of each allele in a population. **33.** The frequencies of phenotypes for a single-gene trait are best expressed as a bar graph because variations in the gene lead to only a few distinct phenotypes. The frequencies of phenotypes for a polygenic trait, however, are usually best expressed as a bell-shaped curve because one polygenic trait can have many possible phenotypes. **34.** The founder effect is a type of genetic drift in which allele frequencies change as a result of the migration of a small subgroup of a population that carries alleles in different relative frequencies than the population from which it came. The founder effect results in a change in allele frequencies due to chance rather than to natural selection. **35.** In the gene pool of a large population, the allele frequencies will remain the same when (1) random mating occurs, so each individual has the same chance of mating; (2) no alleles are added or removed through migration or mutation; and (3) natural selection does not favor any particular phenotype.

Chapter 16—Test B

Multiple Choice 1. D **2.** D **3.** B **4.** A **5.** A **6.** B **7.** C **8.** D **9.** B **10.** D **11.** C **12.** D **13.** B **14.** D **15.** D **Completion 16.** alleles **17.** single-gene **18.** polygenic **19.** Hardy-Weinberg **20.** reproductively **Short Answer 21.** Yes, the term *population* is defined in terms of species. A population is a collection of individuals of the same species in a given area. **22.** The two main sources of genetic variation are mutations and the gene shuffling that results from sexual reproduction. Gene shuffling is the more common of the two. **23.** Stabilizing selection would keep the center of the curve in its current position but narrow the overall graph. **24.** These individuals may carry alleles in different relative frequencies from those of the original population. **25.** Three mechanisms for reproductive isolation are behavioral isolation, geographic isolation, and temporal isolation. Populations of frogs that have different mating calls are an example of behavioral isolation. **Using Science Skills 26.** Figure 16-1 shows periods of mating activity for seven frog species. Four species mate in the early spring; the others, in the late spring or summer. **27.** Frog mating nearly stops in early to mid-May. **28.** Bullfrogs typically mate after the other species shown, an example of temporal isolation. Bullfrogs might also use different mating calls, an example of behavioral isolation. **29.** The two species must be behaviorally isolated because they are not isolated either temporally or geographically. **30.** Since temperatures in the south are typically warmer than in the north, frog mating seasons might also occur earlier. In that case, the curves likely would shift to the left.

Chapter 17—Test A

Multiple Choice **1.** C **2.** C **3.** C **4.** A **5.** C **6.** D **7.** D **8.** B **9.** D **10.** D **11.** C **12.** D **13.** B **14.** C **15.** B **Completion** **16.** extinct **17.** nitrogen **18.** DNA **19.** Microfossils **20.** mass extinction **Short Answer** **21.** The formation of any fossil depends on a precise combination of conditions; therefore, many organisms die without leaving a trace. **22.** Rock layers with major changes in fossil animals and plants mark where one division of geologic time ends and another begins. These divisions have been found to vary in length. **23.** During the Cenozoic Era, mammals evolved adaptations that enabled them to live in a variety of environments. **24.** Proteinoid microspheres have selectively permeable membranes and a means of storing and releasing energy. Several hypotheses suggest that similar structures could have evolved into living cells. **25.** Free oxygen was produced by photosynthetic bacteria that lived in the sea. **Using Science Skills** **26.** The half-life of carbon-14 is 5730 years. Accept answers within a reasonable range. **27.** three **28.** about 23,000 years old **29.** Carbon-14 dating would be more accurate on the younger fossils. The remaining fraction of carbon-14 in a 100,000-year-old sample would be very small and probably difficult to measure precisely. **30.** The amount of nitrogen gas produced by radioactive decay would decline throughout the period of decay, with the largest amount being produced at the beginning. This is when the greatest proportion of carbon-14 decays, with progressively smaller fractions decaying (and correspondingly smaller amounts of nitrogen gas produced) over time. **Essay** **31.** Radioactive dating involves using a radioactive isotope in a sample to determine its age. A radioactive element decays at a steady rate called half-life. Researchers measure the amount of an isotope, such as carbon-14, to find the fraction that has decayed. That information, along with the isotope's half-life, reveals the sample's age. **32.** Miller and Urey filled a flask with hydrogen, methane, ammonia, and water, heated it, and then passed electric sparks through the mixture to simulate lightning. Over a few days, several amino acids accumulated. Later experiments based on more current knowledge of Earth's early atmosphere have also produced organic compounds, including cytosine and uracil, two RNA bases. **33.** The endosymbiotic theory proposes that eukaryotic cells arose from living communities formed by several prokaryotic organisms. This began when some prokaryotic cells began evolving internal cell membranes, forming the ancestors of eukaryotes. Later, other prokaryotic organisms entered the ancestral eukaryotes and began living inside them, resulting in a symbiotic relationship between these eukaryotes and the prokaryotes they carried. Early prokaryotes that could use oxygen to generate ATP were the forerunners of modern mitochondria. Photosynthetic prokaryotes evolved into chloroplasts. **34.** Life arose, and the first living things were simple anaerobic organisms. Later, photosynthetic organisms evolved and added oxygen to Earth's atmosphere. Still later, aerobic life forms evolved, and the first eukaryotes appeared. Some eukaryotic forms evolved into multicellular organisms that became increasingly complex. Life existed only in the sea. **35.** A large asteroid probably struck Earth during the Cretaceous. An impact of this size would have thrown large amounts of dust and water into the atmosphere, changing the global climate and causing the mass extinction that ended the dinosaurs. However, many scientists now think that most mass extinctions did not have a single major cause, but were produced by multiple factors. For the Cretaceous extinction, those factors may have included intense volcanic activity, continental movement, and changes in sea levels, all in addition to the impact of a large asteroid.

Chapter 17—Test B

Multiple Choice **1.** D **2.** C **3.** A **4.** C **5.** C **6.** B **7.** A **8.** C **9.** A **10.** D **11.** A **12.** C **13.** B **14.** A **15.** B **Completion** **16.** fossil **17.** periods **18.** amino acids **19.** endosymbiotic **20.** Coevolution **Short Answer** **21.** Two of the following: the organisms' structure, what they ate, what ate them, the environment in which they lived, how long ago they lived, the order in which they evolved, what other organisms were similar to them, and how they changed over time. **22.** Relative dating allows paleontologists to estimate a fossil's age compared with that of other fossils. **23.** Oceans did not exist because the surface of Earth was extremely hot. **24.** Both are selectively permeable membranes through which water can pass. **25.** All of the dinosaurs died. **Using Science Skills** **26.** Twelve periods are shown in Figure 17-1. The Vendian Period is the earliest. **27.** mass extinction; Permian Period **28.** 30 million years **29.** 65 million years **30.** Cretaceous

Chapter 18—Test A

Multiple Choice **1.** C **2.** A **3.** D **4.** B **5.** C **6.** D **7.** A **8.** A **9.** C **10.** C **11.** A **12.** D **13.** C **14.** A **15.** C **Completion** **16.** phylum **17.** genus **18.** derived character **19.** molecular clock **20.** Bacteria **Short Answer** **21.** The name might be different in different locations or different languages. **22.** Their scientific names show that they both belong to the same genus. **23.** The animals share some common features, such as similar body plans and internal functions. **24.** segmentation and a molted exoskeleton **25.** the ability to "read" the information coded in DNA **Using Science Skills** **26.** Bacteria have cell walls with peptidoglycan, while Archaea cell walls lack peptidoglycan. **27.** No; it could belong to Protista or Plantae. **28.** No; the table gives no information about the numbers of species in kingdoms. **29.** Yes; it could only be a protist because the protists are the only eukaryotic unicellular organisms that have chloroplasts. **30.** The presence or absence of a nucleus is more important. All three domains contain both autotrophs and heterotrophs. **Essay** **31.** Descriptive names were very long, and they were not standardized among all scientists. Binomial names are brief and standard. **32.** Traditional classification places organisms into categories based mainly upon similarities in body structure. Evolutionary classification, on the other hand, places organisms into categories that represent lines of evolutionary descent, rather than just physical similarities. **33.** Neutral mutations, which have no effect on phenotype, are not subject to selection pressures. Because neutral mutations can accumulate in regions of the DNA of different species at about the same rate, a comparison of such DNA sequences in two species can indicate how long they have been evolving independently. **34.** When scientists realized that microorganisms differed from plants and animals, they grouped them all in the kingdom Protista. Later, the fungi (which includes microscopic yeasts) were placed in their own kingdom. Later still, prokaryotes were separated from single-celled eukaryotes and placed in their own kingdom, Monera. Monera have been divided into two kingdoms, Eubacteria and Archaebacteria. **35.** Scientists have performed comparative studies of a small subunit of ribosomal RNA present in all living things. Using these data to establish a molecular clock, scientists have grouped modern organisms according to how long they have been evolving independently, resulting in three domains.

Chapter 18—Test B

Multiple Choice **1.** B **2.** C **3.** A **4.** D **5.** D **6.** B **7.** D **8.** C **9.** C **10.** A **11.** C **12.** B **13.** B **14.** C **15.** C **Completion** **16.** scientific name **17.** binomial **18.** species **19.** plants **20.** Eukarya **Short Answer** **21.** Evolutionary classification groups organisms by evolutionary descent; traditional classification focuses on body structure. **22.** A **23.** B; a cladogram **24.** Organisms that were microscopic were placed in the kingdom Protista. **25.** the presence of a nucleus **Using Science Skills** **26.** the kingdom **27.** Yes; each general category contains the categories shown beneath **28.** No; the genus *Ursus* does not contain any of the categories above it. **29.** The species has the least diversity. From genus upward, each category has more diversity than the one before it. **30.** the genus

Unit 5—Test A

Multiple Choice **1.** C **2.** B **3.** A **4.** B **5.** C **6.** A **7.** B **8.** C **9.** D **10.** D **11.** A **12.** B **13.** B **14.** A **15.** C **Completion** **16.** Thomas Malthus **17.** relative frequency **18.** Quaternary **19.** eukaryotic cells **20.** molecular clock **Short Answer** **21.** Lamarck proposed that by selective use or disuse of organs, organisms acquired or lost certain traits during their lifetime. These traits could then be passed on to their offspring. Over time, this process led to change in a species. **22.** Reproductive isolation has occurred when the members of two populations cannot interbreed and produce fertile offspring. **23.** The distribution of phenotypes for a polygenic trait often fits a bell-shaped curve. **24.** The fossil record shows how different groups of organisms, including species, have changed over time. **25.** DNA analysis has revealed that American vultures are more closely related to storks than to African vultures. **Using Science Skills** **26.** As beak size increases, bird survival also increases. **27.** Through time, the curve would shift toward the larger-beak end. **28.** Directional selection appears to be occurring in the bird population. **29.** No birds survive with beaks 8 mm or less in size. **30.** If birds with smaller beaks had greater survival, the curve would slope down, rather than up, as beak size increased. **Essay** **31.** Directional selection, in which individuals at one end of the distribution have highest fitness, leads to the curve's shifting toward the high-fitness end. Stabilizing selection, in which individuals near the center of the distribution have highest fitness, leads to the curve's narrowing around the center. Disruptive selection, in which individuals at both ends of the curve have higher fitness than individuals near the middle, leads to the curve's developing two peaks at either end and a low spot in the middle. **32.** Hutton's and Lyell's ideas about geological change and the ancient age of Earth helped Darwin realize that life could change and that there had been enough time for changes in life forms to occur. **33.** Most fossils form in sedimentary rock. As layers of sediment build up over time, dead organisms may sink to the bottom and become buried. The weight of layers of sediment, along with chemical activity, turns the dead organisms into rock. Most organisms are not preserved as fossils because the formation of fossils depends on a precise combination of conditions. **34.** The domain Bacteria corresponds to the kingdom Eubacteria. The domain Archaea corresponds to the kingdom Archaebacteria. The domain Eukarya includes the kingdoms Protista, Fungi, Plantae, and Animalia. **35.** Darwin presented evidence for evolution from the fossil record, the geographical distribution of living species, homologous structures of living organisms, and similarities in early development, or embryology.

Unit 5—Test B

Multiple Choice **1.** B **2.** C **3.** A **4.** D **5.** A **6.** C **7.** D **8.** B **9.** D **10.** A **11.** C **12.** D **13.** B **14.** C **15.** D **Completion** **16.** gene pool **17.** homologous **18.** punctuated equilibrium **19.** phylogeny **20.** species **Short Answer** **21.** The principle of descent with modification is Darwin's idea that each living species has descended, with changes, from other species over time. **22.** In genetic terms, evolution is any change in the relative frequency of alleles in a population. **23.** The evolution of photosynthetic bacteria, which produced oxygen as an end product of photosynthesis, led to the accumulation of oxygen in Earth's atmosphere. **24.** Coevolution is the process by which two species evolve in response to changes in each other over time. **25.** Cladograms, unlike systems of classification based on visible similarities, are based on derived characters, which show evolutionary relationships. **Using Science Skills** **26.** Red lizards died out, brown lizards became less common, and black lizards became more common. **27.** Natural selection could cause the lizard population to change in this way if red lizards had lowest fitness, brown lizards intermediate fitness, and black lizards highest fitness. **28.** Assuming that lizard color is a single-gene trait, the change in color could not be an example of directional selection, because this type of selection acts on polygenic traits. **29.** This type of random genetic change is called the founder effect. It is a type of genetic drift. **30.** Green lizards would become more common and both brown lizards and black lizards would become less common.

Chapter 19—Test A

Multiple Choice **1.** D **2.** A **3.** D **4.** C **5.** A **6.** B **7.** D **8.** B **9.** A **10.** B **11.** C **12.** B **13.** A **14.** A **15.** B **Completion** **16.** bacillus (rod-shaped bacteria) **17.** nitrogen fixation **18.** prophage **19.** tail fiber **20.** tail sheath **Short Answer** **21.** Prokaryotes are identified by (1) their shape, (2) the chemical nature of their cell walls, (3) the way they move, and (4) the way they obtain energy. **22.** Accept any two of the following roles: carrying out photosynthesis, performing nitrogen fixation, decomposing dead matter and recycling nutrients. **23.** A virus is a particle of nucleic acid, protein, and, in some cases, lipids that can reproduce only by infecting living cells. A typical virus is composed of a core of either DNA or RNA surrounded by a protein coat called a capsid. **24.** Living things are made up of cells, and they are able to live independently. Viruses are not cellular and must infect a living cell to reproduce. Also, unlike cells, viruses don't grow and develop, don't obtain and use energy, and don't respond to the environment. **25.** Three methods used to control bacteria growth are sterilization by heat, use of disinfectants, and correct food storage and processing. **Using Science Skills** **26.** Sample answer: The student has controlled all the variables except the solution that each disk is soaked in. She is probably trying to test the effectiveness of different disinfectants against *E. coli* bacteria. **27.** The disk soaked in distilled water is the control. **28.** Sample answer: The student can measure the width of the area around each disk where bacteria have not grown to determine the effectiveness of the solution in which each disk was soaked. **29.** Disinfectant 1 shows the largest area where growth has been inhibited; thus, it is the most effective disinfectant tested against *E. coli* growth. **30.** Sample answer: Disinfectants act differently to combat bacterial growth. Some disinfectants are more effective at controlling one species of bacteria than another. **Essay** **31.** Although eubacteria and archaebacteria are both small, have cell walls, and lack nuclei, they differ in the chemical composition of their cell walls and membrane lipids. Archaebacteria also contain key gene sequences that are more like those of eukaryotes than those of eubacteria. Scientists reason from this that archaebacteria may be the ancestors of eukaryotes. **32.**

Binary fission is an asexual form of reproduction that occurs in bacteria. When a bacterium has grown so that it has nearly doubled in size, it replicates its DNA and divides in half. In conjugation, a hollow bridge forms between two bacterial cells, and genes move from one cell to another. **33.** In a lytic infection, a bacteriophage injects DNA into a bacterium and takes over the bacterium's metabolism, causing synthesis of new bacteriophage proteins and nucleic acids. Bacteriophage particles assemble within the bacterium and then lyse the bacterial cell wall, causing the bacterium to burst and releasing bacteriophage particles that can attack other cells. In a lysogenic infection, a bacteriophage injects DNA into a bacterium, and the bacteriophage DNA inserts itself into the bacterial chromosomes. The embedded bacteriophage DNA, called a prophage, may replicate with the bacterium for many generations. **34.** Bacteria cause disease in one of two general ways. Some damage the tissues of the infected organism by breaking them down for food. For example, the bacteria that cause tuberculosis break down lung tissue. Other bacteria release toxins that harm the body. For example, food poisoning and the characteristic rash of scarlet fever are caused by toxins that bacteria have released in the body. **35.** Bacterial infections are caused by bacteria either breaking down the body's tissues for food or releasing harmful toxins into the body. A viral infection results when viruses infect body cells, take over their cellular functioning, and begin replicating themselves. In the process, viruses can destroy body cells.

Chapter 19—Test B

Multiple Choice **1.** B **2.** C **3.** D **4.** C **5.** C **6.** D **7.** D **8.** B **9.** D **10.** B **11.** C **12.** B **13.** A **14.** A **15.** D **Completion** **16.** conjugation **17.** nitrogen fixation **18.** capsid **19.** vaccine **20.** viroids **Short Answer** **21.** A: cell wall, B: cell membrane, C: DNA, D: flagellum **22.** The kingdoms are Archaebacteria and Eubacteria. Accept either of the following differences in the two groups: (1) Archaebacteria lack an important carbohydrate found in the cell walls of eubacteria. (2) DNA sequences of key archaebacterial genes are more like those of eukaryotes than those of eubacteria. **23.** A pathogen is a disease-causing agent. **24.** Some bacteria damage the cells and tissues of the infected organism directly by breaking down the cells for food. Other bacteria release toxins that travel throughout the body interfering with the normal activity of the host. **25.** Three methods are sterilization by heat, use of disinfectants, and correct food storage and processing. **Using Science Skills** **26.** the lysogenic cycle **27.** during the lytic cycle **28.** E **29.** C **30.** A

Chapter 20—Test A

Multiple Choice **1.** D **2.** A **3.** B **4.** C **5.** A **6.** B **7.** C **8.** C **9.** D **10.** B **11.** B **12.** D **13.** C **14.** C **15.** C **Completion** **16.** pseudopods **17.** dinoflagellates **18.** chlorophyll *a* **19.** photosynthesis **20.** oomycetes (or water molds) **Short Answer** **21.** Members of the phylum Sporozoa do not move on their own. **22.** Chlorophyll and accessory pigments allow algae to harvest and use the energy from sunlight. **23.** "Red tides" are algal blooms of the dinoflagellates *Gonyaulax* and *Karenia.* They are dangerous because eating shellfish from water infected with red tide can cause serious illness, paralysis, and even death in humans and fishes. **24.** Both are large multicellular organisms that are so similar they are difficult to tell apart. The sporophyte produces haploid spores, while the gametophyte produces haploid gametes. **25.** Cells of cellular slime molds are separated by cell membranes during every phase of the mold's life cycle. Cells of acellular slime molds fuse to form a large structure with many nuclei. **Using Science Skills** **26.** malaria; the sporozoan *Plasmodium* **27.** Malaria would be most common in environments where the mosquito that carries the protist is common. Some students may know that malaria is common in tropical environments. **28.** A mosquito has picked up the protist and in turn, injected it (in the form of sporozoites) into a human host. The sporozoites have started to grow in the human liver. **29.** The protists infect red blood cells and ultimately cause them to burst, releasing gametes. **30.** Increased symptoms are associated with the period of blood-cell rupture. **Essay** **31.** African sleeping sickness is caused by the animallike protist *Trypanosoma.* The protists are spread through the bite of the tsetse fly. African sleeping sickness is characterized by chills and rashes. If the protist infects nerve cells, it results in severe damage to the nervous system. Some victims lose consciousness, lapsing into a fatal sleep. **32.** A typical euglena is about 50 micrometers in length. Two flagella emerge from a gullet at one end of the organism, and the longer of these spins in a pattern that pulls the organism rapidly through the water. Near the gullet end of the organism is a cluster of reddish pigment known as the eyespot. The euglena is covered by an intricate cell membrane called a pellicle, which is folded into a series of ribbonlike ridges supported by microtubules. **33.** The cells of *Volvox* are arranged in hollow spheres of between 500 to 50,000 cells, each connected to others by strands of cytoplasm. *Volvox* colonies can coordinate their movement with some cells using their flagella to push while others pull. Most cells in a *Volvox* colony are identical—except for a few gamete-producing cells specialized for reproduction. *Ulva,* on the other hand, is composed of several specialized cell types. For example, some cells form a structure called a holdfast, which attaches the alga to a rock. **34.** During much of their life cycle, cellular slime molds are unicellular and both their appearance and behavior are similar to that of animallike protists. When they aggregate, however, cellular slime molds act like multicellular organisms—they migrate together and produce a single fruiting body as if they were a single organism. **35.** Ecosystems would be littered with the bodies of dead animals and plants. Materials would remain tied up in dead bodies and would not reenter the ecosystem to be used by other living things. Some students may further predict that eventually raw materials for new living things might be depleted to the extent that no more new living things could be produced, and life on Earth would end.

Chapter 20—Test B

Multiple Choice **1.** B **2.** C **3.** A **4.** B **5.** B **6.** D **7.** B **8.** A **9.** A **10.** D **11.** C **12.** B **13.** A **14.** A **15.** A **Completion** **16.** protists **17.** malaria **18.** red blood **19.** sunlight **20.** heterotrophs **Short Answer** **21.** *Trichonympha* breaks down the cellulose in wood, making it possible for termites to eat wood. **22.** Unicellular algae make up a portion of the phytoplankton—the population of small, photosynthetic organisms found near the surface of the ocean. **23.** Accept answers such as: to produce chemicals that are used to maintain proper health and to produce chemicals used in industry (plastics, waxes, transistors, deodorants, paints, lubricants, artificial wood). **24.** funguslike protists that play key roles in recycling organic material **25.** The water mold *Phytophthora infestans* infected and destroyed the potato crops in Ireland. **Using Science Skills** **26.** a single-celled organism **27.** Structure A traps food particles in the organism's environment, allowing them to enter the organism's body. **28.** Excess water likely would build up inside the organism. **29.** Structure C is a "working library" of genetic information, which contains multiple copies of most of the genes that the organism uses for its day-to-day existence. **30.** The organism likely could not move normally through its environment.

Chapter 21—Test A

Multiple Choice **1.** A **2.** C **3.** A **4.** C **5.** A **6.** B **7.** C **8.** B **9.** A **10.** B **11.** C **12.** A **13.** A **14.** D **15.** B **Completion** **16.** chitin **17.** haploid **18.** club **19.** enzymes **20.** wheat **Short Answer** **21.** The hyphae on the left have cross walls dividing them into individual cells, while the hyphae on the right have no cross walls and thus contain many nuclei in a single chamber. **22.** Rhizoids anchor the fungus, secrete digestive enzymes, and absorb nutrients. **23.** An untrained person cannot tell the difference between poisonous and nonpoisonous mushrooms and should never eat wild fungi. **24.** Fungi are most important as decomposers. Without decomposers, organic molecules would remain tied up in the bodies of dead organisms. **25.** Lichens gradually break down the rocks on which they grow. **Using Science Skills** **26.** opposite mating types (+ and –) **27.** Structure B, the zygospore, probably would become dormant until its environmental conditions improved. **28.** Zygomycota **29.** II **30.** The zygospore is 2N and is formed by the fusion of gametes of opposite mating types. **Essay** **31.** During the greater part of their life cycle, the nuclei of most fungi are haploid. Diploid nuclei form during sexual reproduction. Shortly after the nuclei fuse, however, meiosis occurs and produces haploid nuclei that dominate the remainder of the life cycle. **32.** The four phyla are Zygomycota (common molds), Ascomycota (sac fungi), Basidiomycota (club fungi), and Deuteromycota (imperfect fungi). Classification is based on their pattern of sexual reproduction. **33.** The fruiting body of a mushroom begins as a mass of growing hyphae that forms a button, or thick bulge, at the soil's surface. The button expands into a fruiting body that consists of a cap at the top of a stalk. The cap is composed of tightly packed hyphae, and the underside of the cap is composed of gills—thin blades of tissues lined with basidia that produce basidiospores. A mushroom can spring up overnight as the cells of the hyphae enlarge by rapidly taking in water. **34.** Some helpful fungi produce the antibiotic penicillin, make bread rise, are good to eat, and break down dead organisms. Harmful fungi cause ringworm, athlete's foot, and other human and animal diseases, and they cause many serious plant diseases that destroy food crops. **35.** Some bacteria are beneficial and keep other organisms such as fungi from growing out of control. Killing the body's normal bacteria can result in an overgrowth of yeasts, which cause thrush and reproductive tract infections.

Chapter 21—Test B

Multiple Choice **1.** B **2.** A **3.** C **4.** D **5.** D **6.** D **7.** C **8.** A **9.** C **10.** B **11.** D **12.** C **13.** B **14.** C **15.** D **Completion** **16.** Asexual **17.** caps **18.** sexual

19. mycorrhizae **20.** lichen **Short Answer 21.** Individual threads of fungi are hyphae; together they form a mycelium. **22.** Two nuclei are found in each cell; cells are separated by cross walls. **23.** baking and brewing **24.** The fruiting body of basidiomycetes is shaped like a club. **25.** Examples include any food crop disease caused by fungi, such as rusts or mildews. **Using Science Skills 26.** One fruiting body is shown. **27.** Both are composed of hyphae. **28.** the mycelium **29.** No; a mycelium could be completely buried and thus be undetectable. **30.** It is a reproductive structure.

Unit 6—Test A

Multiple Choice 1. B **2.** D **3.** B **4.** A **5.** A **6.** C **7.** C **8.** D **9.** C **10.** B **11.** D **12.** B **13.** A **14.** D **15.** C **Completion 16.** cocci **17.** retroviruses **18.** antibiotics **19.** phytoplankton **20.** imperfect **Short Answer 21.** Archaebacteria lack the peptidoglycan of eubacteria and also have different membrane lipids. The DNA sequences of key archaebacterial genes are more like those of eukaryotes than those of eubacteria. **22.** Antibiotics are compounds that block the growth and reproduction of bacteria. Viruses are not bacteria, and viruses have few of the characteristics of living things. **23.** Like land plants, green algae have cellulose in their cell walls, contain chlorophyll *a* and *b,* and store food in the form of starch. **24.** Alternation of generations is the switching back and forth between haploid and diploid stages during an organism's life cycle. **25.** They recycle nutrients by breaking down the bodies and wastes of other organisms. **Using Science Skills 26.** Figure 2 shows the life cycle of a bread mold, which is in the phylum Zygomycota and the kingdom Fungi. **27.** The figure depicts the life cycle of a common mold. Section I shows the process of asexual reproduction, and section II shows the process of sexual reproduction. **28.** The two structures are hyphae called stolons, and they are of opposite mating types. **29.** That structure is a zygospore. The opposite mating types fused to form gametangia. Haploid gametes in the gametangia then fused to form zygotes, which developed into zygospores. **30.** When conditions are favorable, the zygospore germinates and undergoes meiosis, and then haploid spores are released. This sexual process produces new combinations of genetic information that may help the organism meet changing environmental conditions. **Essay 31.** A typical eubacterium is a unicellular organism with a cell wall and a cell membrane. It lacks a nucleus; its DNA is found within the cytoplasm. The cell moves with its flagella. **32.** In a lytic infection, a bacteriophage injects DNA into a bacterium and then takes over the bacterium's metabolism, causing synthesis of new bacteriophage proteins and nucleic acids, which assemble into viral particles. The bacteriophage then lyses the bacterium's cell wall, releasing the bacteriophage particles. **33.** Malaria is caused by the sporozoan *Plasmodium.* Symptoms include severe chills and fever. The sporozoan is spread through bites of the female *Anopheles* mosquito. **34.** The bodies of multicellular fungi are composed of thin filaments called hyphae. Many hyphae are tangled together into a thick mass called a mycelium, which can absorb food. A reproductive structure called a fruiting body grows above ground from the mycelium in the soil beneath it. **35.** Mycorrhizae are mutualistic relationships of plant roots and fungi. The fungi help the plant roots absorb water and minerals from the soil, which are essential for the growth of many plants.

Unit 6—Test B

Multiple Choice 1. B **2.** B **3.** D **4.** C **5.** C **6.** A **7.** D **8.** A **9.** C **10.** B **11.** C **12.** B **13.** D **14.** B **15.** C **Completion 16.** archaebacteria **17.** antibiotics **18.** pseudopodia **19.** plants **20.** hyphae **Short Answer 21.** Sterilization by heat, use of disinfectants, and correct food storage and processing **22.** A typical virus is composed of a core of DNA or RNA surrounded by a protein coat. **23.** The basis for that classification is the way in which different groups obtain nutrition. **24.** Many algae switch back and forth between haploid and diploid stages during their life cycles. **25.** The four main groups are the zygomycetes, or common molds; the ascomycetes, or sac fungi; the basidomycetes, or club fungi; and the deuteromycetes, or imperfect fungi. **Using Science Skills 26.** unicellular **27.** Structure A represents a cell wall. **28.** The DNA is found free within the cytoplasm and not within a nucleus. **29.** The flagellum is a structure used in movement. **30.** The organism is a prokaryote, or a bacterium. Its characteristics include a cell wall, a cell membrane, DNA but no nucleus, and flagella used in movement.

Chapter 22—Test A

Multiple Choice 1. C **2.** A **3.** B **4.** B **5.** B **6.** B **7.** C **8.** B **9.** B **10.** C **11.** D **12.** D **13.** D **14.** C **15.** C **Completion 16.** flowering plants **17.** vascular tissue **18.** gametophytes **19.** seed coat **20.** dicot **Short Answer 21.** Bryophytes are considered successful because they are abundant in various habitats in different parts of the world. **22.** Like roots, rhizoids anchor plants in the ground and absorb water and minerals from the soil. Unlike roots, rhizoids do not have vascular tissue. **23.** The roots, stems, and leaves are considered to be true because they contain vascular tissue. **24.** Land environments became much drier, prompting the replacement of moss and fern species with seed plants. **25.** A living fossil is a living species that has changed very little over millions of years, so it has characteristics that are very similar to those of its fossil ancestors. **Using Science Skills** **26.** Structures F, G, H, I, and J are part of the gametophyte stage of a fern. **27.** Structures C, which are spores, are released from structure B, which is a sporangium. **28.** Structures G and J produce different kinds of gametes. Structure G, an archegonium, produces eggs. Structure J, an antheridium, produces sperm. **29.** Structures C will become gametophytes. **30.** Structure E, the mature sporophyte, grows from structure F, the gametophyte. **Essay** **31.** Plants and multicellular green algae are both eukaryotes. They have identical cell walls and photosynthetic pigments. They also have similar reproductive cycles. These similarities suggest that plants evolved from an organism much like the multicellular green algae living today. **32.** When a sperm fertilizes an egg in an archegonium of a gametophyte, a zygote is formed in the archegonium. The zygote is a sporophyte that continues to grow directly out of the gametophyte. **33.** Vascular tissue allows water and dissolved nutrients to move throughout the plant body more efficiently than by osmosis alone. Vascular tissue also provides support. As a result, plants with vascular tissue do not have to grow close to the ground and can become large in size. **34.** Animals transfer the male gametophytes (pollen grains) of seed plants to the female reproductive structure. They also help to disperse seeds by picking up seeds on their fur or feathers or by eating seeds and then passing them out of their bodies. **35.** Both angiosperms and gymnosperms are vascular plants that produce seeds. The gametophytes of angiosperms and gymnosperms grow and mature within the sporophyte. In gymnosperms, the gametophytes are in cones. In angiosperms, the gametophytes are in flowers. In both gymnosperms and angiosperms, the male gametophyte is located in a pollen grain. The pollen grain is transferred to the female reproductive structure by wind or animals. In gymnosperms, the seeds that result from pollination are formed on the surfaces of cones. In angiosperms, the seeds are formed in flowers. In angiosperms, a protective tissue called an ovary covers a seed. The ovary develops into a fruit.

Chapter 22—Test B

Multiple Choice **1.** D **2.** D **3.** B **4.** A **5.** D **6.** D **7.** B **8.** C **9.** D **10.** C **11.** A **12.** D **13.** C **14.** C **15.** A **Completion** **16.** cell walls **17.** gases (or carbon dioxide and oxygen) **18.** vascular tissue **19.** spores **20.** phloem **Short Answer** **21.** Plants require sunlight, water and minerals, gas exchange, and movement of water and nutrients through the plant body. **22.** A fern gametophyte is a heart-shaped green structure with rhizoids. On the gametophyte's underside are antheridia and archegonia. **23.** A pollen grain is a tiny structure produced by seed plants that contains the male gametophyte. **24.** Fruit is a thick wall of tissue that surrounds an angiosperm seed. **25.** A monocot has one cotyledon; a dicot has two. **Using Science Skills** **26.** A corn seed has one seed leaf. **27.** a dicot, because it has branching veins **28.** The vascular bundles are scattered throughout the stem. **29.** the iris, because it has six floral parts **30.** a taproot

Chapter 23—Test A

Multiple Choice **1.** D **2.** B **3.** D **4.** C **5.** C **6.** D **7.** D **8.** B **9.** D **10.** D **11.** A **12.** A **13.** D **14.** B **15.** C **Completion** **16.** sieve tube **17.** osmosis **18.** ground elements **19.** spongy mesophyll **20.** transpiration pull **Short Answer** **21.** Water moves up the xylem from the roots to the leaves. Carbohydrates and other materials move up and down the phloem, throughout the plant. **22.** Fibrous roots spread out near the surface of the soil. Taproots grow deep into the soil. **23.** Structure A is the vascular cylinder, which is made up of xylem and phloem tissues. **24.** Like a root, a stem consists of dermal, vascular, ground, and meristematic tissues. **25.** As sugar concentrations increase in the phloem, water moves by osmosis from xylem into the phloem, increasing the pressure in the phloem, which forces the sugars to flow through it. **Using Science Skills** **26.** open, structure E **27.** The spaces connect with the stomata, allowing gases to be exchanged between the mesophyll cells and the atmosphere. **28.** Structure F is a leaf vein. Xylem and phloem tissues make up the leaf vein. **29.** Structure A (the cuticle) and structure D (the epidermis) form a waterproof barrier. **30.** The cells that make up the leaf vein, structure F, do not contain chlorophyll. **Essay** **31.** From the soil, roots absorb water, which the leaves use for photosynthesis, and nutrients, which the leaves need for growth. The stem transports the water and nutrients from the roots to the leaves. The stem also produces leaves and holds them up to the sun, allowing them to absorb sunlight for photosynthesis. **32.** The main function of parenchyma cells is photosynthesis. Collenchyma and sclerenchyma cells function in providing support to the plant. **33.** The cell membranes of root hairs and other cells in the root epidermis contain active transport proteins. These proteins use ATP to pump mineral ions from the soil into the plant. The high concentration of mineral ions in the plant cells causes water molecules to move into the plant by osmosis. **34.** Leaves absorb light and carry out most of the photosynthesis in a plant. Water is lost from leaves in a process called transpiration. It is replaced by water drawn into the leaves through xylem tissue, causing transpiration pull. Leaves take in carbon dioxide, and give off oxygen during photosynthesis. **35.** When a plant is watered, it takes in the water through its roots. The water then travels to the leaves. There, the water moves by osmosis into the guard cells, increasing the pressure in the cells. The increased pressure causes the thin outer walls of the cells to become curved, pulling the thick inner walls of the guard cells away from each other. This movement of the inner walls of the guard cells causes the stomata to open.

Chapter 23—Test B

Multiple Choice **1.** B **2.** A **3.** B **4.** C **5.** A **6.** A **7.** C **8.** B **9.** D **10.** C **11.** B **12.** D **13.** A **14.** D **15.** C **Completion** **16.** stems **17.** fibrous **18.** vascular cylinder **19.** buds **20.** spongy mesophyll **Short Answer** **21.** dermal, vascular, and ground tissues **22.** water and minerals **23.** In monocot stems, the vascular bundles are scattered, whereas in dicot stems the vascular bundles are arranged in a ring. **24.** Water rises from the roots to the top of a tree by root pressure, capillary action, and transpiration pull. **25.** *Sink* refers to a region of a plant that is using or storing sugars. **Using Science Skills** **26.** The arrows represent the movement of water through a plant. **27.** Xylem tissue transports water through a plant. **28.** Capillary action causes water to move upward in part B. **29.** Transpiration is helping to bring water to the top of the plant. **30.** Root pressure is helping to move water upward through part A.

Chapter 24—Test A

Multiple Choice **1.** A **2.** D **3.** B **4.** C **5.** C **6.** B **7.** D **8.** A **9.** D **10.** A **11.** B **12.** B **13.** B **14.** D **15.** D **Completion** **16.** pollen cones **17.** outer seed coat **18.** style **19.** seed **20.** risen (or increased) **Short Answer** **21.** It is a sticky material that helps pollen grains stay on the female cone, or seed cone. **22.** Because it has winglike structures that are easily carried in the wind, seed B is typically dispersed by wind. **23.** Germinating seeds absorb water, which causes the endosperm to swell, cracking open the seed coat and allowing the young root to emerge. **24.** from horizontal stems, from plantlets, and from underground roots **25.** plant propagation by cuttings, grafting, or budding **Using Science Skills** **26.** the pollen tubes **27.** There are eight nuclei in the embryo sac. They arose from a haploid cell, which underwent mitosis. **28.** The anther, ovary, and seedling are diploid. **29.** The zygote develops into the embryo. **30.** Because flowers are attractive to animals, the pollen in Figure 24-5 will likely be dispersed by an animal feeding on the flower's pollen or nectar. **Essay** **31.** The term refers to the fact that all plants have a life cycle in which a diploid sporophyte generation alternates with a haploid gametophyte generation. In seed plants, the sporophyte is the visible part of the plant and the gametophytes are hidden deep within the tissues of the sporophyte. In gymnosperms, the gametophytes are found inside cones. In angiosperms, they are found inside flowers. **32.** After fertilization, nutrients flow into the flower tissue and support the development of the growing embryo within the seed. As the seed matures, the ovary wall thickens and joins with other parts of the flower stem. These structures together form a fruit that encloses the seed(s). **33.** The seeds of some pine species are enclosed in sealed cones, which open only after being exposed to the heat of a forest fire. When the cones open, the seeds come out of dormancy and germinate. This process allows the pines to quickly reclaim the forest floor after a fire. **34.** Grafting and budding can reproduce seedless plants and plants that do not have strong root systems. A piece of stem or a lateral bud is cut from the parent plant and attached to another plant. The two plants then grow together. When stems are used, the process is called grafting. When buds are used, the process is called budding. **35.** The efficiency of agriculture has been improved through the selective breeding of crop plants and improvements in farming techniques. Some of the most important improvements have been the use of pesticides and fertilizers.

Chapter 24—Test B

Multiple Choice **1.** A **2.** B **3.** C **4.** D **5.** A **6.** C **7.** A **8.** A **9.** B **10.** C **11.** A **12.** B **13.** C **14.** C **15.** C **Completion** **16.** anther **17.** tube **18.** ovary **19.** water **20.** selective breeding **Short Answer** **21.** sepals, petals, stamens, and carpels **22.** temperature and moisture **23.** If the parent plant is knocked over or if the plantlets fall into the soil, they can take root and grow. **24.** Horticulturists use plant propagation to make many identical plants or to produce offspring from seedless plants. **25.** Although there have been yearly fluctuations, the overall yield of corn has increased. **Using Science Skills** **26.** the style **27.** J **28.** F and H (or G) are male parts; A, B, C, and D (or E) are female parts. **29.** F **30.** D

Chapter 25—Test A

Multiple Choice **1.** D **2.** D **3.** C **4.** B **5.** B **6.** A **7.** D **8.** D **9.** A **10.** D **11.** B **12.** D **13.** C **14.** D **15.** D **Completion** **16.** receptor **17.** growth **18.** deciduous **19.** air roots **20.** adapted **Short Answer** **21.** Unlike animals, plants continue to grow throughout their lives. **22.** Photosynthetic pathways are turned off, nutrients are transported from the leaves to the roots, and the leaves are sealed off from the rest of the plant. **23.** The cactus has reduced leaves in the form of thin, sharp spines. **24.** They both grow on other plants but epiphytes, unlike parasites, gather their own moisture and produce their own food. **25.** Some of the defensive chemicals that plants produce are used by humans as medicines because of their powerful effects. These include those used in aspirin and codeine. **Using Science Skills** **26.** Chrysanthemum. A chrysanthemum blooms and produces seeds in the fall. Cytokinin is produced in developing seeds. **27.** Phototropism and photoperiodism are both plant responses to light. The plants in Figure 25-3 will both grow toward the light, but will flower at different times according to their own photoperiods. **28.** The chrysanthemum will bloom, but the bearded iris will not. The chrysanthemum is a short-day plant, and the beaded iris is a long-day plant. **29.** Chrysanthemums are adapted to blooming only when they have a long period of uninterrupted darkness. **30.** Auxins cause chrysanthemums to bend toward light, to have stems that grow upright and roots that grow downward, and to have a tall thin growth pattern (apical dominance). **Essay** **31.** Fruit tissues produce ethylene, a hormone that ripens fruit. Growers often use ethylene to ripen fruit after harvest. Sealing fruit in a bag would hold in the ethylene the fruit produces and would help the fruit ripen faster. **32.** The tendency to grow toward light is phototropism. The growth response toward or against the force of gravity is gravitropism. Both behaviors are regulated and controlled by varying concentrations of auxins that are produced in the plant's meristems. **33.** If a gardener snipped off the tip of a growing plant, the apical meristem would be removed. Auxin from the apical meristem would stop inhibiting growth of the lateral meristems. The buds would begin to develop new branches, and the plant would take on a new shape that was rounder and fuller. **34.** Even though water lilies are aquatic plants, their root cells require oxygen from the air. Water lilies have the adaptation of large, open spaces in their petioles that reach all the way down to the stems. These structures allow oxygen into the plant, which diffuses down to the roots. **35.** Pitcher plants are carnivorous plants that trap insects in their leaves, where the insects drown and are digested. The insect is used for nutrients. A parasite obtains the nutrients it needs from the tissues of a host plant. These plants are alike in that they both obtain nutrients from other organisms.

Chapter 25—Test B

Multiple Choice **1.** C **2.** D **3.** D **4.** A **5.** C **6.** A **7.** A **8.** D **9.** A **10.** D **11.** B **12.** C **13.** C **14.** D **15.** A **Completion** **16.** herbicides **17.** ethylene **18.** phototropism **19.** thigmotropism **20.** cells **Short Answer** **21.** Answers may include any two of the following: light, temperature, moisture, gravity. **22.** A plant hormone is a chemical substance that controls and regulates a plant's pattern of growth and development. **23.** They demonstrated that an oak seedling grows toward light unless the growing tip of the seedling is removed or covered. **24.** Hollyhocks are long-day plants. **25.** An abscission layer is a layer of cells that form at the petiole to seal the leaf off from the rest of the plant. **Using Science Skills** **26.** The hormone produced in area D, the apical meristem, will move through the plant and stimulate growth in area B, the stem. **27.** Auxins accumulate in the cells on the side of the root next to the rock and inhibit their growth. Then, the cells on the top side of the root grow longer, so the root bends down. **28.** If it is a long-day, flowering plant, structure D, the apical meristem, will produce flowers in the fall. **29.** An increase in ethylene in the plant stimulates an abscission layer to form at area C, causing the leaf to drop off the plant. **30.** Yes; all plants are adapted to their natural habitat or they wouldn't grow there.

Unit 7—Test A

Multiple Choice **1.** A **2.** A **3.** D **4.** C **5.** B **6.** C **7.** B **8.** C **9.** A **10.** A **11.** D **12.** B **13.** B **14.** D **15.** C **Completion** **16.** gametophyte; sporophyte **17.** vascular tissue **18.** roots **19.** ovary walls **20.** tropisms **Short Answer** **21.** Plants need sunlight, water and minerals, gas exchange, and the transport of water and nutrients throughout the plant body. **22.** Monocots have seeds with one cotyledon, parallel leaf veins, floral parts in multiples of three, vascular bundles scattered throughout stems, and fibrous roots. Dicots have two cotyledons, branched leaf veins, floral parts in multiples of four or five, vascular bundles arranged in a ring, and taproots. **23.** Meristematic tissue is the only plant tissue that produces new cells by mitosis. It is responsible for plant growth. **24.** The dandelion (labeled A) is dispersed by wind. The cherries (labeled B) are dispersed by animals. **25.** Many plants defend themselves against insect attack by producing chemicals that have powerful effects on animals. **Using Science Skills** **26.** Leaves absorb light and carry out photosynthesis. **27.** No, a wilting plant would have closed stomata. The stomata in the leaf in Figure 3 are open. **28.** The structure labeled F transports water and the products of photosynthesis. **29.** As cold weather approaches, chlorophyll production stops. When light destroys the remaining green pigment, the leaves change color. Photosynthesis stops. Enzymes extract nutrients from the broken-down chlorophyll, which are then stored in other parts of the plant. The cells labeled B and C begin shriveling up as the leaf prepares to drop from the plant. **30.** If the leaf were from a salt-tolerant plant, it would have specialized cells that pump salt out of the plant and onto the leaf surface, where it is washed off by the rain. **Essay** **31.** Both bryophytes and ferns reproduce by spores. Both also require water for fertilization. Gymnosperms and angiosperms both reproduce with seeds. The reproductive structures in gymnosperms are cones. The reproductive structures in angiosperms are flowers. **32.** In bryophytes, the gametophyte is the dominant stage of the life cycle. The sporophyte depends on the gametophyte for water and nutrients. In ferns, the sporophyte is the dominant stage of the life cycle. The tiny fern gametophyte grows independently of the sporophyte. **33.** The three main plant organs are roots, stems, and leaves. These organs are made up of three types of tissues: dermal tissue, vascular tissue, and ground tissue. Dermal tissue consists of epidermal cells. Vascular tissue is composed of xylem tissue and phloem tissue. Xylem is composed of tracheids and vessel elements. Phloem is composed of sieve tube elements and companion cells. Ground tissue is composed of three cell types: parenchyma, collenchyma, and sclerenchyma. A fourth type of tissue, meristematic tissue, is found only at the growing tips of stems and roots or on the sides of stems that produce secondary growth. **34.** Angiosperms have adaptations to entice mobile animals to spread their pollen and seed away from the parent plant. Flowers have colorful petals and nutrients to attract pollinators. The animal pollinators are attracted to the source of nectar and pollen provided by the flower. While feeding from one flower to the next, the animals also pollinate the flowers. Many angiosperm seeds are surrounded by protective fleshy fruits that are eaten by many different animals. The seeds travel unharmed through the animal's digestive system and are deposited in feces far from the parent plant. Other seeds cling to animal fur when animals brush past the plant. Later, the seed drops off the fur in a new location. **35.** Auxins accumulate on the shaded sides of stems, causing cell elongation, and the plant grows toward light. They accumulate at the base of stems, and the stems grow against the force of gravity. Auxins accumulate on the lower sides of roots, causing the roots to grow downward. Auxins prevent lateral bud growth, so the plant grows tall and thin rather than bushy. Auxins also stimulate fruit tissues to release ethylene, which stimulates fruit ripening. Auxin production in leaves drops as the days get shorter, setting off a series of events that prepare the plant for winter dormancy.

Unit 7—Test B

Multiple Choice **1.** D **2.** B **3.** C **4.** B **5.** C **6.** A **7.** A **8.** B **9.** D **10.** B **11.** C **12.** A **13.** C **14.** D **15.** A **Completion** **16.** cell walls **17.** dicot **18.** xylem; phloem **19.** fruit **20.** thigmotropism **Short Answer** **21.** The vascular system allowed water and nutrients to move through a plant more efficiently. This transport system enabled plants to grow much taller. **22.** Adaptations that allow seed plants to reproduce without water include flowers or cones, the transfer of sperm by pollination, and the protection of embryos in seeds. **23.** Stomata on the undersides of leaves open and close in response to changes in water pressure within the guard cells that surround the opening. When the guard cells are swollen with water, the stomata are open. When the guard cells lose water, the stomata close, limiting further water loss from the leaf. **24.** The structure labeled F is an anther. Meiosis occurs within the anther to form the haploid male gametophytes—pollen grains. **25.** As cold weather approaches, deciduous plants turn off photosynthetic pathways, transport materials from leaves to roots, and seal leaves off from the rest of the plant. **Using Science Skills** **26.** The endodermis is the structure labeled C. It completely encloses the root's vascular system. Once water passes through the endodermis into the vascular tissue, it is trapped there because of the waxy Casparian strip that surrounds the cells of the endodermis. This causes root pressure. **27.** The apical meristem is labeled I. The apical meristem produces new cells for root growth. **28.** The root in Figure 3 is a taproot. Taproots are characteristic of dicots. **29.** Epidermal tissue, or the epidermis, is labeled D. Vascular tissue, labeled H, includes both xylem, labeled G, and phloem, labeled F. Ground tissue, or the cortex, is labeled B. **30.** Auxins cause the root to grow downward. Cytokinins are produced in growing roots and stimulate cell division.

Chapter 26—Test A

Multiple Choice **1.** B **2.** D **3.** C **4.** B **5.** B **6.** B **7.** D **8.** C **9.** A **10.** C **11.** B **12.** B **13.** B **14.** A **15.** B **Completion** **16.** vertebrates **17.** feedback inhibition **18.** anus **19.** gastrovascular cavity **20.** polyp **Short Answer** **21.** A protostome is an animal whose mouth is formed from the blastopore. A deuterostome is an animal whose anus is formed from the blastopore. **22.** Complex animals tend to have high levels of cell specialization and internal body organization, bilateral symmetry, a front end or head with sense organs, and a body cavity. **23.** Cephalization is the concentration of sense organs and nerve cells at the front end of the body. An advantage is that an animal with cephalization can respond to the environment more quickly and in more complex ways than a simpler animal can. **24.** Scyphozoans, such as jellyfishes, live their lives primarily as medusas. A medusa has a motile, bell-shaped body with its mouth on the bottom. **25.** As colonies of coral polyps grow, they secrete an underlying skeleton of calcium carbonate that can build up for thousands of years, forming a coral reef. **Using Science Skills** **26.** Sponges are placed in the phylum Porifera, which means "pore-bearers." This name is appropriate because sponges have tiny openings, or pores, all over their bodies. Structure D is an example of a sponge pore. **27.** A; The osculum is the large hole through which water exits the central cavity of a sponge. **28.** F; Archaeocytes complete the digestive process and transport digested food throughout the sponge. **29.** structure C, which is a choanocyte **30.** Structure B is a spicule, which functions like a lens or a magnifying glass. Spicules focus and direct incoming sunlight to cells lying below the surface of the sponge, where symbiotic organisms carry out photosynthesis. **Essay** **31.** During early development, the cells of most animal embryos differentiate into three germ layers. The endoderm, the innermost germ layer, gives rise to the linings of the digestive tract and much of the respiratory system. The mesoderm, the middle layer, gives rise to muscles and much of the circulatory, reproductive, and excretory systems. The ectoderm, the outermost layer, gives rise to sense organs, nerves, and the outer layer of skin. **32.** Sample answer: A dog exhibits bilateral symmetry, with distinct right and left sides. A dog has pronounced cephalization, and its anterior end is called its head, where the sense organs for sight, smell, hearing, and taste are concentrated. At the posterior end a dog has a tail. A dog's dorsal side is called its back, and its ventral side is called its belly. **33.** A body cavity is important because it provides a space in which internal organs can be suspended, allows for specialized regions to develop, and provides room for internal organs to grow and expand. In some animals, body cavities contain fluids involved in circulation, feeding, and excretion. **34.** Gemmules are groups of archaeocytes surrounded by a tough layer of spicules. Gemmules can survive long periods of freezing temperature and drought. When conditions again become favorable, gemmules can grow into new sponges. In budding, part of the sponge breaks off the parent plant, settles onto the sea floor, and grows into a new sponge. **35.** Cnidarians are predatory carnivores. They are able to capture prey by paralyzing it with stinging nematocysts. Once a cnidarian's prey is paralyzed, it uses its tentacles to push the prey through its mouth and into its gastrovascular cavity, where it is digested.

Chapter 26—Test B

Multiple Choice **1.** C **2.** D **3.** A **4.** A **5.** D **6.** A **7.** B **8.** B **9.** C **10.** C **11.** C **12.** D **13.** A **14.** A **15.** C **Completion** **16.** invertebrates **17.** feedback **18.** deuterostomes **19.** larva **20.** cnidocytes **Short Answer** **21.** feeding, respiration, circulation, excretion, response, movement, and reproduction **22.** The endoderm is the innermost layer, the mesoderm is the middle layer, and the ectoderm is the outermost layer. **23.** Cephalization is the concentration of sense organs and nerve cells at the front end of an animal's body. **24.** The movement of water through a sponge provides a simple mechanism for feeding, respiration, circulation, and excretion. **25.** A polyp has a cylindrical body, tentacles, and a mouth that opens upward. **Using Science Skills** **26.** B **27.** swimming larva **28.** asexually **29.** medusa **30.** Hydras spend their entire lives as polyps. Figure 26-1 includes a medusa stage.

Chapter 27—Test A

Multiple Choice **1.** A **2.** C **3.** D **4.** B **5.** B **6.** C **7.** B **8.** D **9.** C **10.** C **11.** B **12.** D **13.** B **14.** B **15.** A **Completion** **16.** mesoderm **17.** hermaphrodite **18.** gizzard **19.** polychaetes **20.** foot **Short Answer** **21.** by removing remnants of feces that might contain ascarid eggs **22.** A roundworm has a pseudocoelom, which is a cavity formed between the endoderm and mesoderm layers but only partially lined with mesoderm tissue. Annelids have a true coelom, a cavity lined with tissue derived from the mesoderm. **23.** Earthworm feces fertilize the soil; earthworms also mix and aerate the soil, providing passageways for roots and water and allowing the growth of beneficial bacteria. **24.** As water passes through the mantle cavity, oxygen and carbon dioxide diffuse over the animal's gills. **25.** Students should mention any three examples for each class:

Gastropoda, which includes pond snails, land slugs, sea butterflies, sea hares, limpets, and nudibranchs; Bivalvia, which includes clams, oysters, mussels, and scallops; and Cephalopoda, which includes octopi, squids, cuttlefishes, and nautiluses. **Using Science Skills** **26.** The organism is an earthworm, of the class Oligochaeta and the phylum Annelid. **27.** Structure A is a clitellum. When eggs are ready for fertilization, it secretes a mucous ring into which eggs and sperm are released. Fertilization takes place in the mucous ring. **28.** Structure B is the dorsal blood vessel, through which blood moves from head to tail. **29.** The structures labeled D are setae, which help an earthworm prevent slippage as it moves. **30.** Structure C is the brain. Attached to the brain is structure E, a nerve cord. Both are parts of a well-developed nervous system. **Essay** **31.** A blood fluke matures and reproduces sexually in the blood vessels of human intestines. Fluke embryos are released and pass out of the human body with feces. If they get into the water, embryos develop into swimming larvae that infect freshwater snails, the intermediate host. In snails, flukes reproduce asexually, and new larvae are released into the water. They then infect humans, the primary host, by burrowing through the skin. **32.** Many flatworms rely on diffusion to remove wastes from their bodies. Some flatworms have a network of flame cells that empties metabolic wastes such as ammonia and urea from the body into the outside environment. In annelids, digestive waste passes out through the anus at the end of the digestive tract. Cellular waste containing nitrogen is eliminated by nephridia, which are excretory organs that filter fluid in the coelom. **33.** A trochophore is a free-swimming larva. Since both mollusks and annelids have trochophore larvae, those groups may be closely related. **34.** They circulate water through their body cavities using currents created by cilia on their gills. Food particles in the water are trapped by mucus and cilia on the gills. **35.** In an open circulatory system, blood is pumped through vessels by a simple heart. Blood eventually leaves the vessels and works its way through different sinuses, which are large saclike spaces. The blood passes from the sinuses to the gills, where oxygen and carbon dioxide are exchanged, and then back to the heart. Mollusks have an open circulatory system. In a closed circulatory system, blood is contained within a network of blood vessels. Annelids have a closed circulatory system.

Chapter 27—Test B

Multiple Choice **1.** B **2.** A **3.** B **4.** A **5.** A **6.** A **7.** B **8.** B **9.** C **10.** C **11.** A **12.** C **13.** A **14.** A **15.** B **Completion** **16.** flatworms **17.** carnivores **18.** closed **19.** mesoderm **20.** tentacles **Short Answer** **21.** Flatworms are soft, flattened worms with tissues, organ systems, three germ layers, bilateral symmetry, and cephalization. **22.** Snails are intermediate hosts, in which the blood fluke reproduces asexually. **23.** The inner tube is the digestive tract, and the outer tube is the body wall. **24.** Hookworms develop in the soil and can enter the body through the unprotected skin of a human. **25.** foot, mantle, shell, and visceral mass **Using Science Skills** **26.** Platyhelminthes, or flatworms **27.** D, pharynx **28.** B, ganglia **29.** A, eyespot **30.** Its body is thin, making most of its cells close to the external environment.

Chapter 28—Test A

Multiple Choice **1.** C **2.** C **3.** A **4.** B **5.** A **6.** D **7.** C **8.** D **9.** B **10.** D **11.** A **12.** A **13.** C **14.** D **15.** C **Completion** **16.** nerve cord **17.** crustaceans **18.** pedipalps **19.** chelicerae **20.** water vascular **Short Answer** **21.** insect: head, thorax, abdomen; chelicerate: cephalothorax, abdomen **22.** Spiders use book lungs, and horseshoe crabs use book gills. **23.** Insects pollinate crops; produce commercially valuable products such as silk, wax, and honey; and are eaten by humans. **24.** 1 eggs, 2 larva, 3 pupa, 4 adult **25.** Echinoderms are bilaterally symmetrical as larvae but radially symmetrical as adults. **Using Science Skills** **26.** The animal has radial symmetry demonstrated by the centrally located ring canal, the arms, and internal structures, which are arranged around the central body. **27.** the stomach; To feed, a sea star pushes its stomach out through its mouth, pours out enzymes, and digests its food. When the sea star has finished feeding, it pulls its stomach back into its mouth. **28.** Structure 2 is the madreporite, structure 3 is the ring canal, and structure 4 is a radial canal. They are all parts of the water vascular system. **29.** tube feet; when water is pushed into them, they expand. When water is pulled out, the cups on their ends shrink, creating a partial vacuum that holds them to whatever they are touching and allows the animal to walk. **30.** The tube feet use suction to pull open shelled prey. **Essay** **31.** Skin glands digest the inner part of the exoskeleton, and other glands secrete a new skeleton. The animal pulls itself out of what remains of the original skeleton. While the new exoskeleton is still soft, the animal fills with air or fluids to allow room for growth before the next molting. **32.** Spiders have two body sections, whereas insects have three. Spiders have four pairs of legs, whereas insects have three. Spiders lack antennae, whereas insects have one pair. Spiders have two pairs of appendages that are used as mouthparts, whereas insects have three pairs, including a pair of mandibles. **33.** Both centipedes and millipedes are uniramians. They both have jaws, one pair of antennae, unbranched appendages, and a highly segmented body. Both live under rocks. Centipedes have one pair of legs on most body segments, whereas millipedes have two pairs. Centipedes eat other animals, whereas millipedes eat dead and decaying plant material. **34.** The bee runs forward in a straight line while waggling her abdomen. She then circles around one way, runs in a straight line again, and circles around the other way. The farther away the food is, the longer the bee takes to perform the straight run and the more she waggles. The direction of the straight run indicates the direction of the food. **35.** Sperm are produced in testes, and eggs are produced in ovaries. Both types of gametes are shed into the water, where fertilization takes place. The larvae, which have bilateral symmetry, swim around for some time and then swim to the ocean bottom, where they develop into radially symmetrical adults.

Chapter 28—Test B

Multiple Choice **1.** A **2.** D **3.** C **4.** B **5.** C **6.** A **7.** A **8.** C **9.** A **10.** B **11.** D **12.** B **13.** C **14.** B **15.** D **Completion** **16.** segments **17.** chitin **18.** fangs **19.** three **20.** castes **Short Answer** **21.** a segmented body, a tough exoskeleton, and jointed appendages **22.** crustaceans or chelicerates **23.** Mites and ticks are often parasites; they may cause itching or rashes and can transmit bacteria that cause diseases. **24.** complete metamorphosis **25.** Each piece will grow into a new animal, as long as it contains a portion of the central part of the body. **Using Science Skills** **26.** 6 mm (12 mm – 6 mm) during the first molt and by 8 mm (20 mm – 12 mm) during the second molt **27.** While its new exoskeleton was still soft, the arthropod expanded its body by filling with air or fluids. **28.** 12 mm **29.** The exoskeleton of an arthropod is inflexible, so an arthropod cannot grow between molts. **30.** 14 days

Chapter 29—Test A

Multiple Choice **1.** C **2.** C **3.** D **4.** C **5.** D **6.** A **7.** B **8.** B **9.** D **10.** B **11.** B **12.** D **13.** D **14.** C **15.** B **Completion** **16.** segmentation **17.** germ **18.** aquatic **19.** water **20.** hydrostatic **Short Answer** **21.** Some animals of the Cambrian period had shells, skeletons, and other hard body parts, which are well preserved in fossils. **22.** Accept any three of the following: flatworms, roundworms, annelids, mollusks, and arthropods. **23.** Cephalization allows an animal to respond to the environment more quickly and in more sophisticated ways. **24.** Gases diffuse most efficiently across a moist membrane. **25.** A—open, B—closed **Using Science Skills** **26.** radial symmetry **27.** flatworms, roundworms, mollusks, annelids, arthropods, and echinoderms **28.** sponges **29.** mollusks and arthropods **30.** protostome development **Essay** **31.** Early animals had little internal specialization. They carried out essential functions using specialized cells or simple tissues. As animals became larger and more complex, specialized cells joined together to form tissues, organs, and organ systems. **32.** Cnidarians and echinoderms have radial symmetry; mollusks have bilateral symmetry. Cnidarians have no body cavity or cephalization; mollusks and echinoderms have a true coelom and cephalization. **33.** Terrestrial invertebrates must conserve body water while removing nitrogenous wastes. Urea is much less toxic than ammonia, so it can be concentrated by the excretory system without posing a danger to the animal. Concentrating urea conserves body water. **34.** Cnidarians have nerve nets, which consist of individual nerve cells that form a netlike arrangement throughout the animal's body. Flatworms have a few small clumps of nerve tissue, or ganglia, in the head. Cephalopod mollusks have ganglia that are organized into a brain, which controls and coordinates the nervous system. **35.** Hydrostatic skeletons are fluid-filled body cavities surrounded by muscles. Exoskeletons are external skeletons, such as arthropods' hard body covering made of chitin. Endoskeletons are structural supports located inside the body.

Chapter 29—Test B

Multiple Choice **1.** B **2.** D **3.** A **4.** A **5.** C **6.** C **7.** D **8.** C **9.** A **10.** D **11.** A **12.** D **13.** C **14.** A **15.** A **Completion** **16.** Trace **17.** pseudocoelom **18.** blastopore **19.** flame cell **20.** nerve nets **Short Answer** **21.** the opening of a blastula **22.** an animal that has a true coelom, or body cavity lined completely with mesoderm **23.** digestion of food inside cells **24.** saclike organs that convert ammonia into uric acid **25.** Ammonia is toxic. **Using Science Skills** **26.** gastrovascular cavity **27.** cnidarians and flatworms **28.** digestive tract **29.** Accept any two of the following: roundworms, annelids, mollusks, arthropods, and echinoderms. **30.** Food enters the digestive tract through the mouth; passes through the pharynx, crop, gizzard, and intestine; and undigested food leaves through the anus.

Unit 8—Test A

Multiple Choice **1.** C **2.** D **3.** D **4.** B **5.** C **6.** A **7.** B **8.** A **9.** C **10.** B **11.** A **12.** D **13.** D **14.** C **15.** B **Completion** **16.** invertebrates **17.** fission **18.** siphon **19.** molting **20.** cephalization **Short**

Answer **21.** Complex animals tend to have high levels of cell specialization and internal body organization, bilateral body symmetry, cephalization, and a body cavity. **22.** A polyp has a sessile, cylindrical body with armlike tentacles, with the mouth pointing upward. A medusa has a motile, bell-shaped body with the mouth on the bottom. **23.** Internal fertilization takes place inside the animal's body. External fertilization takes place outside the body. **24.** For each subphylum, students should mention any two groups or organisms. Crustacea: crabs, shrimps, lobsters, crayfishes, barnacles; Chelicerata: spiders, horseshoe crabs; Uniramia: insects, centipedes, millipedes. **25.** Echinoderms have a closer evolutionary relationship to chordates than to mollusks. Both echinoderms and chordates are deuterstomes, whereas mollusks are protostomes. **Using Science Skills** **26.** The process is complete metamorphosis. Bees, moths, and beetles undergo this process. **27.** Drawing 1 represents the insect's eggs, which are formed through internal fertilization during sexual reproduction. **28.** Drawing 2 represents a larva, which looks and acts nothing like the parents. Larvae typically feed voraciously and grow rapidly. They molt several times but change little in appearance. **29.** Drawing 3 represents a pupa. During the pupal stage, the body is completely remodeled inside and out. **30.** The other form, called incomplete metamorphosis, is much less dramatic. The immature form, called a nymph, looks very much like an adult, and there is a gradual change from nymph to adult. **Essay** **31.** In a developing embryo, the blastopore leads into a central tube that runs the length of the embryo. The tube becomes the digestive tract and is formed in one of two ways. A protostome is an animal whose mouth is formed from the blastopore. A deuterstome is an animal whose anus is formed from the blastopore. Protostomes include flatworms, roundworms, annelids, mollusks, and arthropods. Deuterstomes include echinoderms. **32.** During the life cycle, a parasitic organism lives in multiple hosts. In the primary host, the organism reproduces sexually. In the intermediate host, the organism reproduces asexually. **33.** The evolution of arthropods has led to fewer body segments and highly specialized appendages for feeding, movement, and other functions. A typical primitive arthropod was composed of many identical segments. Most living arthropods have only two or three segments, and modern arthropod appendages include antennae, claws, walking legs, wings, flippers, mouthparts, and tails. **34.** A hydrostatic skeleton consists of muscles surrounding a fluid-filled body cavity that supports the muscles. When the muscles contract, they push against the body cavity, causing the body to change shape. An exoskeleton is a tough external covering. An endoskeleton is an internal skeleton. **35.** Terrestrial invertebrates must conserve water while removing nitrogenous wastes from the body. To do this, many animals convert ammonia into a compound called urea, which is less toxic than ammonia. Urea is eliminated from the body in urine, which is highly concentrated. Some insects and arachnids have Malphighian tubules, saclike organs that convert ammonia into uric acid. The uric acid and digestive wastes combine to form a thick paste that leaves the body through the rectum. This paste contains little water.

Unit 8—Test B

Multiple Choice **1.** B **2.** D **3.** B **4.** C **5.** C **6.** A **7.** D **8.** A **9.** B **10.** B **11.** D **12.** A **13.** C **14.** B **15.** C **Completion** **16.** larva **17.** flatworm **18.** bivalves **19.** cephalothorax **20.** radial **Short Answer** **21.** polyp and medusa **22.** foot, mantle, shell, visceral mass **23.** An open circulatory system is one in which blood is pumped through vessels into sinuses. **24.** An insect's three body sections are the head, the thorax, and the abdomen. The legs are attached to the thorax. **25.** A coelom is a fluid-filled body cavity that is lined with tissue derived from the mesoderm. Invertebrates that have a true coelom include annelids, mollusks, arthropods, and echinoderms. **Using Science Skills** **26.** earthworm; Annelida **27.** clitellum; reproduction **28.** C; brain **29.** closed circulatory system **30.** The structures are setae, which prevent slipping as the animal moves.

Chapter 30—Test A

Multiple Choice **1.** C **2.** D **3.** D **4.** B **5.** A **6.** D **7.** C **8.** D **9.** B **10.** A **11.** D **12.** A **13.** D **14.** C **15.** D **Completion** **16.** pharyngeal pouches **17.** gills **18.** amphibian **19.** drying out **20.** right atrium **Short Answer** **21.** The front end grows into a brain. **22.** bodies that were armored with bony plates **23.** The pyloric ceca are fingerlike pouches located near the stomach. They secrete digestive enzymes and absorb nutrients from digested food. **24.** Anadromous fishes spend most of their lives in the ocean but migrate to fresh water to breed, whereas catadromous fishes live their lives in fresh water but migrate to the ocean to breed. **25.** A—cartilage, B—bone, C—fibers and cartilage **Using Science Skills** **26.** A is the sinus venosus, B is the atrium, C is the ventricle, and D is the bulbus arteriosus. **27.** It is poor in oxygen. **28.** C, the ventricle, is the actual pumping portion. **29.** the aorta **30.** the gills **Essay** **31.** An adult lancelet is a fishlike animal with a hollow nerve cord, a notochord, and segmented muscles that run the length of the body. It has a definite head region and a tail. An adult tunicate has a body that is covered by a tough, nonliving tunic. The tunic is flattened at the base, where the animal is typically attached to a solid surface, and it has two large openings called siphons. An adult tunicate does not have a hollow nerve cord, a notochord, segmented muscles, a head region, or a tail. **32.** Because jaws could hold teeth and muscles, jaws made it possible for these fishes to chew on plants and other animals. Thus, fishes with jaws could eat a much wider variety of food than could jawless fishes, which were limited to eating small particles of food that they filtered out of the water or sucked up from the sea floor. Jaws also allowed fishes to defend themselves by biting. **33.** The main parts are the olfactory bulbs, cerebrum, optic lobes, cerebellum, and medulla oblongata. The olfactory bulbs and cerebrum are both involved with the sense of smell, or olfaction. The optic lobes process information from the eyes. The cerebellum coordinates body movements. The medulla oblongata controls the functioning of many internal organs. **34.** The number of living species of amphibians in the world is decreasing. Possible explanations for the decline include environmental threats, such as decreasing habitat, depletion of the ozone layer, acid rain, water pollution, fungal infections, introduced aquatic predators, and an increasing human population. **35.** Adult amphibians have a double-loop circulatory system and a heart with three chambers. One loop carries oxygen-poor blood from the ventricle to the lungs and takes oxygen-rich blood from the lungs to the left atrium. The left atrium empties this blood into the ventricle. The other loop transports oxygen-rich blood from the ventricle to the rest of the body and oxygen-poor blood from the body to the right atrium. The right atrium also empties its blood into the ventricle.

Chapter 30—Test B

Multiple Choice **1.** B **2.** A **3.** A **4.** A **5.** D **6.** A **7.** B **8.** A **9.** C **10.** D **11.** C **12.** B **13.** A **14.** B **15.** C **Completion** **16.** vertebrae **17.** cartilage **18.** smell **19.** Amphibians **20.** three **Short Answer** **21.** a dorsal, hollow nerve cord; a notochord; pharyngeal pouches; and a tail that extends beyond the anus **22.** Lancelets move in a fishlike motion by contracting muscles on either side of the body. **23.** A, a shark, is a cartilaginous fish (class Chondrichthyes), B is a bony fish (class Osteichthyes), and C is a jawless fish (a lamprey). **24.** Accept any two: strong bones in the limbs and limb girdles; lungs and breathing tubes to breathe air; ribs form a bony cage that support and protect internal organs against the pull of gravity. **25.** salamanders, frogs and toads, and caecilians **Using Science Skills** **26.** the notochord **27.** the dorsal hollow nerve cord **28.** the tail **29.** pharyngeal pouches **30.** They develop slits that connect to the outside of the body. The slits may then develop into gills.

Chapter 31—Test A

Multiple Choice **1.** D **2.** A **3.** D **4.** A **5.** C **6.** A **7.** D **8.** B **9.** C **10.** B **11.** B **12.** C **13.** A **14.** C **15.** C **Completion** **16.** extinction **17.** amniotic **18.** energy **19.** four **20.** magnetic **Short Answer** **21.** A reptile's skin does not grow when the rest of the reptile grows. Reptiles overcome this disadvantage by shedding their skin periodically as they increase in size. **22.** The lungs of reptiles have more gas-exchange area than those of amphibians. **23.** Student answers may include: Many reptiles in this order (turtles and tortoises) can pull their head, legs, and tail into their shells to protect themselves. They also have jaws that are often powerful enough to deliver a damaging bite. **24.** Structure A (the collarbone) consists of bones that are fused together. It is therefore more rigid than the corresponding structure in other vertebrates. **25.** In flying birds, Structure B (the sternum or breastbone) is enlarged and has a long keel that provides an attachment site for the large chest muscles that are necessary for flight. **Using Science Skills** **26.** The bird labeled A probably eats insects because it has a short, fine bill that can pick insects off surfaces or catch them as they fly. **27.** The long, thin bill is adapted to probing soft mud for worms and other invertebrates. **28.** The bird labeled C probably eats meat because its strong, hooked bill can tear and shred prey. **29.** This bird would not be successful at eating insects because its large bill would not be fine enough to pick insects off leaves and branches. **30.** The bird labeled E probably eats seeds because its short, thick bill can be used for breaking open seeds. **Essay** **31.** Dinosaurs ranged in size from small to enormous. Some walked on two legs, and others walked on four. Some ate leafy plants, and others hunted in large herds. Some dinosaurs lived in small family groups and cared for their young in nests. Certain dinosaurs may even have had feathers. **32.** Both lizards and alligators have a double-loop circulatory system and a heart with two atria. One of the loops sends blood to the lungs and returns it to the left atrium. The other loop sends blood to the rest of the body and returns that blood to the right atrium. In a lizard, there is a single ventricle with a partial internal wall that helps separate oxygen-rich and oxygen-poor blood. In an alligator, there are two ventricles, so oxygen-rich and oxygen-poor blood remain completely separated. **33.** *Archaeopteryx* looked like a small, running dinosaur, but it had well-developed feathers covering most of its body. The presence of feathers supports the classification of *Archaeopteryx* as a bird. However, *Archaeopteryx* also had teeth in its beak, a bony tail, and toes and claws on its wings, all of which are features that modern birds lack. Taken together, the current evidence indicates that *Archaeopteryx* was a transitional species with characteristics of both dinosaurs and birds. **34.** When a bird inhales, most air first enters large air sacs in the body cavity and bones. The inhaled air then flows to the lungs and travels through the lungs in a series of small tubes. This system ensures that air flows into the air sacs and out through the lungs in a single direction, which constantly exposes the lungs to oxygen-rich air. This constant exposure helps birds maintain a high metabolic rate and provides the oxygen necessary for high-altitude flight. **35.** Hummingbirds and other nectar-eating birds transfer pollen from flower to flower as they feed, thereby pollinating flowers. Fruit-eating birds swallow seeds but may not digest them, so their droppings disperse seeds over great distances. Insect-eating birds, such as swallows and chimney swifts, catch great numbers of mosquitoes and other insects, thereby helping control insect populations. Many other examples would be acceptable.

Chapter 31—Test B

Multiple Choice **1.** B **2.** B **3.** C **4.** C **5.** C **6.** C **7.** D **8.** B **9.** A **10.** D **11.** B **12.** D **13.** A **14.** D **15.** B **Completion** **16.** cloaca **17.** plastron **18.** collarbone (or wishbone) **19.** sternum (or breast bone) **20.** seeds **Short Answer** **21.** Uric acid is much less toxic than ammonia, so it does not have to be diluted with as much water. Therefore, water is conserved in the bodies of land reptiles. **22.** Students can choose one of the following two threats. Many reptiles are in danger because their habitats have been and are being destroyed. Humans also hunt reptiles for food, to sell as pets, and for their skins. **23.** Birds are endotherms, which means that they generate their own body heat through their metabolism by eating food. **24.** One explanation is that birds evolved directly from dinosaurs. A second explanation is that both dinosaurs and birds evolved from a common ancestor. **25.** Possible answers include hooked bills, large wingspans, or sharp talons. **Using Science Skills** **26.** the amnion; It is a fluid-filled sac that surrounds and cushions the developing embryo. **27.** the allantois; It stores the waste produced by the embryo and also serves as a respiratory organ. **28.** the chorion; It regulates the transport of oxygen and carbon dioxide between the embryo and the surface of the egg. **29.** the yolk sac; It contains yolk, the food supply for the embryo. **30.** the shell; It is soft and leathery in most reptile eggs.

Chapter 32—Test A

Multiple Choice **1.** B **2.** A **3.** A **4.** C **5.** B **6.** C **7.** C **8.** B **9.** B **10.** C **11.** A **12.** B **13.** B **14.** C **15.** A **Completion** **16.** mammary glands **17.** insects **18.** placenta **19.** bipedal **20.** anthropoids **Short Answer** **21.** Mammals have a lower jaw consisting of a teeth-bearing bone connected by a joint directly to the skull, complex teeth that are replaced just once in a lifetime, and distinctive features of the limbs and the backbone. **22.** The rumen is a stomach chamber that stores and processes newly swallowed plant food. Bacteria in the rumen break down the hard-to-digest cellulose of plant tissues. Carnivores do not have rumens because digestive enzymes alone can quickly digest meat. **23.** Mammals maintain homeostasis by regulating their body heat and excreting or retaining water with the kidneys. **24.** After the embryo is born, it crawls across its mother's fur and attaches to a nipple. In most marsupials, the nipples are located in a pouch called the marsupium. The young marsupial continues to grow and develop while attached to the nipple, until it is large enough to survive on its own. **25.** Binocular vision provides depth perception, which is important for judging the locations of tree branches. **Using Science Skills** **26.** artiodactyls **27.** perissodactyls **28.** The bones are slender, reducing the wing's weight. The "fingers" are spread out to support flaps of skin that form a wing. **29.** The bones are much thicker than the bat's. **30.** The long, slender snout is adapted to probe into the ground, looking for ants. **Essay** **31.** Mammals have a high rate of metabolism, which helps them generate body heat. Mammals also have body hair and subcutaneous fat, which help them conserve heat. Many mammals have sweat glands that cool the body when the sweat evaporates from the skin. Mammals that lack sweat glands often pant when they need to rid themselves of excess heat. **32.** Specialized teeth and digestive systems process food more quickly and more thoroughly so that mammals can get the most nutrients from their food. This is important because mammals require a lot of energy to maintain their high metabolic rate. This high metabolic rate generates body heat. **33.** Monotremes have a cloaca, a structure into which the reproductive and urinary systems open, as do reptiles. Like most reptiles, monotremes also lay eggs. Like other mammals, monotremes nourish their young with milk produced by mammary glands. **34.** Both groups of monkeys are anthropoids. New World monkeys are found in Central and South America. They live almost entirely in trees and have long, flexible arms and long, prehensile tails. Old World Monkeys are found in Africa and Asia. They also spend time in trees but lack prehensile tails. **35.** Paleontologists once thought that hominid evolution ocurred in a simple, straight-line transformation of one species into another. Now, however, they think that hominid evolution occurred as a series of complex adaptive radiations that produced a large number of different species. Paleontologists changed their way of thinking as the result of many recent fossil finds.

Chapter 32—Test B

Multiple Choice **1.** D **2.** B **3.** D **4.** D **5.** A **6.** B **7.** D **8.** A **9.** C **10.** B **11.** D **12.** D **13.** D **14.** A **15.** B **Completion** **16.** hair **17.** duckbill platypus **18.** cerebrum (or cerebral cortex) **19.** New World **20.** opposable **Short Answer** **21.** canines, incisors, molars, and premolars **22.** The diaphragm pulls the bottom of the chest cavity downward, increasing the volume of the cavity and drawing air into the lungs. **23.** the way mammals reproduce and develop **24.** A marsupium is a pouch on the outside of a female marsupial's body in which embryos complete their development. **25.** A prosimian is a type of primate that is generally small and nocturnal and has large eyes that are adapted to seeing in the dark. **Using Science Skills** **26.** II is the skull of a carnivore, and I is the skull of an herbivore. **27.** molars and premolars; They are used to crush and grind tough plants. **28.** incisors; They are used for cutting, gnawing, grooming and to grasp and tear vegetation. **29.** canines; They are used for piercing, gripping, slicing, and tearing flesh from prey. **30.** In the herbivore (mammal I), the jaws generally move from side to side. In the carnivore (mammal II), the jaws usually move up and down.

Chapter 33—Test A

Multiple Choice **1.** D **2.** A **3.** D **4.** A **5.** A **6.** A **7.** D **8.** B **9.** D **10.** B **11.** A **12.** A **13.** C **14.** B **15.** A **Completion** **16.** adaptive radiations **17.** ectotherm **18.** filter feeder **19.** alveoli **20.** cerebrum **Short Answer** **21.** The ecological conditions probably were similar. **22.** Animal B is the ectotherm because ectotherms have lower metabolic rates than endotherms. **23.** Tunicates have survived to the present day, so their organ systems are well equipped to perform the essential functions of life. **24.** The digestive tracts of carnivores produce enzymes that help digest meat, while the digestive tracts of herbivores often have colonies of bacteria that help digest cellulose fibers in plant tissues. **25.** Larval tunicates swim with a fishlike movement of their muscular tails, whereas adult tunicates either swim by using their siphons or remain attached to a surface. **Using Science Skills** **26.** Vessels C and D contain oxygen-rich blood, and vessels A and B contain oxygen-poor blood. **27.** E and H are atria, F and G are ventricles. **28.** to capillaries in the lungs and the rest of the body **29.** Crocodilians, birds, and mammals have a circulatory system that is represented by diagram I. Noncrocodilian reptiles have a circulatory system that is represented by diagram II. **30.** The heart in diagram I has two separate ventricles, whereas the heart in diagram II has a single, partially divided ventricle. Hence, mixing of oxygen-rich and oxygen-poor blood does not occur in the heart shown in diagram I, but some mixing does occur in diagram II. **Essay** **31.** Many of the chemical reactions that are important in metabolism are affected by temperature. Therefore, essential life functions can be carried out most efficiently when an animal's body temperature is within a particular operating range. For example, if the temperature is too low, muscles may contract slowly. **32.** One hypothesis is that dinosaurs were endotherms. A second hypothesis is that endothermy evolved long after the appearance of the dinosaurs. Evidence suggests that endothermy evolved once along the evolutionary line of reptiles that led to birds and once along the evolutionary line of reptiles that led to mammals. **33.** In mammals, inhaling brings oxygen-rich air into the lungs through the trachea and many small branches, and exhaling forces oxygen-poor air out of the lungs in the reverse direction through the same passageways. Consequently, there is always stale, oxygen-poor air trapped in the lungs. In birds, a system of tubes within the lungs and air sacs attached to the lungs ensures that air flows in only one direction through the lungs. Thus, the gas-exchange surfaces are constantly in contact with fresh air, and stale air is never trapped in the lungs. **34.** The hearts of fishes have two chambers: one atrium and one ventricle. The hearts of amphibians and all reptiles except crocodilians have three chambers: two atria and one ventricle. In those reptiles, however, the ventricle contains a partial partition. The hearts of crocodilians, birds, and mammals have four chambers: two atria and two ventricles. **35.** Tunicates, aquatic amphibians, and most fishes excrete ammonia. Mammals, land amphibians, and cartilaginous fishes convert ammonia into urea, which is excreted. Reptiles and birds change ammonia into uric acid. Tunicates also produce some uric acid, but it is stored within the body and released only when the animal dies.

Chapter 33—Test B

Multiple Choice **1.** C **2.** B **3.** B **4.** C **5.** B **6.** B **7.** B **8.** C **9.** A **10.** C **11.** C **12.** A **13.** B **14.** B **15.** D **Completion** **16.** lancelets **17.** ectothermy **18.** alveoli **19.** double **20.** viviparous **Short Answer** **21.** a notochord, a dorsal hollow nerve cord, a tail that extends posterior to the anus, and pharyngeal pouches **22.** a source of heat for the body, a way to conserve that heat, and a method of eliminating excess heat when necessary **23.** One loop carries blood between the heart and the lungs, while the other loop carries blood between the heart and the rest of the body. **24.**

Vertebrate kidneys excrete nitrogenous wastes and regulate the amount of water in the body. **25.** The optic lobes are involved in vision, and the olfactory bulbs are involved in the sense of smell. **Using Science Skills 26.** the litopterns **27.** during the Early Cenozoic **28.** Accept any two: mastodons, condylarths, creodonts, litopterns. **29.** proboscideans **30.** chiropterans

Chapter 34—Test A

Multiple Choice 1. A **2.** D **3.** C **4.** C **5.** C **6.** B **7.** B **8.** C **9.** C **10.** B **11.** A **12.** D **13.** C **14.** B **15.** A **Completion 16.** stimulus **17.** habituation **18.** threatening (harmful) **19.** classical conditioning **20.** communication **Short Answer 21.** Innate behaviors help an animal survive. Many innate behaviors, such as a spider's building a web or the suckling of a newborn mammal, help the animal to get food. Others, such as nest building, help an animal to reproduce. Without these behaviors, the animal would not survive. **22.** In habituation, an animal decreases or stops its response to a repetitive stimulus that is nonthreatening or unrewarding, which allows the animal to spend its time and energy more efficiently. **23.** Young geese have an innate urge to follow the first moving object they see, but they must learn from experience what object to follow. **24.** When food or other resources are scarce, dormancy and migration allow animals to survive by becoming inactive or by moving to another place where resources are more plentiful. **25.** The members of a society usually share a large proportion of one another's genes, so helping other members survive increases the chance that an animal's own genes will be passed to the next generation. **Using Science Skills 26.** innate **27.** the female **28.** the male, because the male must guide the female to the nest **29.** The male swims zigzag to the female, swims toward the nest, points to the nest, trembles, and nudges the female. **30.** The female gives a head-up display, swims head-up toward the male, enters the nest, lays eggs in the nest, and leaves the nest. **Essay 31.** Some behaviors can be inherited by an animal's offspring because these behaviors are influenced by genes. Sometimes a new adaptive behavior that is influenced by genes helps an individual to survive and reproduce better than individuals that do not have the behavior. After many generations of natural selection, most individuals in the population will have the new adaptive behavior. **32.** In classical conditioning, an animal learns to make a mental connection between a stimulus and some kind of reward or punishment. The stimulus is one that the animal did not previously associate with the reward or punishment. In operant conditioning, an animal learns through trial and error to produce a certain behavior to receive a reward or avoid punishment. **33.** A Skinner box is a box in which an animal is placed to test for the ability to learn by operant conditioning. The box contains a button or lever that, when pressed, delivers a food reward. After the animal is rewarded several times, it learns that it can receive food whenever it presses the button or lever. At that point, the animal has learned by operant conditioning how to obtain food. **34.** Territories contain resources, such as food and water, that enable the animal to survive and reproduce. By claiming a territory, an animal keeps other animals away. This allows the animal to use all the resources in the territory. **35.** Communication can occur whenever one individual passes information to another. Animals may use visual, sound, touch, or chemical signals to communicate with one another. Animals use these signals to communicate such things as warnings, readiness to mate, and the boundaries of a territory.

Chapter 34—Test B

Multiple Choice 1. D **2.** C **3.** B **4.** A **5.** A **6.** D **7.** A **8.** C **9.** C **10.** A **11.** B **12.** B **13.** D **14.** A **15.** C **Completion 16.** response **17.** Ivan Pavlov **18.** insight **19.** aggression **20.** language **Short Answer 21.** the sense organs, the nervous system, and the muscles **22.** An innate behavior is a behavior that appears in fully functional form the first time it is performed and does not require previous experience with the stimulus that triggers it. **23.** habituation, classical conditioning, operant conditioning, and insight learning **24.** Courtship helps individuals attract, locate, and mate with other members of the same species and can help animals identify healthy mates. **25.** visual, sound, touch, and chemical signals **Using Science Skills 26.** innate behavior **27.** classical conditioning **28.** Food and a ringing bell are the two stimuli. The response is salivation by the dog. **29.** habituation, operant conditioning, and insight learning **30.** imprinting

Unit 9—Test A

Multiple Choice 1. C **2.** D **3.** B **4.** A **5.** D **6.** B **7.** B **8.** D **9.** C **10.** A **11.** D **12.** A **13.** C **14.** B **15.** D **Completion 16.** amphibian **17.** four-chambered heart **18.** Triassic Period **19.** invertebrate **20.** response; stimulus **Short Answer 21.** Both tunicates and lancelets have a hollow nerve cord, a notochord, pharyngeal pouches, and a tail at some stage of their life cycle. **22.** Amphibians require water for reproduction and for the development of their fishlike embryos. **23.** Some adaptations that enable birds to fly include highly efficient digestive, respiratory, and circulatory systems; aerodynamic feathers and wings; strong, lightweight bones; and strong chest muscles. **24.** Primates have binocular vision, a well-developed cerebrum, relatively long fingers and toes, and arms that can rotate around their shoulder joints. **25.** When an animal's senses have detected an external stimulus, the information is passed along nerve cells to the brain. The brain and other parts of the nervous system process the information and direct body muscles to produce the response. **Using Science Skills** **26.** The cow has the largest stomach. **27.** The shark is a carnivore. The cow and the pigeon are herbivores. Herbivores have long intestines to help digest plant tissues. Carnivores have short digestive tracts. **28.** Because birds and mammals are endotherms, they require large amounts of food to supply the energy for their high metabolic rate. Their more complex digestive systems break food down more efficiently. Ectothermic fishes do not require as much energy from food because they have a much lower metabolic rate. **29.** The pigeon has a crop and a gizzard. The crop stores and moistens food before it moves toward the stomach. The gizzard is a muscular part of the stomach that breaks down food by crushing and grinding it. **30.** Cows have flat-edged incisors to grasp and tear plants and broad, flattened molars to grind the food. **Essay** **31.** Like in the evolution of other chordate groups, a series of complex adaptive radiations produced a large number of species. These adaptive radiations occurred in response to new environmental conditions or the appearance of new adaptations. **32.** Humans are the only animals known to use language. Language is a complicated form of communication that combines sounds, symbols, and gestures according to sets of rules about word order and meaning. Humans are capable of learning and using language because of their well-developed brain. The cerebrum is greatly enlarged, compared with that of other vertebrates, and includes a well-developed cerebral cortex. Binocular vision enables humans to use visual cues and gestures as part of language. **33.** Aquatic vertebrates, such as fishes and larval amphibians, use gills for respiration. Adult amphibians, reptiles, birds, and mammals use lungs to breathe. As you move from amphibians to mammals, lungs become more complex and efficient at obtaining oxygen from air. Reptilian lungs often are divided into chambers that increase the surface area for gas exchange. Mammals have thousands of bubblelike alveoli that produce a very large area for gas exchange. Birds have the most efficient respiratory system. Their system of air sacs and tubes in the lungs enables a one-way air flow. This ensures that the lungs are in constant contact with oxygen-rich air. **34.** The chemical reactions for essential life functions are carried out most efficiently when an animal's internal body temperature is within a certain range. Animals control their body temperature either with their behavior (ectotherms) or by generating their own heat and conserving it with body insulation (endotherms). **35.** The behavior of an animal determines whether it will get food and other resources, find a mate, and get protection from predators. If an animal has an adaptive behavior that makes it better suited to get any of the things it needs, it will have a better chance of surviving and passing its genes to offspring. If an animal has a behavior that prevents it from getting any of the things it needs, the animal will probably not survive.

Unit 9—Test B

Multiple Choice **1.** C **2.** B **3.** D **4.** B **5.** A **6.** C **7.** A **8.** A **9.** D **10.** B **11.** C **12.** D **13.** A **14.** B **15.** C **Completion** **16.** jaws and paired fins **17.** shell **18.** opposable thumb **19.** endotherm **20.** stimulus; response **Short Answer** **21.** This chordate, a fish, is adapted to an aquatic lifestyle. It has fins and gills, adaptations for moving and getting oxygen in water. **22.** The embryos of both birds and reptiles develop within amniotic eggs. Both excrete nitrogenous wastes as uric acid and have a cloaca. Both have similar bones that support the front and hind limbs. **23.** Mammalian kidneys filter urea from the blood, as well as excrete excess water or retain needed water. **24.** Adaptive radiations occurred with a rapid increase in the number and diversity of species as they adapted to new conditions. Convergent evolution occurred when unrelated species evolved similar adaptations to similar environments. **25.** Members of a society are often closely related, and therefore, share a large proportion of each other's genes. Helping a relative survive increases the chance that the genes an individual shares with the relative will be passed along to offspring. **Using Science Skills** **26.** Diagram III illustrates a single-loop circulatory system. **27.** The circulatory system in diagram III supports gills. The circulatory systems in diagrams I and II support lungs. **28.** Vessels C and D contain oxygen-rich blood. Vessels A and B contain oxygen-poor blood. **29.** The heart in diagram II has a ventricle with a partition. Three-chambered hearts, such as this one, are found in all reptiles except crocodiles and alligators. **30.** The circulatory system in diagram I is most efficient at delivering oxygen because oxygen-rich blood is always completely separated from oxygen-poor blood. And, the double pump of the four-chambered heart pumps blood through the body more forcefully and quickly.

Chapter 35—Test A

Multiple Choice **1.** D **2.** B **3.** D **4.** C **5.** A **6.** B **7.** D **8.** C **9.** A **10.** A **11.** C **12.** B **13.** A **14.** A **15.** B **Completion** **16.** potential **17.** nerves **18.** direction **19.** Homeostasis **20.** nodes **Short Answer** **21.** A tissue is a group of similar cells that perform a single function. An organ is a group of tissues that work together to perform a complex function. **22.** Fetal alcohol syndrome is caused by consuming alcohol while pregnant. The effect of this syndrome is a range of birth defects, such as heart defects, malformed faces, delayed growth, and poor motor development. **23.** Sensory receptors react to the tap on the knee and stimulate the sensory neuron. The sensory neuron relays the signal to an interneuron within the spinal cord. A signal is then sent to the motor neuron, which in turn stimulates the muscle to extend the knee. **24.** Both rods and cones are photoreceptors in the retina. Rods are sensitive to light, but do not distinguish colors. Cones are less sensitive to light than rods, but cones respond to light of different colors. **25.** The spinal cord is the main communications link between the brain and the rest of the body, transmitting information, just as a telephone line carries many calls at once. **Using Science Skills** **26.** vesicles **27.** synaptic cleft **28.** The impulse travels from the axon to the adjacent neuron. **29.** The neurotransmitters may be broken down by enzymes, or taken up and recycled by the axon terminal. **30.** muscle cell or gland cell **Essay** **31.** The sympathetic nervous system and the parasympathetic nervous system have opposite effects on the same organ system. When the sympathetic system speeds up an activity, the parasympathetic system slows down the same activity. **32.** The cerebrum controls voluntary activities, intelligence, learning, and judgment. The cerebellum controls the action of muscles. The brain stem regulates the flow of information between the brain and the rest of the body. The brain stem controls blood pressure, heart rate, breathing, and swallowing. The thalamus receives messages from the sense organs and sends them to the cerebrum. The hypothalamus recognizes sensations of hunger, thirst, fatigue, anger, and body temperatures. **33.** The five general categories are pain receptors, thermoreceptors, mechanoreceptors, chemoreceptors, and photoreceptors. Pain receptors are located throughout the body except in the brain. Thermoreceptors are located in the skin, body core, and hypothalamus. Mechanoreceptors are found in the skin, skeletal muscles, and inner ears. Chemoreceptors are found in the nose and taste buds. Photoreceptors are found in the eyes. **34.** Stimulants increase the release of neurotransmitters at some synapses in the brain, which then leads to a feeling of energy and well-being. However, when the effects of the stimulants wear off, the brain's supply of neurotransmitters has been depleted. The user quickly falls into fatigue and depression, which can impair his or her ability to drive and lead to accidents. **35.** The difference in electrical charge across the cell membrane of a resting neuron is the resting potential. Once an impulse begins, positive ions rush across the cell membrane, reversing the charge difference. The inside of the membrane gains a positive charge, and the outside of the membrane gains a negative charge. This rapid reversal of charges is called the action potential. As the impulse passes, the positively charged ions flow out of the cell, and the resting potential of the membrane is reestablished.

Chapter 35—Test B

Multiple Choice **1.** C **2.** B **3.** C **4.** D **5.** A **6.** A **7.** A **8.** A **9.** A **10.** D **11.** C **12.** A **13.** B **14.** D **15.** C **Completion** **16.** peripheral **17.** feedback **18.** mechanoreceptors **19.** interneurons **20.** alcohol **Short Answer** **21.** Dendrites carry impulses from the environment or from other neurons toward the cell body. Axons carry impulses away from the cell body. **22.** Opiates mimic natural chemicals in the brain known as endorphins, which normally help to overcome sensations of pain. **23.** the cerebrum, the cerebellum, the brain stem, the thalamus, and the hypothalamus **24.** vision, hearing, smell, taste, and touch **25.** the process by which organisms maintain a relatively constant internal environment **Using Science Skills** **26.** spinal cord **27.** structure A, the cerebrum **28.** hypothalamus **29.** structure B **30.** medulla oblongata and pons

Chapter 36—Test A

Multiple Choice **1.** C **2.** C **3.** B **4.** A **5.** B **6.** A **7.** D **8.** D **9.** B **10.** B **11.** A **12.** A **13.** C **14.** B **15.** D **Completion** **16.** spongy bone **17.** movable **18.** sliding-filament model **19.** neuromuscular junction **20.** blood vessels **Short Answer** **21.** The two major parts are the axial skeleton and the appendicular skeleton. The axial skeleton consists of the skull, vertebral column, and rib cage. The appendicular skeleton is made of the arms and legs and the bones of the pelvis and shoulder area. **22.** Cartilage is found in parts of the body where flexibility is needed, such as the tip of the nose and the external ear. It is also located where bones are attached to the sternum, which allows the rib cage to move during breathing. **23.** The small sac of synovial fluid is called a bursa; it reduces the friction between the bones of a joint and also acts as a tiny shock absorber. **24.** The left hand would probably be stronger because the person would be using it much more than the right hand, and exercising muscles causes them to get stronger. **25.** Keratin and melanin are both produced in the epidermis. Keratin is a tough, fibrous protein. Melanin is a dark brown pigment. **Using Science Skills** **26.** yellow marrow and red marrow **27.** The cell lines the Haversian canal, so it is an osteoclast, which breaks down bone, or an osteoblast, which produces bone. **28.** The Haversian canal contains blood vessels and nerves. **29.** The periosteum contains blood vessels that provide oxygen and nutrients to the bone. **30.** Haversian canals, veins, arteries, osteoclasts, and osteoblasts **Essay 31.** Compact bone is found beneath the periosteum. It is a dense bone, although not solid. Running through compact bone is a network of tubes called Haversian canals that contain blood vessels and nerves. Spongy bone is less dense than compact bone and is found in long bones and in the middle of short, flat bones. It is strong and organized in a latticework structure, which adds strength to the bone without adding mass. **32.** The three types of joints are immovable joints, slightly movable joints, and freely movable joints. The places where the bones in the skull meet are immovable joints. Joints between the two bones of the lower leg and the joints between adjacent vertebrae are examples of slightly movable joints. Freely movable joints include ball-and-socket joints, hinge joints, pivot joints, and saddle joints. **33.** Skeletal muscle tissue is striated, is generally attached to the bones of the skeleton, and is under voluntary control. It is spindle-shaped and contains many nuclei. Cardiac muscle tissue is striated, contains one or two nuclei, and is not under direct control of the central nervous system. Cardiac muscle is found only in the heart. Smooth tissue has spindle-shaped cells that have a single nucleus and are not striated, and it is generally not under control of the central nervous system. **34.** Regular exercise increases muscle size and strength. Exercise also helps body systems to become more efficient, including organs such as the heart and lungs. **35.** Hair is produced at the base of hair follicles from columns of cells that are filled with keratin and then die. Hair follicles, made up of clusters of such cells, are anchored in the dermis. Cells multiply rapidly in the base of the follicle, causing the hair to grow longer.

Chapter 36—Test B

Multiple Choice 1. C **2.** B **3.** C **4.** C **5.** B **6.** D **7.** D **8.** C **9.** A **10.** A **11.** C **12.** D **13.** A **14.** C **15.** D **Completion 16.** axial **17.** ribs **18.** ossification **19.** melanin **20.** integumentary **Short Answer 21.** Answers should include any three of the following five functions: supporting the body, protecting internal organs, allowing movement, storing mineral reserves, providing a site for blood cell formation. **22.** Bones are a solid network of living cells and protein fibers that are surrounded by deposits of calcium salts. **23.** hinge **24.** The three types of muscles are skeletal muscle, smooth muscle, and cardiac muscle. **25.** The integumentary system serves as a barrier against infection and injury, helps to regulate body temperature, removes waste products from the body, and provides protection against ultraviolet radiation from the sun. **Using Science Skills 26.** A **27.** The figure shows that there are no blood vessels in the outer layer of the skin, so a slight scratch will probably not be deep enough to break blood vessels. **28.** dermis **29.** Structure F is an opening similar to that for sweat gland D. **30.** This gland produces oil that keeps the epidermis flexible and waterproof.

Chapter 37—Test A

Multiple Choice 1. B **2.** A **3.** A **4.** B **5.** C **6.** B **7.** D **8.** B **9.** C **10.** B **11.** C **12.** D **13.** A **14.** C **15.** B **Completion 16.** B **17.** blood pressure **18.** atherosclerosis **19.** water **20.** hemophilia **Short Answer 21.** The ribs are lowered. **22.** Red blood cells transport oxygen. White blood cells perform a variety of protective functions, including guarding against infection, fighting parasites, and attacking bacteria. Platelets help in the clotting process. **23.** The air passes from the outside into the body through the nose or mouth, passes through the pharynx and the trachea, and then passes into the bronchi, bronchioles, and alveoli within the lungs. **24.** The chest cavity is largest at inhalation. **25.** Smoking can cause chronic bronchitis, emphysema, and lung cancer. In chronic bronchitis, the bronchi become swollen and clogged with mucus; emphysema is a lack of elasticity in the tissues of the lungs; and lung cancer is a deadly disease that can spread to other parts of the body. **Using Science Skills 26.** The inferior vena cava brings oxygen-poor blood from the lower part of the body to the right atrium of the heart. **27.** These are valves that prevent blood from flowing backward in the heart. **28.** A, the aorta **29.** The valves close. **30.** It carries oxygen-poor blood from the heart to the lungs. **Essay 31.** (a) The muscular walls of the large arteries help them withstand blood pressure. (b) The valves in the veins help to prevent the backward flow of blood in them. (c) The valves in the heart prevent the backward flow of blood in the heart. **32.** The body regulates blood pressure in two ways. First, sensory receptors measure blood pressure. When blood pressure is too low, the sensory receptors send impulses to the medulla oblongata, which stimulates the autonomic nervous system to contract the muscles in blood vessel walls. When blood pressure is too high, the same system helps to decrease blood pressure. Second, the kidneys remove water from the blood, lowering blood volume. The loss of fluid lowers the blood pressure. **33.** Some white blood cells engulf and digest foreign cells. Others slip out of capillary walls and attack invading organisms in the tissues of the body. Some produce antibodies that fight infection. **34.** The lymphatic system is a network of vessels, nodes, and organs that collect fluid leaking from the bloodstream and return it to the circulatory system. Lymph nodes act as filters, trapping bacteria and other microorganisms that cause disease. The organs produce specialized white blood cells, which protect the body from infection. Lymph vessels also aid in nutrient absorption. They absorb fats from the digestive tract and carry fat-soluble vitamins to the blood. **35.** Cellular respiration is the release of energy from the breakdown of food in the presence of oxygen. The second type of respiration is the exchange of oxygen and carbon dioxide between the lungs and the environment.

Chapter 37—Test B

Multiple Choice **1.** A **2.** D **3.** D **4.** C **5.** B **6.** B **7.** A **8.** A **9.** A **10.** B **11.** C **12.** B **13.** A **14.** A **15.** B **Completion** **16.** blood **17.** lung **18.** hemoglobin **19.** bronchi **20.** nervous **Short Answer** **21.** The circulatory system consists of the heart, a series of blood vessels, and the blood that flows through them. **22.** Arteries carry blood away from the heart. Capillaries bring nutrients and oxygen to the tissues and absorb carbon dioxide and waste products. Veins carry blood toward the heart. **23.** Blood pressure is the force of the blood on the walls of arteries. **24.** Blood plasma is a straw-colored fluid that makes up 55 percent of the volume of blood. It is 90 percent water and 10 percent dissolved gases, salts, nutrients, enzymes, hormones, waste products, and plasma proteins. **25.** The human respiratory system brings about the exchange of oxygen and carbon dioxide. **Using Science Skills** **26.** C **27.** trachea, windpipe **28.** A, alveoli **29.** F, epiglottis **30.** D, bronchus

Chapter 38—Test A

Multiple Choice **1.** B **2.** D **3.** D **4.** B **5.** D **6.** C **7.** B **8.** D **9.** D **10.** B **11.** C **12.** B **13.** A **14.** B **15.** A **Completion** **16.** minerals **17.** water **18.** mechanical **19.** pepsin **20.** blood **Short Answer** **21.** Accept any two of the following functions: to protect body organs; insulate the body; store energy; produce cell membranes, myelin sheaths, and certain hormones; help body absorb fat-soluble vitamins. **22.** A calorie is the amount of heat needed to raise the temperature of one gram of water by one Celsius degree. A Calorie is equal to 1000 calories, or 1 kilocalorie (kcal). **23.** a tool made by nutritionists to aid people in selecting a healthful diet **24.** The ureter carries urine from the kidneys to the urinary bladder. **25.** The urinary bladder stores liquid waste, or urine. **Using Science Skills** **26.** the loop of Henle; this is where water is conserved and the volume of urine is minimized **27.** D, the capillaries **28.** This structure, a collecting duct, collects impurities filtered from the blood and transports them to the ureter. **29.** filtration and reabsorption **30.** kidney **Essay** **31.** The body loses water through sweat. As the water in sweat evaporates, it cools the body. Water in the form of vapor is also lost with every breath exhaled. More water is lost in the urine. **32.** The kidneys remove the wastes of cellular metabolism from the blood and regulate the concentrations of substances found in the body fluids. The ureter transports urine from each kidney to the urinary bladder. The urinary bladder stores the urine until it is excreted by the urethra out of the body. **33.** The digestive system includes the mouth, pharynx, esophagus, stomach, small intestine, and large intestine. Several major accessory structures, including the salivary glands, the pancreas, the gall bladder, and the liver, add secretions to the digestive system. **34.** An increased intake of water causes the concentration of water in the blood to increase. As the amount of water in the blood increases, the rate of water reabsorption in the kidneys decreases. Thus, less water is returned to the blood, and the excess water is excreted as urine. An increased intake of salt causes the level of salt in the blood to rise. The kidneys respond by returning less salt to the blood by reabsorption. The excess salt is excreted in urine. **35.** Dialysis is a process in which a machine is connected to a patient to filter out wastes from the patient's blood. In one type of dialysis, blood is removed from the body through a tube inserted in the arm and pumped through special tubing that acts like nephrons. Tiny pores in the tubing allow salts and small molecules, including nitrogen wastes, to pass through. Wastes diffuse out of the blood into the fluid-filled chamber, allowing purified blood to be returned to the body.

Chapter 38—Test B

Multiple Choice **1.** C **2.** C **3.** B **4.** A **5.** C **6.** D **7.** C **8.** C **9.** A **10.** B **11.** A **12.** A **13.** D **14.** C **15.** C **Completion** **16.** nutrition **17.** peristalsis **18.** ulcer **19.** nephrons **20.** homeostasis **Short Answer** **21.** water, carbohydrates, fats, proteins, vitamins, and minerals **22.** When the chewed-up food reaches the stomach, the stomach muscles contract to churn and mix stomach fluids and food, gradually producing a mixture known as chyme. **23.** The function of the digestive system is to help convert foods into simpler molecules that can be absorbed and used by the cells of the body. **24.** They maintain homeostasis by regulating the water content of the blood, and therefore, blood volume, by maintaining blood pH, and by removing waste products from the blood. **25.** The person can have a kidney transplant or use a kidney dialysis machine. **Using Science Skills** **26.** A **27.** D **28.** C, liver **29.** F, pancreas **30.** G, small intestine

Chapter 39—Test A

Multiple Choice **1.** B **2.** A **3.** B **4.** A **5.** D **6.** A **7.** C **8.** C **9.** D **10.** D **11.** B **12.** D **13.** D **14.** C **15.** A

Completion **16.** endocrine, bloodstream **17.** ovulation **18.** scrotum **19.** gastrulation, endoderm **20.** placenta **Short Answer** **21.** Endocrine glands (such as the pituitary, parathyroid, adrenal, or thyroid glands) secrete hormones into the bloodstream, while exocrine glands (such as those that secrete sweat, tears, and digestive juices) pass their secretions through ducts. **22.** Prostaglandins are hormonelike substances that are produced by a wide range of cells. Unlike hormones, they usually affect only nearby cells and tissues. **23.** Students should list a feedback mechanism, such as the feedback inhibition that controls metabolism or the one that controls water balance in the body. **24.** Sperm are produced in the seminiferous tubules. From there, they travel through the epididymis, vas deferens, and urethra. **25.** Once a sperm nucleus has entered the egg, the cell membrane of the egg cell changes, preventing other sperm from entering the cell. **Using Science Skills** **26.** It shows the feedback mechanism that controls the thyroid gland. **27.** High thyroxine levels inhibit the hypothalamus and the anterior pituitary so that less TRH and TSH are released. **28.** thyroxine **29.** Hyperthyroidism is a condition that occurs when the thyroid gland produces too much thyroxine; thus, any change that shows an overproduction of thyroxine is a correct response. Hyperthyroidism is characterized by nervousness, elevated body temperature, increased metabolic rate, increased blood pressure, and weight loss. **30.** The hypothalamus is sensitive to both temperature and the level of thyroxine in the blood. When body temperature decreases, such as on a cold day, the hypothalamus will produce more TRH even if the level of thyroxine does not stimulate it to do so. TRH will stimulate the anterior pituitary to produce more TSH and more thyroxine will be released into the bloodstream, speeding up metabolism and increasing body temperature. **Essay** **31.** Complementary hormone action occurs when two hormones with opposite effects act to regulate the body's internal environment. An example is the regulation of blood calcium levels. When calcium levels in the blood are too high, calcitonin is released to help reduce the level. When calcium levels in the blood are too low, parathyroid hormone is released to increase the level. **32.** The hormones released by the pituitary gland regulate many bodily functions. In addition, they control the actions of several endocrine glands. The hypothalamus controls the release of these pituitary glands. **33.** Nonsteroid hormones generally do not enter the target cell. Rather, they bind to receptors on the cell membrane. This action activates enzymes on the inner surface of the cell membrane. This enzyme activates secondary messengers that carry the message of the hormone inside the cell. Once released, these second messengers can activate or inhibit a wide range of other cell activities. **34.** Puberty is a period of rapid growth and sexual maturation during which the reproductive system becomes fully functional. Before this time, neither testes nor ovaries are capable of producing active reproductive cells. At the completion of puberty, however, the male and female gonads are fully developed. **35.** In males, the sex hormones allow for the normal production of sperm as well as bring about an increase of facial hair, increase in body size, and deepening of the voice. In females, the sex hormones bring about the development of the female reproductive system, the widening of the hips, and development of the breasts.

Chapter 39—Test B

Multiple Choice **1.** B **2.** C **3.** C **4.** A **5.** B **6.** A **7.** C **8.** B **9.** D **10.** D **11.** C **12.** C **13.** D **14.** A **15.** D **Completion** **16.** Hormones **17.** prostaglandins **18.** Puberty **19.** follicle **20.** placenta **Short Answer** **21.** The endocrine system releases hormones, which broadcast messages throughout the body. **22.** These are the parathyroid glands, which regulate the amount of calcium in the blood. **23.** The four phases are as follows: follicular phase, ovulation, luteal phase, menstruation. **24.** The germ layers that result from gastrulation are called endoderm, mesoderm, and ectoderm. **25.** Any one of the following: (1) During infancy, a baby learns to walk and talk. (2) During childhood, language and motor coordination are perfected. (3) With puberty begins adolescence and sexual maturity. (4) Signs of aging begin to appear and childbearing ends in adulthood. **Using Science Skills** **26.** Diagram I is the female reproductive system. Diagram II is the male reproductive system. **27.** The ovary—structure A—releases eggs. **28.** Fertilization usually occurs in a Fallopian tube. Structure F indicates a Fallopian tube. **29.** Sperm are formed in the testes, structure K. **30.** Sperm exit the male body through the urethra, structure L.

Chapter 40—Test A

Multiple Choice **1.** D **2.** C **3.** C **4.** B **5.** B **6.** B **7.** B **8.** C **9.** B **10.** C **11.** A **12.** C **13.** A **14.** A **15.** A **Completion** **16.** infectious **17.** vectors **18.** Antibiotics **19.** pathogens **20.** AIDS **Short Answer** **21.** Interferon, which interrupts viral replication, is produced by virus-infected cells and helps slow down viral infections; thus, it is not effective at stopping nonviral pathogens. **22.** Allergies result when antigens from allergens bind to mast cells, which release histamines. **23.** When the immune system mistakes its own body's cells for pathogens, an autoimmune disease results. **24.** HIV attacks helper T cells, reducing their number in the blood, weakening the immune system, and making the body susceptible to rare infections that it is usually able to resist. **25.** Antigens are substances that trigger an immune response. Antibodies are immune system proteins that recognize and bind to antigens. **Using Science Skills** **26.** The purpose of this culture is to isolate and grow the pathogen that is responsible for the cow's illness. **27.** A healthy mouse is being injected with the pathogen. **28.** The purpose of this culture is to isolate and grow the pathogen that is responsible for the mouse's illness and to compare it with the pathogen that caused the cow's illness. **29.** A healthy mouse would become as ill as the first mouse and the cow because it would be infected with the same pathogens. **30.** Sample answer: The microorganism isolated in A is the cause of the cow's (as well as the mouse's) disease. **Essay** **31.** Hereditary diseases, such as hemophilia, are caused by heredity, or genes passed from parents to children. Materials in the environment can also cause diseases. For example, cigarette smoke causes cancer, and exposure to the sun can cause skin cancer. Infectious diseases are caused by pathogens, such as bacteria or viruses. **32.** In active immunity, the body produces its own antibodies against a particular antigen. In passive immunity, a person is given antibodies obtained from the blood of either another person or an animal. Active immunity is a type of immunity that may develop as a result of having had a disease, and the disease rarely recurs. Passive immunity, on the other hand, is only temporary and usually does not last for more than several weeks. **33.** The inflammatory response is a nonspecific defense reaction to tissue damage caused by an injury or infection. When a wound occurs, nearby blood vessels expand and white blood cells leak into the infected tissues. Many of these white blood cells are phagocytes, which engulf and destroy pathogens. The infected tissue may swell and feel painful. The area may also be red and feel warm to the touch. **34.** The three main causes of cancer are viruses, radiation, and chemicals. Radiation causes cancer by producing mutations in DNA. If these mutations occur in genes that control cell growth, a normal cell may become a cancer cell. Chemicals, or carcinogens, also produce cancer by causing mutations in DNA. **35.** Both types of immunity are specific defenses. In cell-mediated immunity, killer T cells attack cells infected by pathogens. In humoral immunity, B cells produce antibodies that disable pathogens in the blood and other body fluids.

Chapter 40—Test B

Multiple Choice **1.** B **2.** C **3.** D **4.** C **5.** B **6.** A **7.** C **8.** B **9.** C **10.** C **11.** C **12.** D **13.** C **14.** C **15.** D **Completion** **16.** pathogen **17.** bacteria **18.** cell-mediated **19.** vaccine **20.** Humoral **Short Answer** **21.** Diseases can be inherited, caused by materials in the environment (such as asbestos or tobacco smoke), and produced by pathogens (such as bacteria and viruses). **22.** Antibiotics kill bacteria by interfering with the cellular processes of the bacteria. **23.** As long as the skin is not broken, it forms a barrier against invading pathogens; in addition, oil and sweat glands in the skin produce an acidic environment that kills bacteria on the surface of the skin. **24.** Allergies occur when the immune system overreacts to certain antigens, causing sneezing, runny nose, and watery eyes. **25.** A benign tumor is one that is not cancerous and will not spread throughout the body; a malignant tumor is cancerous and is likely to spread to other parts of the body. **Using Science Skills** **26.** an antigen (or pathogen) and an antibody **27.** Antigen-binding sites **28.** Structure A—an antigen—triggers the production of antibodies. **29.** humoral immunity **30.** Antigen-binding sites allow an antibody to bind to two antigens, destroying the pathogens.

Unit 10—Test A

Multiple Choice **1.** B **2.** C **3.** D **4.** B **5.** A **6.** C **7.** D **8.** A **9.** B **10.** D **11.** C **12.** C **13.** A **14.** B **15.** D **Completion** **16.** threshold **17.** compact **18.** bronchus **19.** water **20.** B cells **Short Answer** **21.** The dermis contains collagen fibers, blood vessels, nerve endings, glands, sensory receptors, smooth muscles, and hair follicles. **22.** The pancreas produces enzymes that break down carbohydrates, proteins, lipids, and nucleic acids in the small intestine. The pancreas also produces sodium bicarbonate, a base that neutralizes stomach acid so that these enzymes can be effective. **23.** A carcinogen is a chemical compound that is known to cause cancer. Examples include aflatoxin, chloroform, benzene, and chemicals found in tobacco smoke. **24.** When platelets come into contact with the edges of a broken blood vessel, they become sticky and cluster together around the wound. The platelets then release proteins called clotting factors, which start a series of chemical reactions that lead to the formation of a clot. **25.** Nutrients are absorbed across the inside surface of the small intestine. Villi are tiny projections on the surface of the small intestine that greatly increase the intestine's surface area and ability to absorb nutrients. **Using Science Skills** **26.** A cell-mediated immune response is shown in the diagram. **27.** Structure B is a T cell. In Step 1, the T cell binds to the macrophage and is activated to become a helper T cell. **28.** Structure B is a helper T cell. It activates killer T cells and B cells. **29.** Structure C is a killer T cell. In Step 3, the killer T cell binds to the infected cell, disrupting its cell membrane and destroying it. **30.** Cell-mediated immunity, shown in the diagram, involves killer T cells attacking cells infected with pathogens. Humoral immunity, in contrast, involves B cells producing antibodies that disable pathogens in the blood. Both processes are similar in their being specific defenses in the immune response. **Essay** **31.** The endocrine system communicates with hormones, which travel throughout the body via the bloodstream. Hormonal communication is relatively slow but can reach cells everywhere in the body. The nervous system communicates with nerve impulses, which travel through a system of interconnected neurons. Nervous communication is very rapid but limited to parts of the body that are interconnected by nerves. **32.** The levels of organization are cells, tissues, organs, and organ systems. Possible examples include: red blood cell (cell); blood (tissue); heart (organ); and cardiovascular system (organ system). **33.** The skeleton of an embryo is composed almost entirely of cartilage. Beginning around seven months before birth, ossification begins to take place. Bone tissue forms as osteoblasts secrete mineral deposits that replace the cartilage in developing bones. The growth of cartilage at the ends, or growth plates, of long bones causes the bones to lengthen. Gradually this new growth of cartilage is also replaced by bone tissue, and the bones become larger and stronger. By adulthood, the cartilage in the growth plates is replaced by bone and the bones become completely ossified. **34.** The thyroid gland controls metabolism. Its hormone, thyroxine, stimulates cells to increase their metabolic activity. When the hypothalamus senses that the level of thyroxine in the blood is low, it secretes thyroid-releasing hormone (TRH). TRH stimulates the pituitary to produce thyroid-stimulating hormone (TSH), which stimulates the thyroid to produce thyroxine. The opposite events occur if the level of thyroxine in the blood is high. **35.** The body's nonspecific defenses are the skin, inflammatory response, fever, and interferons. The skin is a physical barrier that few pathogens can penetrate. Sweat and skin oils also kill many bacteria. The inflammatory response occurs in reaction to tissue damage. It brings many white blood cells to the site to fight infection. Fever raises the body temperature, which slows the growth of many pathogens. Interferons are proteins produced by cells infected with virus. Interferons help block viral replication.

Unit 10—Test B

Multiple Choice **1.** B **2.** C **3.** D **4.** A **5.** C **6.** B **7.** B **8.** C **9.** A **10.** D **11.** C **12.** C **13.** B **14.** D **15.** C **Completion** **16.** reflex **17.** periosteum **18.** capillaries **19.** bile **20.** autoimmune **Short Answer** **21.** Stimulant and depressant drugs have opposite effects on the nervous system. Stimulants increase and depressants decrease the actions controlled by the nervous system, such as heart rate. **22.** Ligaments hold bones together in joints. Tendons attach skeletal muscles to bones and make them work like levers around the joints. **23.** Atherosclerosis is a buildup of fat deposits on the inner walls of arteries. It increases the risk of cardiovascular diseases such as heart attack and stroke. **24.** The teeth contribute to mechanical digestion by tearing and grinding food into smaller pieces. The enzyme amylase in saliva begins the chemical digestion of starch. **25.** Infectious diseases can be spread by physical contact, including coughing and sneezing; by contaminated food or water; and by infected animals. **Using Science Skills** **26.** Diagram I shows the action of a steroid hormone. Diagram II shows the action of a nonsteroid hormone. **27.** In Diagram I, structure A is the receptor. It is found only in target cells. It binds to the hormone to form structure B, a hormone-receptor complex, which enters the nucleus of the cell and binds to DNA. **28.** The hormone in diagram I directly affects gene expression by turning on or off whole sets of genes. **29.** In Diagram II, structure A is the receptor. It is found only on the cell membranes of target cells. The receptor binds to the hormone and activates an enzyme on the inner surface of the cell membrane. This enzyme activates structure B, a secondary messenger, which carries the message of the hormone inside the cell. **30.** The steroid hormone in Diagram I is made of lipids, so it can cross cell membranes. The nonsteroid hormone in Diagram II is made of proteins, so it cannot cross cell membranes.

Final Exam—Level A

Multiple Choice **1.** A **2.** D **3.** C **4.** B **5.** C **6.** A **7.** D **8.** C **9.** D **10.** A **11.** A **12.** B **13.** D **14.** D **15.** C **Completion** **16.** slows down (stops) **17.** transcription **18.** fungi **19.** fruit **20.** mollusk **Short Answer** **21.** Grow two similar plants in the same environmental conditions. Give each the same amount of water. However, add fertilizer to the water of one plant. The controlled variables are temperature, location, soil, type of plant, and water. The manipulated variable is fertilizer. The responding variable is plant growth. **22.** Plant roots absorb precipitation that seeps into the ground. Plant leaves return water to the air through transpiration. **23.** Cells divide because the larger a cell becomes, the more demands the cell places on its DNA. Also, the cell has more trouble moving enough nutrients and wastes across the cell membrane. **24.** Genetic diversity is important to breeders because the more genetic diversity there is, the more genetic traits there are to choose from, and it is more likely that breeders will find the trait they are interested in. **25.** Scientists think that human evolution occurred in a series of complex adaptive radiations that produced a large number of different species. **Using Science Skills** **26.** A is the promoter site. It is the site on the DNA molecule where RNA polymerase binds to start transcription. **27.** The repressor blocks the movement of RNA polymerase. **28.** When the repressor protein is bound to lactose, its shape changes so that it cannot bind to the operator. When lactose is not present, the shape of the repressor protein allows it to bind to the operator. **29.** The *lac* genes are on when lactose binds to the repressor protein. **30.** The *lac* operon controls when the *lac* genes are turned on. This is useful to the cell so that it doesn't waste resources to make proteins that it doesn't need. **Essay** **31.** With a climate change, abiotic factors such as temperature, precipitation, and humidity would change. Biotic factors, or all the living things in the ecosystem, would die out or emigrate if they could not adapt to the climate changes. Some living things would adapt to the climate changes and remain in the ecosystem. Other living things might immigrate to the area because the new climatic conditions are optimal for their survival. **32.** Plants are the main land autotrophs that use the energy from sunlight to produce food, or carbohydrates. They also add oxygen to the air as a result of photosynthesis. Plants make up the first trophic level in terrestrial food chains. All consumers rely on plants for energy. Without plants, almost all consumers on land could also become extinct. Consumers in aquatic environments may not be affected because green algae are the main autotrophs in freshwater ecosystems and in the oceans. **33.** The human insulin gene is cut from a sequence of human DNA with restriction enzymes. The fragments are separated by gel electrophoresis, and the human insulin gene is isolated. A bacterial plasmid is cut with the same restriction enzyme. The human DNA fragment and the cut plasmid are mixed together, and the sticky ends of each molecule recombine with each other. The recombinant plasmids are transformed into bacterial cells. Transformed bacteria are separated from bacteria that do not contain a plasmid by growing the bacteria on plates that contain a certain antibiotic. The plasmid carries a gene resistant to an antibiotic; so bacterial cells containing the plasmid will grow in the presence of that antibiotic. **34.** Changes in DNA can cause mutations in the sequence of a gene, which in turn changes the action of its protein product. Some mutations are beneficial in that they increase the fitness of individuals that carry them. Individuals with increased fitness are better suited to their environment and better able to survive and reproduce. Over time, most members of a population will carry the mutant allele and exhibit the mutant trait. **35.** Muscle cells make up smooth muscle tissue. Smooth muscle tissue, along with epithelial tissue, nervous tissue, and connective tissue, makes up the stomach. The stomach is an organ that works with other organs in an organ system called the digestive system.

Final Exam—Level B

Multiple Choice **1.** B **2.** C **3.** C **4.** A **5.** A **6.** B **7.** D **8.** D **9.** C **10.** B **11.** A **12.** C **13.** D **14.** B **15.** D **Completion** **16.** abiotic **17.** interphase (S phase) **18.** natural selection **19.** radial **20.** endotherms **Short Answer** **21.** Plants are living things because they are made up of cells, they reproduce, they contain DNA, they grow and develop, they use energy from the sun and nutrients from the soil, they respond to their environment, they maintain a stable internal environment, and they change over time. **22.** As genetic diversity decreases, there are fewer traits that an organism can "choose" from to adapt to changing conditions. If an organism cannot adapt to changing conditions, it will eventually die off. **23.** Photosynthesis uses the energy of sunlight to convert the reactants, carbon dioxide and water, into the products, sugars and oxygen. **24.** The two main sources of genetic variation are mutations and the shuffling of alleles that occurs from sexual reproduction. **25.** A behavior helps an organism survive and reproduce by helping it get resources and find a healthy mate. **Using Science Skills** **26.** The plant cell is Cell I. **27.** These cells are eukaryotic cells because they contain genetic material within a nucleus. **28.** Structure A is a mitochondrion. It functions in cellular respiration. **29.** The nucleus is labeled D. It codes for proteins. The proteins make up cell structures and regulate cell processes. **30.** Cell II would shrink because water would move out of the cell. Water would diffuse from the cell because the concentration of water is higher inside the cell than outside the cell.